The Biology of DNA Tumor Viruses

With an Introduction by Renato Dulbecco

Milestones in Current Research

Collections of Influential Contributions in Key Areas of Science

Series Editor: Samuel Schiminovich

Volume 1: THE BIOLOGY OF DNA TUMOR VIRUSES
With an Introduction by Renato Dulbecco

The Biology of DNA Tumor Viruses

With an Introduction by Renato Dulbecco

Edited by
Samuel Schiminovich

Springer Science+Business Media, LLC

Library of Congress Cataloging in Publication Data

Main entry under title:

The Biology of DNA tumor viruses.

(Milestones in current research; v. 1)
Includes index.
1. Oncogenic viruses. I. Schiminovich, Samuel. II. Title: DNA tumor viruses.
[DNLM: 1. Oncogenic viruses. 2. Polyoma virus. 3. SV40 virus. 4. DNA, neoplasm.
QZ202 B616]
RC268.57.B56 616.9'92'0194 76-25874

ISBN 978-1-4684-6972-1 ISBN 978-1-4684-6970-7 (eBook)
DOI 10.1007/978-1-4684-6970-7

Originally published by Plenum Press, New York in 1976
Softcover reprint of the hardcover 1st edition 1976
A Division of Plenum Publishing Corporation
227 West 17th Street, New York, N.Y. 10011

ACKNOWLEDGMENTS

The permission of the copyright owners to reprint the following material is gratefully acknowledged.

Aloni, Y., Winocour, E., and Sachs, L., Characterization of the simian virus 40-specific RNA in virus-yielding and transformed cells, *Journal of Molecular Biology,* 1968, **31**:415-429.

Benjamin, T.L., Virus-specific RNA in cells productively infected or transformed by polyoma virus, *Journal of Molecular Biology,* 1966, **16**:359-373.

Black, P.H., Rowe, W.P., Turner, H.C., and Huebner, R.J., A specific complement-fixing antigen present in SV40 tumor and transformed cells, *Proceedings of the National Academy of Sciences,* 1963, **50**:1148-1156.

Dulbecco, R., and Eckhart, W., Temperature-dependent properties of cells transformed by a thermosensitive mutant of polyoma virus, *Proceedings of the National Academy of Sciences,* 1970, **67**: 1775-1781.

Dulbecco, R., Hartwell, L.H., and Vogt, M., Induction of cellular DNA synthesis by polyoma virus, *Proceedings of the National Academy of Sciences,* 1965, **53**:403-410.

Dulbecco, R., and Stoker, M., Abortive transformation by the Tsa mutant of polyoma virus, *Nature,* 1969, **223**:397-398.

Dulbecco, R., and Vogt, M., Evidence for a ring structure of polyoma virus DNA, *Proceedings of the National Academy of Sciences,* 1963, **50**:236-243.

Fried, M., Cell-transforming ability of a temperature-sensitive mutant of polyoma virus, *Proceedings of the National Academy of Sciences,* 1965, **53**:486-491.

Gerber, P., Studies on the transfer of subviral infectivity from SV40-induced hamster tumor cells to indicator cells, *Virology,* 1966, **28**:501-509.

Khoury, G., Martin, M.A., Lee, T.N.H., Danna, K.J., and Nathans, D., A map of simian virus 40 transcription sites expressed in productively infected cells, *Journal of Molecular Biology,* 1973, **78**: 377-389.

Koprowski, H., Jensen, F.C., and Steplewski, Z., Activation of production of infectious tumor virus SV40 in heterokaryon cultures, *Proceedings of the National Academy of Sciences,* 1967, **58**:127-133.

Lindstrom, D.M., and Dulbecco, R., Strand orientation of simian virus 40 transcription in productively infected cells, *Proceedings of the National Academy of Sciences,* 1972, **69**:1517-1520.

Oda, K., and Dulbecco, R., Regulation of transcription of the SV40 DNA in productively infected and in transformed cells, *Proceedings of the National Academy of Sciences,* 1968, **60**:525-532.

Pollack, R.E., Green, H., and Todaro, G.J., Growth control in cultured cells: Selection of sublines with increased sensitivity to contact inhibition and decreased tumor-producing ability, *Proceedings of the National Academy of Sciences,* 1968, **60**:126-133.

Pope, J.H., and Rowe, W.P., Detection of specific antigen in SV40-transformed cells by immunofluorescence, *The Journal of Experimental Medicine,* 1964, **120**:121-128.

Radloff, R., Bauer, W., and Vinograd, J., A dye-buoyant-density method for the detection and isolation of closed circular duplex DNA: The closed circular DNA in HeLa cells, *Proceedings of the National Academy of Sciences,* 1967, **57**:1514-1521.

Rapp, F., Kitahara, T., Butel, J.S., and Melnick, J.L., Synthesis of SV40 tumor antigen during replication of simian papovavirus (SV40), *Proceedings of the National Academy of Sciences,* 1964, **52**: 1138-1142.

Renger, H.C., and Basilico, C., Mutation causing temperature-sensitive expression of cell transformation by a tumor virus, *Proceedings of the National Academy of Sciences,* 1972, **69**:109-114.

Sambrook, J., Westphal, H., Srinivasan, P.R., and Dulbecco, R., The integrated state of viral DNA in SV40-transformed cells, *Proceedings of the National Academy of Sciences,* 1968, **60**:1288-1295.

Sauer, G., and Kidwai, J.R., The transcription of the SV40 genome in productively infected and transformed cells, *Proceedings of the National Academy of Sciences,* 1968, **61**:1256-1263.

Stoker, M., Abortive transformation by polyoma virus, *Nature,* 1968, **218**:234-238.

Vinograd, J., Lebowitz, J., Radloff, R., Watson, R., and Paipis, P., The twisted circular form of polyoma viral DNA, *Proceedings of the National Academy of Sciences,* 1965, **53**:1104-1111.

Wall, R., and Darnell, J.E., Presence of cell and virus specific sequences in the same molecules of nuclear RNA from virus transformed cells, *Nature New Biology,* 1971, **232**:73-76.

Watkins, J.F., and Dulbecco, R., Production of SV40 virus in heterokaryons of transformed and susceptible cells, *Proceedings of the National Academy of Sciences,* 1967, **58**:1396-1403.

Weil, R., and Vinograd, J., The cyclic helix and cyclic coil forms of polyoma viral DNA, *Proceedings of the National Academy of Sciences,* 1963, **50**:730-738.

Westphal, H., and Dulbecco, R., Viral DNA in polyoma- and SV40-transformed cell lines, *Proceedings of the National Academy of Sciences,* 1968, **59**:1158-1165.

FOREWORD

Milestones in Current Research is a series of reprint collections distinguished from other such publications by new concepts in preparation, presentation, and intent. The aim of each volume is to gather for a given field the seminal contributions that have defined and shaped the trends within the most active areas of current research. The selections for each volume and the structure of the book have been determined with the help of a novel technique of bibliographic analysis and have then been presented to an acknowledged scientific authority for minor adjustments and the writing of an Introduction. These introductions will lend historic perspective to the material selected for each volume. The bibliographic analysis used tends to select papers central to the areas of current research within, roughly, the last decade and is a systematic procedure for depicting, delineating, and covering all such areas over a wide spectrum of scientific research.

It is hoped that with this procedure it will be possible to achieve an objectivity, authority, and thoroughness not reached by others and that the timeliness of the volumes will not be limited to just a few years. Each volume should have the permanent value of a historical statement and yet be sufficiently interesting to active researchers in the field as well as to students exploring the quiet way in which the relentless drama of research unfolds in the journal literature. Furthermore, follow-up volumes will periodically update the status of research and maintain the currency of the series as a whole.

While the series may well become a prestige forum where an outstanding segment of the journal literature is reprinted, readers may occasionally miss a paper that they would expect to be included. Such omissions are bound to occur for a number of reasons, including among others the need to keep the volumes to a viable size or the sheer unavailability of the given paper for reproduction due to unusual copyright restrictions.

Our gratitude is extended to all those who in one way or another have helped us to bring these concepts to fruition, in particular to Professors

Ralph Schiller and Howard M. Temin for their enlightening insights and early encouragement, to all copyright holders who have made their articles available for reproduction, and to Dr. Renato Dulbecco for making possible such an auspicious start for the series.

Samuel Schiminovich
Series Editor

CONTENTS

I
Introduction

INTRODUCTION

R. DULBECCO

The oncogenic potency of DNA-containing viruses became widely recognized around 1960, when polyoma virus (PV), simian virus 40 (SV40), and adenoviruses were shown to be oncogenic in animals. It had long been known that papilloma virus induces warts in rabbit skin, which later evolve into carcinomas, but it seemed to be an isolated case; and furthermore the connection between the virus and the later malignant growth was undetermined. These new discoveries gave fresh impetus to research. A related factor for this renewed interest in DNA-containing oncogenic viruses was the widespread knowledge of bacterial lysogeny. In this process a viral genome becomes integrated in the cell's DNA as a prophage, and then functions more or less like a set of cellular genes, often changing the cell in recognizable ways. Hence, bacterial lysogeny was thought by many to be a possible model for viral oncogenesis.

Polyoma virus and SV40 were adopted as model systems in many laboratories because they could be studied well in cell culture, in which they cause cell *transformation*, which is equivalent to cancer induction. These viruses multiply lytically in *permissive* cells, which are then killed (mouse cells for PV, African green monkey cells for SV40) and preferentially induce stable transformation in *non-permissive* cells, which are not killed (e.g., hamster cells for both viruses and mouse cells for SV40). In time it became evident that changes similar to those of stable transformation cells also occur transiently in lytic infection of permissive cells before the cells are killed. This observation suggested the concept that the presence of certain functional viral genes always transforms the cells.

In early studies of these viruses investigators were faced with a novel feature of the viral DNA: that it split into two (and sometimes three) components on sedimentation. At first this behavior was attributed to end-to-end dimerization; but the chromatographic behavior of the DNA soon showed this interpretation to be incorrect. It became clear that the various components must differ in conformation (Paper II-1 of this volume). The initial studies suggested that the fastest component (Component I) was a covalently closed ring-shaped double-stranded molecule — the first of its kind to be discovered — and the second one (Component II), its linear counterpart. Indeed, cyclic molecules were soon recognized by electron microscopy (Paper II-2). Subsequently, however, Component II was also shown to be ring-shaped but nicked; the linear equivalent, when present, constituted the slowest component, Component III (Paper II-3). Extensive studies of this interesting DNA in Vinograd's laboratory made possible the complete elucidation of its properties and opened up important ways for characterizing and purifying the DNA. Thus, it was found that Component I molecules bind less ethidium bromide and therefore have a higher buoyant density than linear DNAs (including the cellular DNA) in equilibrium gradient centrifugation in CsCl + ethidium bromide (Paper II-4). The special features of the cyclic DNA together with the selective extraction procedure devised by Hirt (1967) afforded tools that were extensively used in the subsequent work on the molecular biology of these viruses. For instance, using these technics Sebring et al. (1972) have made the surprising discovery that the polyoma DNA ring replicates without any permanent nick. Hence the replicative intermediate has intermediate characteristics between those of Components I and II and can be isolated in CsCl-ethidium chloride gradients. The parental strands are probably unwound during replication by an "unwindase."

More recently, Danna and Nathans (1971) using restriction enzymes have shown that SV40 DNA can be split into 11 characteristic fragments. This discovery provided a basis for the physical mapping of the genome, which had already proved extremely useful in studies of replication and transcription of the DNA.

The recognition of the cyclic configuration of PV and SV40 DNA sparked a search for a similar DNA in other viruses. A significant development was finding cyclic molecules during the replication of temperate bacteriophage, since it gave a molecular basis to Campbell's suggestion, based on genetic evidence, that the phage DNA in cyclic form becomes integrated in the cellular DNA by a single crossover. These findings with phages reinforced the belief that PV or SV40 DNA must also be integrated in the transformed cells.

A test for integration could not be based on genetic evidence since the genetics of the viruses and of their host cells were essentially unknown; however, a molecular approach was feasible since the viral DNA could be recognized in the transformed cells by nucleic acid hybridization. The required reagent was labeled RNA of high specific activity synthesized *in vitro* by purified *E. coli* RNA polymerase, using the viral DNA as template. Its discriminatory power was adequate to detect one genome per cell corresponding to about one part of viral DNA in a million parts of cellular DNA Paper III-1). Suitable fractionation of the DNA of SV40-transformed cells showed that the sequences able to hybridize with this RNA were indeed present in cellular DNA (Paper III-2). The association was alkali resistant, hence the two DNAs were covalently bound. This important result showed that the viral DNA is integrated as provirus in the transformed cells.

Once the molecular anatomy of transformation was solved, emphasis shifted to functional studies, i.e., how the provirus transforms the cell.

An important development relevant to this problem was the discovery, in Huebner's laboratory, of a virus-specific antigen present in SV40-transformed cells and distinct from the V antigen of the viral particles (Paper IV-1). The antigen was designated as T (tumor) antigen. It could be shown by immunofluorescence that the antigen was in the nuclei of most cells in a transformed culture (Paper IV-2). Soon, it was also found in the nuclei of lytically infected cells (Paper IV-3), confirming that it was a consequence of the expression of some viral function in the cells. An antigen with these properties was also recognized in PV-infected or transformed cells.

The continued presence of the T antigen in transformed cells revealed the activity of some viral genes in them although the transformed cells neither produce virus nor contain V antigen. A possible explanation was that the viral genome in the transformed cells was defective, i.e., it was mutated or deleted in the gene for viral coat protein. The solution of this problem became possible when Gerber (Paper V-1) found that cells of a SV40-induced hamster tumor did not produce virus by themselves but did so when cocultivated with SV40-permissive cells. This happened at the time when the introduction of inactivated Sendai virus as fusing agent made cell fusion in culture an important experimental tool. The hypothesis that viral multiplication occurs after fusion of a transformed and a permissive cell was soon tested and found correct (Paper V-2,3). This result established that the non-permissive transformed cells contain a non-defective provirus, and that their failure to express the V antigen must be due to the regulation of the expression of the proviruses in these cells. How regulation is achieved thus became the main problem.

Considerable insight into this problem was gained by studying the transcription of the viral DNA both in lytic infection and in transformed cells. That the viral DNA is transcribed into specific mRNAs was first shown by Benjamin (Paper VI-1), who labelled cells lytically infected or transformed by polyoma virus with $^{32}PO_4$ and then hybridized the RNA extracted from these cells with viral DNA. As the technology for the selective extraction of various nucleic acids from cells and for nucleic acid hybridization became more sophisticated, it became possible to describe the transcription of the viral DNA in considerable detail. An important tool was competition hybridization, in which the hybridization of a sample of labelled RNA with the viral DNA is competed by increasing amounts of another sample of unlabeled RNA. This and other approaches revealed the pattern of transcription (Paper VI-2,3,4). In lytic infection with SV40, transcription is different in the phase preceding the onset of viral DNA replication (early) and in that following it (late). Early RNA transcribes about 40% of the genome, whereas the whole genome is transcribed in the late phase. Subsequent work suggested that the basis for this regulation is a strand switch, since early and late RNAs are transcribed from opposite strands of the viral DNA (Paper VI-5). The actual direction of transcription could be established by using SV40 DNA fragments produced by restriction endonucleases; since the strands of the fragments are incomplete at the 3' end their orientation can be established after completing the ends with radioactive nucleotides. The result is that the origin of both transcriptions is close to the origin of replication of the DNA (Paper VI-6). The regulation of transcription exists whether the transcribed viral DNA derives from infection with SV40 virus or with a defective adenovirus strain in which a segment of adenovirus DNA is replaced by a segment of the SV40 DNA. (Paper VI-7). Although most of SV40 DNA is in the hybrid, only its early sequences are transcribed. These results suggest that regulation of transcription in productive infection is viral.

In SV40-transformed cells, again a proportion of the DNA is is transcribed, but varies with the cell type. In SV3T3 cells about 40% of the provirus is transcribed but in other cells the transcribed fraction reaches 80%. In all cases the transcription includes more than half of the early sequences; in contrast, the proportion of late sequences is very variable. This distribution apparently satisfies a double need: to express the functions of some early viral genes, which are required for transformation (apparently including the one for the T antigen), and to avoid functions of late genes that would kill the cells. These lethal functions may be related to the V antigen.

The regulation of viral transcription is probably cellular, including both the nature of the transcription factors (such as those supplied by permissive cells when they are fused to transformed cells) and the site of provirus integration. In fact, Wall and Darnell (Paper VI-8) have shown that in SV40-transformed 3T3 cells viral sequences are part of hybrid molecules also containing cellular sequences. It is likely that viral transcription begins at a cellular promoter and is therefore influenced both by the nature of the cellular DNA surrounding the provirus and by the mode of integration of the provirus (i.e., where the circle was opened).

These results began to define the viral gene functions required for transformation: clearly, not more than about one-third of the viral genome is essential. The question was what these functions are. This problem was approached by a study of temperature-sensitive (ts) viral mutants (first of polyoma virus and then of SV40) and of the functional changes observed in virus-infected and transformed cells in respect to their uninfected counterparts.

Let's consider the viral functions first. In addition to the T and V antigens already mentioned, Sjogren et al. (1961) and Habel (1961) discovered the existence of a virus-specific surface antigen, which causes the rejection of transformed cells in animals previously infected by the transforming virus (probably owing to induction of the antigen in cells of the animal) and also present in lytic infection (Irlin , 1967). Another viral function of great interest is that which restarts the replication of the cellular DNA in resting cells before they are killed by the infection (Paper VII-1). At least two functions are necessary to cause a permanent cell transformation: provirus integration, and alteration of the cell growth regulation (the transforming function). Perhaps the transforming function is identical to that causing a transient transformation in permissive lytically infected cells, before they are killed, or the abortive transformation (Paper VII-2) of non-permissive cells.

All these functions are affected by ts mutations. The first one, isolated in polyoma virus by Fried (Paper VII-3) and called ts-a, affects the replication of the viral DNA and perhaps also the appearance of the T antigens. The mutation may also affect integration of the viral DNA in the cellular DNA because the mutation affects the stable but not the abortive transformation of BHK cells (Paper VII-4), in which integration may not be essential. In stable transformation the function of the mutated gene is required only transiently, for an initial event which could be integration.

Of the other ts mutants of polyoma virus, two groups affect the capsid proteins (which constitute the V antigens). A mutant of special interest (ts-3) affects the decreased serum requirement of the transformed cells as well as the induction of cellular DNA synthesis in lytic infection (Paper VII-5), supporting the view that the two phenomena result from the same change. The mutation also affects surface characteristics of the transformed cells demonstrable by agglutination by lectins (carbohydrate-binding proteins).

The results summarized so far allow us to build the following picture of the biology of these small oncogenic viruses. As soon as they enter a cell, whether permissive or non-permissive, they express two early functions. One causes the transformation of the cells, the other the formation of the T antigen. In lytic infection both functions may be required for the replication of the viral DNA. In permissive cells replication of the viral DNA allows the expression of late functions, with production of virus and ultimate death of the cells. In non-permissive cells the viral DNA probably does not replicate; late functions are not expressed, allowing cell survival and transformation. Therefore, cell transformation would be a by-product of a function of the virus developed for the purpose of its own multiplication.

We know that this picture is incomplete, and present research is directed at clearing up much that is obscure and which seems likely to hold very important secrets. When these secrets are revealed they will lead not only to a much better understanding of the viral transforming function, but also of the organization of growth regulation of the cells. Following are some of the unsolved questions:

1. Are the alterations of growth properties the only type of permanent effect the virus can cause? The similarity of properties observed in independently transformed cells may be the result of the selection procedure used in their selection. With other selection cells with different properties might be isolated.

2. Does integration play any additional role besides fixing the viral to the cellular DNA? Does it affect the function of neighbor-

ing cellular genes, and if so, with what consequences?

3. Cell transformation is strongly influenced by cellular mutations (Papers VIII-1,2). What cellular functions are affected by these mutations, and how do they interplay with the viral functions?

4. The ts3 mutation is recognizable only in certain cell types; in others the mutant behaves essentially like wild type. Therefore, some cellular function can replace the viral function affected by the mutation. What is this cellular function?

5. SV40 possesses a "helper" function which enables the synthesis of adenovirus coat proteins and therefore the replication of adenovirus in African green monkey cells (Baum et al., 1967). Does the virus specify a translation factor? Or, does it elicit its production by a cellular gene?

6. The nature of the T antigen and its function in DNA replication and integration are still unknown. Does the gene specify an enddnuclease? Or a DNA-binding protein?

7. The nature of the surface antigen is also unclear. There are at least two kinds of antigens: one, virus-specific, determines the rejection of transformed cells in immune animals. The other, nonspecific, is fetal and does not affect cell rejection (Ting et al., 1972). What is the specification of these antigens? How do they correlate to the changes of growth regulation?

These and other unresolved questions suggest that research with these oncogenic viruses is just beginning its most interesting phase.

Another reason for the continuing interest in these viruses is the finding that SV40 and related strains are fairly widespread in man and produce slow infections of the brain (Padgett et al., 1971). Hence the question arises whether they play a role in the etiology of human cancers.

REFERENCES

Hirt, B., "Selective Extraction of Polyoma CNA from Infected Mouse Cell Cultures." *Journal of Molecular Biology,* 1967, 26:365.

Sebring, E.D.; T.J. Kelley Jr.; M.M. Thoren; and N.P. Salzman, "Structure of replicating simian virus 40 deoxyribonucleic acid molecules," *Journal of Virology,* 1971, 8:478.

Danna, K.J.; and D. Nathans, "Specific Cleavage of simian virus 40 DNA by restriction endonuclease of hemophilus influenzae, *"Proceedings of the National Academy of Sciences,* 1971, 68:2913.

Sjogren, H.O.; I. Hellstrom; and G. Klein, "Transplantation of Polyoma Virus-Induced Tumors in Mice," *Cancer Research,* 1961, 21:329.

Habel, K., "Resistance of polyoma virus immune animals to rtansplanted polyoma tumors, *"Proceedings of the Society of Experimental Biology and Medicine,* 1961, 106:722.

Irlin, I.S., "Immunofluorescent Demonstration of a Specific Surface Antigen in Cells Infected or Transformed by Polyoma Virus," *Virology,* 1967, 32:725.

Baum, S.G.; M.S. Horwitz; and J.V. Maizel, Jr., "Studies of the Mechanism of Enhancement of Human Adenovirus Infection in Monkey Cells by Simian Virus 40," *Journal of Virology,* 1972, 10:211.

Ting, C.; D.H. Lavrin; G. Shiv; and R.B. Herberman, "Expression of Fetal Antigens in Tumor Cells," *Proceedings of the National Academy of Sciences,* 1972, 69:1664.

Padgett, B.L.; D.L. Walker; G.M. ZuRhein; and R.J. Eckroade, "Cultivation of Papova-like Virus from Human Brain with Progressive Multifocal Leukoencephalopathy," *The Lancet,* 1971, 1:1257.

II
Structure of Viral DNA

EVIDENCE FOR A RING STRUCTURE OF POLYOMA VIRUS DNA

R. Dulbecco and M. Vogt

The DNA of polyoma (PY) virus has two interesting features: it is unusually resistant to heat or formamide denaturation,[1] and it has two distinct components, of sedimentation coefficient 14 and 21, respectively.[2] A two-component sedimentation was formerly shown to be a characteristic of the DNA of papilloma virus,[3] which is considered a member of the same group of viruses.

Evidence presented in this article shows that both these features of polyoma virus DNA are the consequence of special configurational properties.

Material and Methods.—Purified polyoma virus was prepared according to Winocour;[4] DNA was extracted from the purified virus with phenol[5] and stored at −70°C. Virus labeled with either P^{32} or H^3 was obtained by adding label (P^{32} orthophosphate or H^3 thymidine) 24 hr after infecting the cultures. Sedimentation studies were carried out by the band sedimentation method:[6] 3 ml of CsCl solution of density 1.50 were placed in a centrifuge tube and covered with paraffin oil; the sample, of a volume 0.05–0.25 ml. was added dropwise on top of the oil layer. The tube was spun in an SW-39 rotor in a model L Spinco ultracentrifuge at 35 K rpm; in most experiments, spinning time was 4 hr. For shorter runs, the tube containing the CsCl solution and oil was spun for 4 hr before adding the sample, in order to preform the gradient. The CsCl solution was buffered either at pH 7.5 with 0.01 *M* Tris, or at alkaline pH (11.0, 11.8, and 12.5) with 0.01 *M* phosphate buffer.[7] For density gradient equilibrium sedimentation, 3 ml of a CsCl solution of density 1.71 or 1.72 at pH 7.5, and 1.73 at pHs 11–12.5 were used. The solution was covered with oil, spun in the SW-39 head of the model L ultracentrifuge at 29 or 30 K for 70–80 hr. In all cases, fractions were collected from the bottom of the tube. Columns of methylated albumin were prepared according to Mandell and Hershey.[8] Infectivity of polyoma DNA was measured as described by Weil.[5] Pancreatic DNAase Worthington 1x crystallized was used. *E. coli* phosphodiesterase (Lehman enzyme)[9] was a gift of Dr. R. L. Sinsheimer. DNAase and phosphodiesterase digestion were carried out according to Fiers and Sinsheimer.[16]

Results.—In brief, the following results were obtained. (1) PY DNA always sediments in two bands, a fast one (*F*) and a slow one (*S*). The *F* and *S* components are stable under a variety of conditions; both are infectious. (2) In the column of methylated albumin the *S* component is eluted at *higher* salt concentration than the *F* component. (3) The following observations were made by density gradient equilibrium centrifugation. At pH 7.5 the bands formed by the *F* and by the *S* component have the same density and do not differ greatly in width; at pH 11.8 most of the *F* component has the density of native DNA, whereas the *S* component has the density of denatured DNA. At pH 12.5 both components are denatured; the *S* component forms a wider band. (4) In band sedimentation at pH 12.5 the *F* component sediments about 2.5 times faster than at pH 7.5; after neutralization and annealing it reacquires the original sedimentation characteristics. The *S* component, on the contrary, migrates at pH 12.5, about the same speed as at pH 7.5. (5) The infectivity of the *F* component is increased several times after treatment at pH 12.5 followed by neutralization; the infectivity of the *S* component is lost after this treatment. (6) Pancreatic DNAase converts the *F* component into the *S* component, with loss of infectivity. The kinetics of conversion is unusual, since it is of first order. These results show that molecules of polyoma virus DNA in different states exist in the *F* and in the *S* component. The data are consistent with the hypothesis that the *F* component is made up of molecules in ring form, and the *S* component of molecules in linear, or open, form. The results will now be given in greater detail, in terms of this hypothesis.

Properties of the two components of different sedimentation velocity: Many preparations of PY DNA, of both small plaque and large plaque type, were examined by band sedimentation in CsCl at pH 7.5. They all contained the *F* and the *S* band. The band pattern observed with this method was found to be similar to that observed with the sucrose gradient method. The major component was the *F* component; it represented 60–80 per cent of the total in DNA preparations which had been extracted only once with phenol, and 90–95 per cent in preparations which had been extracted two or three times.

The ratio of the uncorrected distances traveled by the two components was 1.31 in sucrose gradient with 5×10^{-3} *M* Mg^{++}, 2.5×10^{-2} *M* K^{+}, and 5×10^{-2} *M* Tris buffer, pH 7.5; the ratio was 1.22 in CsCl of density 1.50 at pH 7.5.

Both components were infectious and produced plaques of equal type. The average specific infectivity (ratio: infectivity to OD_{260}) was two to three times larger in the *F* band (Fig. 1). In the *F* band the infectivity was distributed almost uniformly, whereas in the *S* band infectivity was present only in the frontal part. The *S* band is thus heterogeneous; this band is assumed to consist in its frontal part of infectious linear molecules, obtained by opening the ring, and in its trailing part, of smaller noninfectious fragments.

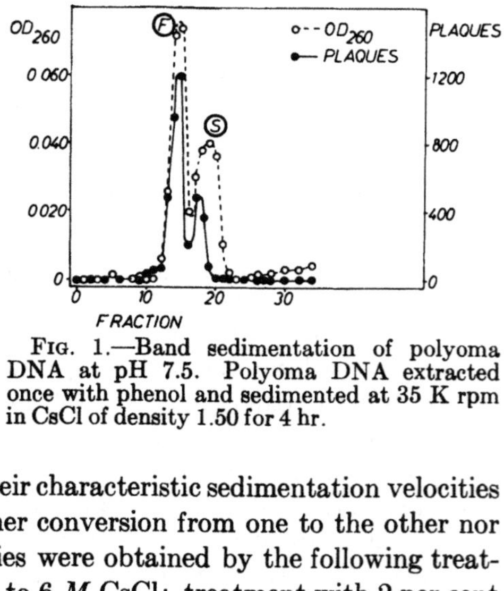

FIG. 1.—Band sedimentation of polyoma DNA at pH 7.5. Polyoma DNA extracted once with phenol and sedimented at 35 K rpm in CsCl of density 1.50 for 4 hr.

Stability: Isolated *F* and *S* components were prepared, either unlabeled or labeled with different labels, by collecting fractions corresponding to the peak part of each band. The *F* and *S* components prepared in this way maintained their characteristic sedimentation velocities after prolonged storage at −70°C. Neither conversion from one to the other nor alterations of their sedimentation properties were obtained by the following treatments: exposure to pHs from 4.5 to 11, or to 6 *M* CsCl; treatment with 2 per cent Na-dodecylsulfate for 30 min at room temperature; annealing separately or together at 70°C for up to 1 hr; chromatography in a column of methylated albumin; treatment with the Lehman enzyme.

Chromatographic properties: P^{32}-labeled isolated *F* and *S* components mixed with unfractionated H^3-labeled PY DNA were chromatographed in the column of methylated albumin (Fig. 2). The *F* component was distributed almost uniformly throughout the whole elution band; the *S* component eluted at higher salt concentration. In turn, the frontal part of the elution band was rich in the fast component, whereas the trailing part was richer in the slow component. The infectivity was distributed almost uniformly throughout the chromatographic band.

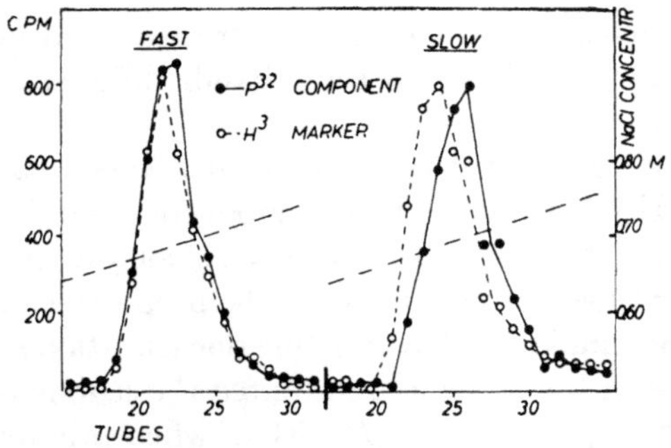

FIG. 2.—Chromatography of F and S components on a column of methylated albumin. 0.5 μg of DNA was used in each run. A column of 1 cm diameter, 11 cm long was used. The DNA was applied in 0.55 M NaCl, 0.05 M phosphate buffer pH 6.9. The elution gradient was linear. Fractions of 2 ml each were collected. To each fraction 0.03 ml of a 4% serum albumin solution and trichloroacetic acid to a final concentration of 5% were added in the cold. After centrifugation the pellets were dissolved in NH_4OH and counted.

These results contradict the hypothesis that the S band is made up of monomer DNA molecules and the F band of end-to-end dimers, as suggested by Watson and Littlefield[3] for the two components of papilloma virus DNA of different sedimentation velocity: the dimers should in fact elute at *higher*, rather than lower, salt concentration. The results are compatible with the ring hypothesis, since the linear molecules may well appear longer to the column than the ring molecules.

Density gradient equilibrium centrifugation: Mixtures of F and S components, labeled with different isotopes, were banded at pHs 7.5, 11, 11.8, and 12.5. At pH 7.5 the two components gave rise to essentially coincident bands at equilibrium (Fig. 3); the band of the S component was wider by a factor of 1.2, perhaps owing to the presence, in this component, of fragments of DNA molecules. The S component reached equilibrium more slowly than the F component. At pH 11 both components had the buoyant density of native PY DNA, although

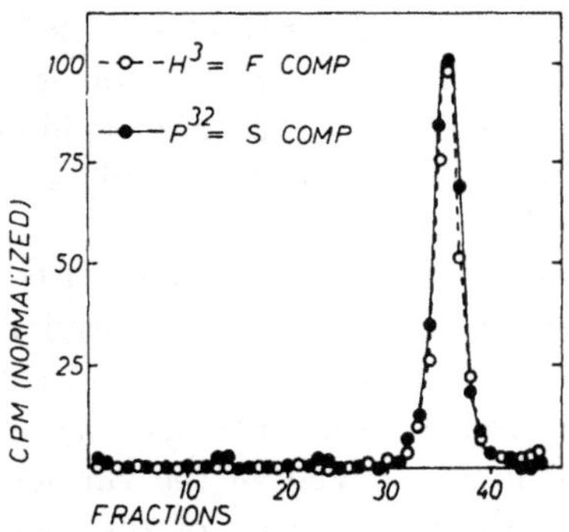

FIG. 3.—Density gradient equilibrium centrifugation of F and S components at pH 7.5. H^3-labeled DNA containing *ca.* 90% F component, and P^{32}-labeled DNA containing *ca.* 80% S component, in amounts of 0.2 μg each, were centrifuged in CsCl of density 1.72 in 0.01 M Tris buffer of pH 7.5 for 80 hr at 29 K rpm.

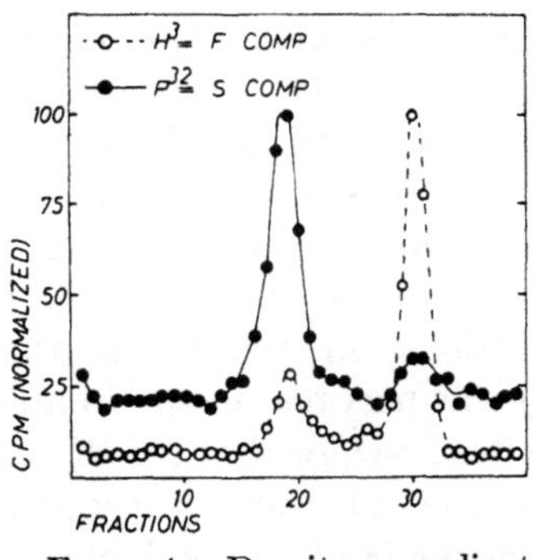

FIG. 4.—Density gradient equilibrium centrifugation of F and S components at pH 11.8. H^3-labeled F and P^{32}-labeled S component in 0.1 M phosphate buffer at pH 11.8 were centrifuged in CsCl solution of density 1.73 in 0.01 M phosphate buffer, pH 11.8, for 80 hr at 30 K rpm.

the band made up of the S component tended to spread toward higher densities, owing to incipient denaturation. At pH 12.5 both components were completely denatured and had a correspondingly higher density;[7] the band of the S component was 1.5 times wider than that of the F component. At pH 11.8 the S component had a denatured density, whereas the F component separated into two bands, with most of the DNA in a band of native density, and the remainder in a band of denatured density (Fig. 4). The proportion of *F* DNA in these bands differed in various preparations.

Band sedimentation at pH 12.5: The *S* component sedimented at a speed not very different from that which it had at pH 7.5, whereas the *F* component gave rise to two bands, one (*SF band*) sedimenting like the *S* component at the same pH, the other (*FF band*) sedimenting about 2.5–2.8 times faster (Fig. 5). The proportion of *F* DNA which appeared in the *FF* band was similar to that resisting denaturation in the equilibrium centrifugation. The DNA of the *FF* band reacquired the sedimentation velocity of *F* DNA when annealed at pH 11.8 in 6 *M* CsCl: when rapidly neutralized, most of the *FF* DNA retained the high sedimentation velocity.

The results obtained at alkaline pH are interpreted as follows: the ring molecules (*F* component) are resistant to denaturation, presumably because they do not have free ends; when denatured, they remain topologically unchanged, i.e., still coiled together, and sediment very rapidly, like incompletely denatured DNA.[11] Upon annealing, they reconstitute the ring molecules characteristic of the *F* component. The linear molecules (*S* component) denature more readily, presumably because they have free ends; upon denaturation they give rise to separate single strands, which sediment at a rate similar to that of the native *S* component. The molecules of the *F* component, which sediment slowly at pH 12.5 (*SF* band) and denature readily, are thought to originate from *F* molecules either with single-chain cuts, or with alkali-labile points, such as pre-existing depurinations. Upon denaturation, such molecules would give rise both to single-stranded rings and single-stranded linear chains, which would move close to each other, like the analogous ϕX forms.[12]

Behavior of the infectivity at alkali pH: Treatment at pH 11.8 or 12.5 abolished the infectivity of the *S* component. The infectivity of the *F* component was not affected by treatment at pH 11.8; after treatment at pH 12.5 and rapid neutralization it was increased about four times. This increase can be probably attributed to increased uptake of partially denatured DNA by the cells.[1]

The two bands produced by the *F* component when sedimented at pH 12.5 were both infectious; in the *SF* band the specific infectivity (ratio: infectivity to radioactivity) was maximal in the frontal and absent in the trailing part of the band.

These results show that the *SF* band is heterogeneous. The heterogeneity is also shown by the different sensitivity to the Lehman enzyme of the various fractions composing the band. The DNA present in the frontal part of the band was resistant to the enzyme, whereas that of the trailing part was partly hydrolyzed. The infectivity was not affected by the enzyme. A possible interpretation of these results is that the infectious DNA present in the frontal part of the *SF* band is made up of single-stranded rings, and the DNA of the trailing part, mostly noninfectious, of single-stranded linear chains. Other interpretations are, however, possible.

DNAase digestion: Samples of the labeled isolated components were treated with the enzyme at constant concentration for various lengths of time. Then the sedimentation pattern of each sample was determined, after mixing it with differentially labeled unfractionated PY DNA as marker. *F component* (Fig. 6): As a function of the time of digestion the height of the *F* band decreased, and a band coincident with the native *S* band was formed. The change affected the height of the *F* band but not its position. No component migrating at a velocity intermediate between

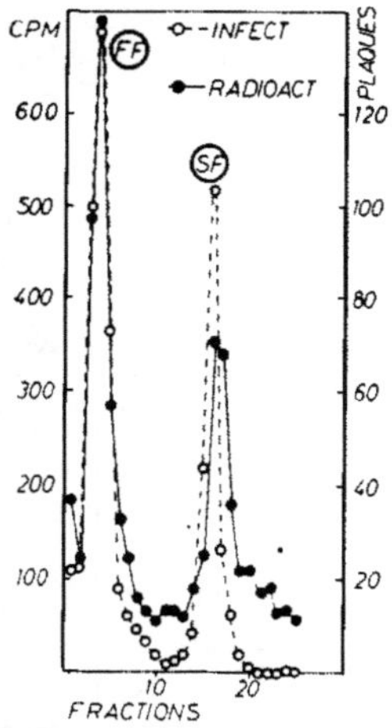

FIG. 5.—Band sedimentation of the F component at pH 12.5. 0.2 μg of H^3-labeled PY DNA, made up almost exclusively of F component, was mixed with 1.5 μg of unlabeled isolated F component, and brought to pH 12.5 with 0.1 M phosphate buffer. The mixture was sedimented at 35 K rpm for two hr in a tube containing CsCl of density 1.50, in 0.01 M phosphate buffer, pH 12.5. The tube had been prespun for 4 hr at 35 K rpm in order to preform the gradient. Both radioactivity and infectivity were determined in all fractions.

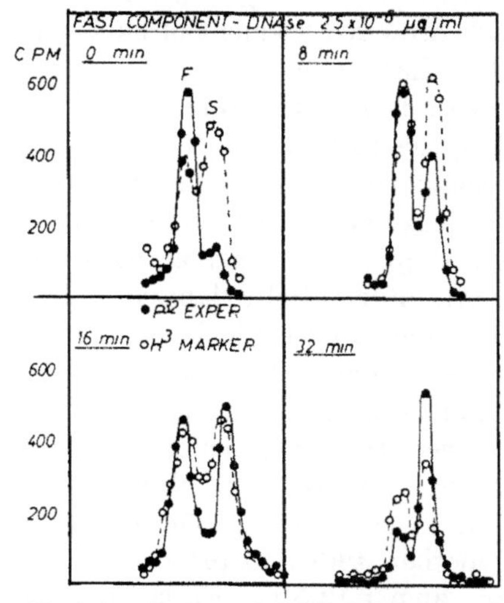

FIG. 6.—Action of DNAase on the F component. Isolated P^{32}-labeled F component DNA (0.5 μg/ml), unfractionated unlabeled PY DNA (4 μg/ml), and DNAase, pre-equilibrated at 37°C, were mixed at time zero and incubated at 37°C for various lengths of time. The incubation medium was: 0.04 M NaCl, 0.01 M Tris buffer, pH 7.5, 0.004 M $MgCl_2$, 5×10^{-5} gr/ml of bovine serum albumin. DNAase action was interrupted by 0.008 M Versene and 0.2 M phosphate buffer of pH 9.0. Each sample was then mixed with unfractionated H^3-labeled PY DNA, and sedimented at 35 K rpm for 4 hr in CsCl of density 1.50.

those of the F and of the S band was formed. In the further study of this phenomenon, a DNAase treatment which converts 37 per cent of the F component into S will be designated as a "converting hit." *S component* (Fig. 7): As a function of the time of digestion the S band at first increased in width, and showed subsequently a progressively lower velocity of sedimentation without generating a new separate band.

The results obtained with the F component can be interpreted as those obtained by Fiers and Sinsheimer[13] with the DNA of phage ϕX 174: that a single scission in the ring molecule converts it to the linear form, thus causing a discrete difference in sedimentation velocity.

Kinetics of DNAase digestion: In repeated experiments the kinetics of disappearance of the F band as a function of time, at constant DNAase concentration, was found of first order. This result suggested at first that the ring contains an inserted single-stranded segment. To search for such a segment, labeled F DNA was exposed to several converting hits with DNAase and then to the Lehman enzyme. No measurable hydrolysis was observed under conditions where 1 per cent hydrolysis could have been detected. We conclude that the DNA of the *F* component does not have a single-stranded segment of sufficient size to explain the kinetics of DNAase action.

In order to interpret the kinetics of *F* → *S* conversion, the kinetics of DNAase action on either component was studied by sedimenting the enzyme-treated DNA both at pH 7.5 and at pH 12.5, together with untreated DNA, differentially labeled. The changes in sedimentation velocity at pH 7.5 reflect the consequences of *complete scissions* of the DNA molecules; the changes observed at pH 12.5 reflect the consequences of *single-chain scissions*. The following observations were made. *F component:* The effect of DNAase, whether studied at pH 7.5 or 12.5, was similar: at pH 12.5 the *FF* component was converted into the *SF* component at exactly the same rate at which the *F* component was converted into the *S* component at pH 7.5. This result shows that DNAase did not produce single-chain cuts in the ring molecules, which would have resulted in faster disappearance of the *FF* component at pH 12.5, compared to the *F* component at pH 7.5. Thus, the enzyme broke both chains at once. *S component* (either native or produced from the action of DNAase on the *F* component): At either pH the band formed was modified in the same qualitative way by DNAase; the band became wider, and then its average velocity of sedimentation progressively decreased, as digestion proceeded. The effect was, however, much more pronounced at pH 12.5.

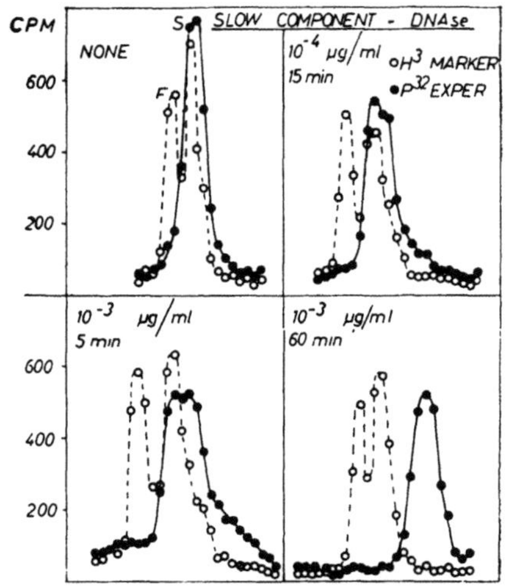

FIG. 7.—DNAase action on the *S* component. Same conditions as in Fig. 6, except that isolated *S* component was used.

To compare the effects observed at the two pHs, the number of scissions was evaluated from the sedimentation behavior of the DNA molecules. When the number of scissions was low, it was evaluated by determining the proportion of material remaining in a position coincident with the *S* band (pH 7.5) or with the *SF* band (pH 12.5). When the number of scissions was higher, it was evaluated from the change in average velocity of sedimentation,[14] with the assumption that the sedimentation velocity is proportional to $MW^{0.37}$ at pH 7.5, and to $MW^{0.5}$ at pH 12.5.[15] In repeated experiments it was found that single-chain cuts, demonstrable at pH 12.5, were produced at a much faster rate than complete scissions, demonstrable at pH 7.5; and furthermore, that complete scissions occurred according to a kinetics which appeared to be of higher order (Fig. 8).

These results show that there is a profound difference in the effect of DNAase on the *F* and on the *S* component.

The interpretation of these results is that the ring molecules are labile for reasons related to their shape. Perhaps the ring form imposes a statistically higher stress on the phosphodiester bonds; hydrolysis of one bond may increase considerably the stress on the complementary bond, making it highly susceptible to hydrolysis by the same enzyme molecule, or to breakage. As soon as the ring opens, the stress disappears.

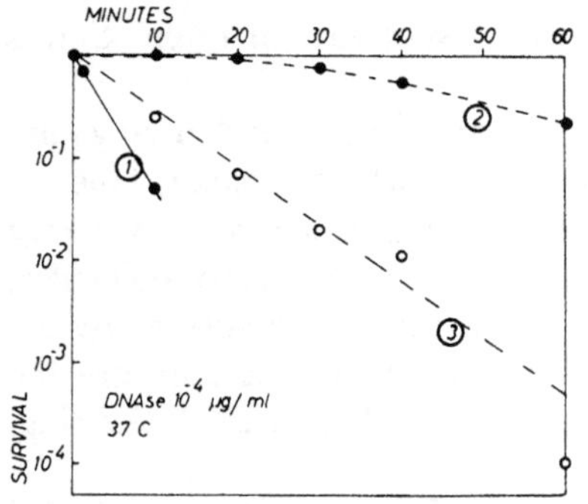

FIG. 8.—Kinetics of DNAase digestion on H^3-labeled PY DNA containing almost exclusively *F* component. All rates were determined within a single experiment, by using a single sample which contained DNA, 15 µg/ml and DNAase, 10^{-4} µg/ml. Survival = proportion of the respective form that remained unchanged. *Curve 1:* kinetics of disappearance of *F* (pH of sedimentation 7.5) or *FF* (pH of sedimentation 12.5). *Curve 2:* kinetics of degradation of *S* (derived from *F*); pH of sedimentation 7.5. *Curve 3:* kinetics of degradation of *SF;* pH of sedimentation 12.5. In Curves 2 and 3 low survivals were calculated from the average number of scissions per molecule, according to the equation: survival = exp (– average number of scissions).

Effect of DNAase on infectivity: These experiments yielded two types of results. The first result is that the infectivity of the DNA of the *F* band decreased as a function of the time of exposure to DNAase at exactly the same rate as the radioactivity. This shows that the infectious molecules have the same properties as the bulk of the DNA molecules. Experiments carried out with pure *F* component showed that not more than 10 per cent of the infectivity lost by the *F* component appeared in the *S* band. Thus, *F* → *S* conversion by DNAase entails loss of infectivity in most, if not all, molecules.

The other result is that the infectivity of the *S* component was more resistant to the action of DNAase than that of the *F* component. The number of hits in the *S* component determined from infectivity loss corresponded to the average number of complete scissions per molecule, as determined from the sedimentation behavior. With increasing exposure to DNAase, as the radioactivity band increased in width, the residual infectivity remained associated with molecules which sedimented like those of the frontal part of the original *S* band.

These results show that in both the *F* and the *S* components loss of infectivity was caused by complete scissions, whereas single-chain breaks were inconsequential. Single-chain cuts had, however, an effect on infectivity, if DNA which had experienced a few DNAase hits was heated at 100°C for 5 min in 0.15 *M* NaCl. This DNA lost all its residual infectivity, in contrast to DNA not exposed to DNAase which increased its infective titer two or three times after heating.[1]

Conclusions.—The results here reported are in agreement with the hypothesis that the DNA of polyoma virus is a ring molecule. Many other possible structures which may account for the existence of two components of different sedimentation velocity are contradicted by the results.

According to the results, the ring should be made up of two complete complementary strands, each one separately closed on itself, plectonemically coiled, and not cross-linked. It is likely that the ends of each polynucleotide chain are connected by a special "linker," as in φX DNA.[13] This is suggested by the presence of infectivity in the native linear molecules, in contrast to its absence in at least most of the linear molecules obtained by DNAase treatment. These results also show that infectivity requires the information of the whole DNA molecule and the integrity of its functional units.

Single-chain breaks, as produced by DNAase in infectious *S* molecules, do not abolish infectivity. Thus, either these breaks are repaired after the DNA enters the cell, or the reading of the DNA molecules for RNA synthesis or replication is undisturbed by the breaks.

The behavior of infectivity at alkali pH suggests that single-stranded rings are infectious; only further experiments may, however, decide whether this is true.

Two results remain to be explained: (1) the difference in sedimentation velocity of the ring and of the linear molecules is larger than for other ring-shaped molecules, such as those of ϕX DNA[12] or λ DNA;[16] (2) the ring is unstable, as suggested by the kinetics of DNAase digestion. It can be tentatively suggested that they are the consequence of the combination of double strandedness and of the size of the ring.

We are grateful to Prof. R. L. Sinsheimer for useful discussions. The competent and dedicated assistance of Miss Maureen Muir is gratefully acknowledged.

This work will be supplemented by studies with the analytical ultracentrifuge.[17]

Aided by grants from the National Institutes of Health, U.S. Public Health Service, the American Cancer Society, and The National Foundation.

[1] Weil, R., these PROCEEDINGS, **49**, 480 (1963).
[2] Crawford, L. V., *Virology*, **19**, 279 (1963).
[3] Watson, J. D., and J. W. Littlefield, *J. Mol. Biol.*, **2**, 161 (1960).
[4] Winocour, E., *Virology*, **19**, 158 (1963).
[5] Weil, R., *Virology*, **14**, 46 (1961).
[6] Vinograd, J., R. Bruner, R. Kent, and J. Weigle, these PROCEEDINGS, **49**, 902 (1963).
[7] Vinograd, J., J. Morris, N. Davidson, and W. F. Dove, Jr., these PROCEEDINGS, **49**, 12 (1963).
[8] Mandell, J. D., and A. D. Hershey, *Anal. Biochem.*, **1**, 66 (1960).
[9] Lehman, I. R., *J. Biol. Chem.* **235**, 1479 (1960).
[10] Fiers, W., and R. L. Sinsheimer, *J. Mol. Biol.*, **5**, 408 (1962).
[11] Freifelder, D., and P. F. Davidson, *Biophys. J.*, **3**, 49 (1963).
[12] Sinsheimer, R. L., *J. Mol. Biol.*, **1**, 43 (1959).
[13] Fiers, W., and R. L. Sinsheimer, *J. Mol. Biol.*, **5**, 424 (1962).
[14] Thomas, C. A., *J. Am. Chem. Soc.*, **78**, 1861 (1956).
[15] Doty, P., J. Marmur, J. Eigner, and C. Schildkraut, these PROCEEDINGS, **46**, 461 (1960).
[16] Hershey, A. D., E. Burgi, and L. Ingraham, these PROCEEDINGS, **49**, 748 (1963).
[17] Weil, R., and J. Vinograd, in preparation.

THE CYCLIC HELIX AND CYCLIC COIL FORMS OF POLYOMA VIRAL DNA

Roger Weil and Jerome Vinograd

The DNA extracted from polyoma virus exhibits certain properties which have not been reported for other viral base-paired DNA's.[1] The DNA renatures monomolecularly. The loss of helical configuration does not impair biological activity. Heating at 100° for 10–20 min followed by rapid cooling does not reduce the infective titer. Dulbecco has given evidence suggesting that a fraction of the polyoma DNA molecules are cyclic.[2] This form would in part account for the above properties.

The results reported in this communication give strong evidence for cyclic polyoma DNA molecules; they also define certain properties of the cyclic and open forms. Some of the evidence reported here is similar to that obtained in an independent investigation by Dulbecco and Vogt.[3]

Materials and Methods.—Isolation and purification of the virus and extraction of the DNA: The virus was isolated and purified by the procedure described by Winocour.[4] The main visible band from the CsCl density gradient was collected, and the DNA extracted by a modified phenol procedure.[1] The homogenization step was omitted. The CsCl, Harshaw Chemical Company, Optical Grade, formed solutions which remained clear upon addition of alkaline buffers. Reagent grade inorganic chemicals and Merck's reagent grade formaldehyde were used.

Analytical ultracentrifugation: Sedimentation velocity analyses were performed in the Spinco analytical ultracentrifuge by band-centrifugation.[5] Boundary-velocity experiments were precluded by the limited supply of the virus. Buoyant density experiments were performed at 44,770 rpm at 25°C and analyzed by methods previously described.[6]

Preparation of the alkaline solutions: The alkaline CsCl solutions were prepared with the buffers described by Vinograd *et al.*[7] The alkaline NaCl solutions were prepared by mixing 1 volume of 4 M NaCl with 3 volumes of 0.11 M KOH.[7a] The pH measured with a small general-purpose Beckman glass electrode was 12.2 ± 0.1. DNA samples were made alkaline in the sample hole by addition of 1 volume of 1.1 M KOH[7a] to 9 volumes of DNA solution, and mixed by drawing the solutions back into the Kel-F tubing several times. The final pH was 12.5 ± 0.2. Pre-incubation for longer periods was performed in 0.032-in. I.D. Kel-F tubing sealed at both ends with 4-mm plugs of silicone grease.

Denaturation and renaturation: These experiments were performed as previously described.[1] Formaldehyde denaturations[8] were performed with 25 μl samples in 11% CH_2O, pH 8.3, ionic strength 0.01, in sealed glass melting-point tubes at 70–75°C 2–3 min followed by chilling at 0°C.

Sonication: The DNA solutions in thin 1-ml nitrocellulose test tubes were sonicated in a 9,000 cps Raytheon Sonicator Model no. S102A at the indicated power level.

Calculations: The molecular weight of anhydrous NaDNA was calculated from the standard deviation of the band.[6] The additional cesium ions and preferential hydration[7] introduced at high pH were included in the calculations. Buoyant densities were measured with *M. lysodeikticus* DNA, $\rho_0{}^0 = 1.732$ gm cm^{-3}, as a reference, a value based on $\rho_0{}^0 = 1.710$ gm cm^{-3} for DNA from *E. coli.*[9] This value disagrees with a recent redetermination by +0.006 gm cm^{-3}.[7]

Results.—Sedimentation analysis of DNA extracted from the virus: Band-centrifugation analyses of the DNA from six different virus batches all showed three components which we refer to as I, II, and III, in order of decreasing sedimentation coefficients. The values, $s^{\circ}_{20,w}$, in 1 M NaCl, 0.01 M Na_2HPO_4, pH 8.0 were

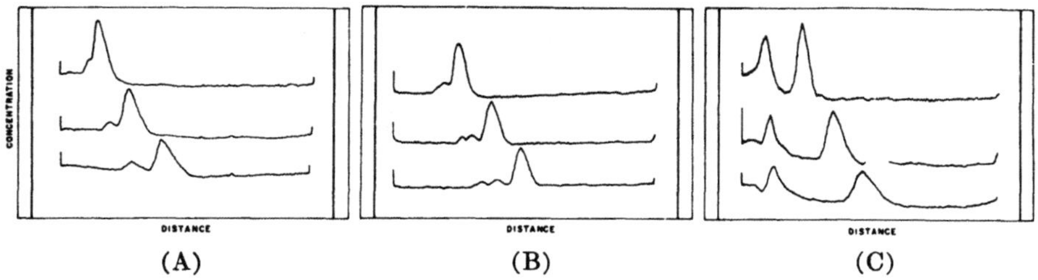

FIG. 1.—Band sedimentation velocity analyses of polyoma DNA in (*A*) neutral NaCl, (*B*) neutral CsCl, (*C*) alkaline NaCl. (*A*) Lamella: 15 μl, 24 μg/ml DNA; bulk solution: 1 *M* NaCl, 0.01 *M* Na_2HPO_4, pH 8.0. Densitometer records are from photographs taken at 16-min intervals. 35,600 rpm, 20°. (*B*) Lamella: 20 μl, 60 μg/ml DNA; bulk solution: CsCl, $\rho = 1.36$, 0.01 *M* Tris, pH 8.1. 16-min intervals, 35,600 rpm, 20°. (*C*) Lamella: 25 μl, 24 μg/ml DNA, pH = 12.75; bulk solution: 1.0 *M* NaCl, 0.10 *M* KOH, pH 12.2, 8-min intervals. 35,600 rpm.

20.3 S ± 0.4 S.D., 15.8 S ± 0.4 S.D., and 14.4 S[10] (Fig. 1*A*). The component composition was determined in 3.5 *M* CsCl (Fig. 1*B*). In this solvent the band profiles are more symmetrical than in 1 *M* NaCl or 1 *M* CsCl. In all preparations I comprised 80–90 per cent of the DNA. Component II varied between 1 and 20 per cent and III between 1 and 10 per cent. DNA released from the virus in the sample hole, pH 11.2 or pH 12.3, gave patterns indistinguishable from those obtained with the phenol extract. The sedimentation pattern at neutral pH did not change upon storage at 4° for periods up to 20 weeks, or upon freezing and thawing. The DNA formed a single band at equilibrium in CsCl with a buoyant density of 1.709 gm cm^{-3}. All preparations formed bands which were skewed on the light side. It is shown later that this skewness is due to III.

The renaturation behavior of polyoma DNA suggested that the strands did not physically separate when heated under conditions in which strand separation normally occurs. To examine this problem we employed the procedure described by Freifelder and Davison.[8] Solutions of polyoma DNA were denatured in CH_2O and examined in sedimentation velocity experiments in 11 per cent CH_2O, 1 *M* NaCl. The DNA now sedimented in two components (80 per cent fast) with a ratio of sedimentation coefficients of 2.2. This result was not compatible with the hypothesis that the fast material represented singly cross-linked double strands, and the slow material single strands[8] because a ratio of $\sqrt{2}$ is to be expected. Strand separation was also examined at high pH, 12.2 ± 0.1, in 1 *M* NaCl. In this solvent, complete strand separation is to be expected.[7] All fresh polyoma DNA preparations, including those stored at −70°C, sedimented in two well-resolved components[11] (about three-fourths fast) with sedimentation coefficients, $\overset{\circ}{s}_{20,w}$ of 53.1 S ± 0.3 S.D. and 14.6 S ± 0.5 S.D. (Fig. 1*C*). We will refer to the alkali components as the 53 S and 15 S DNA's and will show below that these are double-stranded cyclic coils and single-stranded random coils, respectively. That both of these components can derive from I was shown in experiments in which the DNA was so dilute that II and III could not be detected at pH 8; in alkali, however, both the 53 S and 15 S components were observed.

The sedimentation velocity of the 15 S DNA was compared with the sedimentation velocity of φX-174 DNA (16.0 S) in the same solvent, and its molecular weight calculated with the relation for random coils, $S_1/S_2 = (M_1/M_2)^{0.5}$. With

M = 1.7 × 10^6 for ϕX-174 DNA[12] we obtain M = 1.4 × 10^6 for the 15 S component. The same value is obtained if we use single-stranded T-7 DNA as a reference in this calculation, $s^{\circ}_{20,w}$ = 39 S, M = 9.5 × 10^6.[13]

Fresh polyoma DNA forms two well-separated bands in buoyant density experiments in alkaline CsCl pH 12.3. In *band-buoyancy* experiments (band-centrifugation performed with buoyant bulk solutions[5]) (Fig. 2), the 53 S DNA forms a band at a higher density than the 15 S DNA. The buoyant densities are 1.784 and 1.766 gm cm^{-3}, referred to 1.710 for *E. coli* DNA. The molecular weight calculated from several band profiles for the 53 S DNA was 2.4 ± 0.3 × 10^6. The band formed by the 15 S DNA was skewed on the light side because of slowly sedimenting DNA and a small amount of III.

With a ratio of molecular weights $M_{53\,S}/M_{15\,S} \approx 2$, as indicated in the arguments in the above two paragraphs, we expect for random coils a ratio of sedimentation coefficients of 1.4. The high velocity, 53 S, in alkali of the double-stranded DNA has thus to be attributed to a drop in the frictional coefficient by a factor of 2.5 relative to a random coil containing two strands (Fig. 3). We therefore propose that the double-stranded cyclic helix denatures to form a double-stranded cyclic coil in which the turns originally present in the helix are conserved. The topological constraint imposes a compact structure on this molecule (Fig. 3) and gives rise to an unusually high sedimentation coefficient and buoyant density.

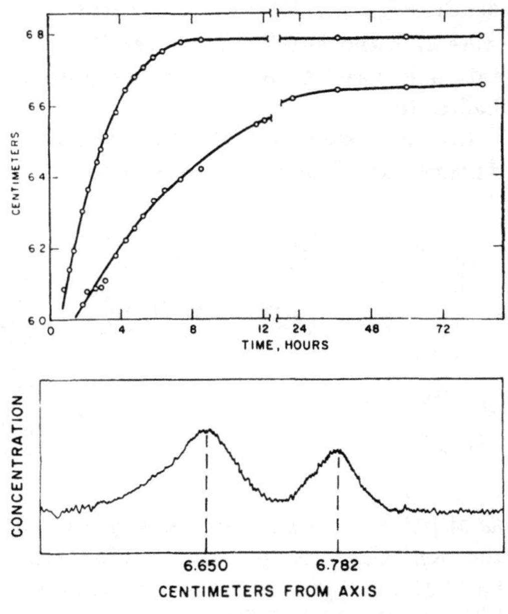

FIG. 2.—Band-buoyancy analysis of polyoma DNA in alkaline CsCl. Lamella: 20 μl, 24 μg/ml DNA, pH 12.75. Bulk solution: CsCl, ρ = 1.76, 0.04 *M* K_3PO_4, pH = 12.35. 44,770 rpm. *Upper:* position of 15 S and 53 S components at various times. *Lower:* densitometer records of bands at 72 hr, 25°.

Several physical and chemical agents can convert the cyclic helix, the linear helix and the double-stranded cyclic coil to single strands, as well as the cyclic helix to the double-stranded cyclic coil and to the linear helix.

(1) *Exposure to high pH:* After the initial rapid conversion of about one fourth of component I into single strands, a further conversion into single-stranded 15 S DNA occurs at a slow rate. For example, after 72 hr incubation at pH 12.1 in 0.01 *M* NaCl, the amount of 53 S DNA decreased to about one third with a concomitant increase in 15 S DNA. Unlike the usual 15 S DNA, this material contained two velocity components, a small fraction, presumably cyclic single strands, sedimenting about 10 per cent faster than the main component.[11] The patterns were similar to those observed in experiments with ϕX-174 DNA, which contained both linear and circular molecules.[14]

(2) *Exposure to high CsCl concentrations:* The sedimentation velocity patterns of polyoma DNA in dilute CsCl solutions are skewed because of the effects of concentration-dependent sedimentation. At a density of 1.35 gm cm^{-3} the bands become symmetrical, presumably because of a reduction in the equivalent hydrodynamic volume.

At still higher concentrations of CsCl, $\rho = 1.5$ and $\rho = 1.6$, the relative amounts of the components changed in two of the three preparations studied. Component I became smaller and II larger. That this conversion is not rapidly reversible was shown in experiments in which polyoma DNA was incubated at 23°C in CsCl, $\rho = 1.72$, 0.006 *M* Na citrate pH 7.0 for 1, 24, and 44 hr, and subsequently analyzed, after dilution, in CsCl, $\rho = 1.35$ gm cm^{-3}. Analyses in neutral CsCl showed that irreversible conversion of I in all three DNA preparations had occurred to the extent of 50, 75, and 90 per cent, respectively. In one of the preparations of DNA it was confirmed that the conversion noted after 44 hr by analysis in neutral CsCl represented a ring cleavage. This was shown by a change in the relative amounts of fast and slow forms in alkaline CsCl, $\rho = 1.36$, 0.04 *M* K_3PO_4, pH = 12.1. The ratio of amounts changed from 0.6 to 0.1. The corrected sedimentation coefficients, $s^0_{20,w}\ \eta_{rel}$, for the double-stranded cyclic coils and the single-stranded coils were 37 S and 11.6 S. These results indicate that preparative equilibrium sedimentation in CsCl at room temperature can lead to a substantial loss of the cyclic helical form.

(3) *Exposure to high CsCl concentration at high pH:* A series of buoyant density experiments were performed to examine the effects of pH. At pH 11,

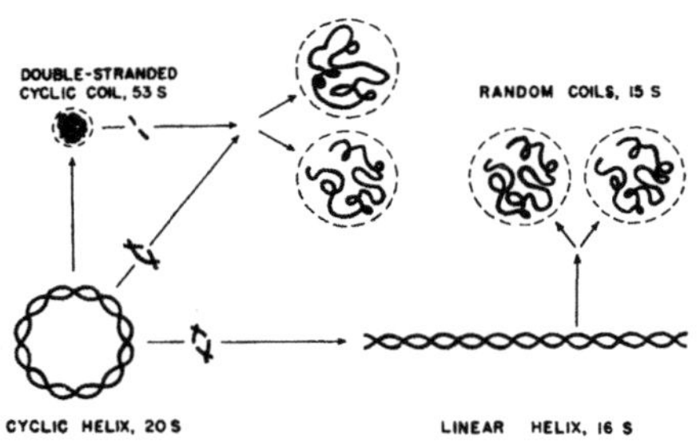

FIG. 3.—Diagrammatic representation of the two forms of polyoma DNA. The double-stranded cyclic helix denatures to form a double-stranded cyclic coil. The linear helix denatures to form two single-stranded random coils. The dashed circles indicate the relative equivalent hydrodynamic diameters. Possible links between chains or special bonds connecting chain ends are not considered. The sedimentation coefficients of the helices and coils were measured in neutral and alkaline NaCl solutions, respectively.

as at pH 8.1, the buoyant density was 1.709 gm cm^{-3}, and the band was skewed at the light side. At pH 11.8 a substantial fraction of the DNA titrated to form a band at 1.766 gm cm^{-3} that was skewed at the light side while the remainder, the intact cyclic helical DNA, formed a symmetrical band at 1.714 gm cm^{-3}.[11] At pH 12.2 the latter material denatured and titrated to form a band of double-stranded cyclic coils at 1.784 gm cm^{-3}. The results show that the cyclic helix is more stable than the linear helix to alkali denaturation. Both at pH 11.8 and 12.2 the relative amounts of the double-stranded cyclic DNA were smaller than the relative amount of 53 S DNA initially present in the preparations. A portion of the double-stranded cyclic DNA was therefore converted to single strands.

(4) *Sonication:* Sonication at 75 per cent full power for 1.5 and 5 min gave the same results. The relative amount and the sedimentation coefficient of I were

essentially unchanged in neutral 1 *M* NaCl (Fig. 4). Components II and III formed a relatively broad band with decreased sedimentation coefficient, 12.2 S. At full power, sonication for 2 and 4 min resulted in the conversion of 50 and 80 per cent, respectively, of I into a slower material, 8.3 S. In previous studies with native DNA's, the resistance to shearing increases with decreasing sedimentation coefficient.[15] We have here the reverse effect, presumably because the extended length of the cyclic form is less than that of the linear form.

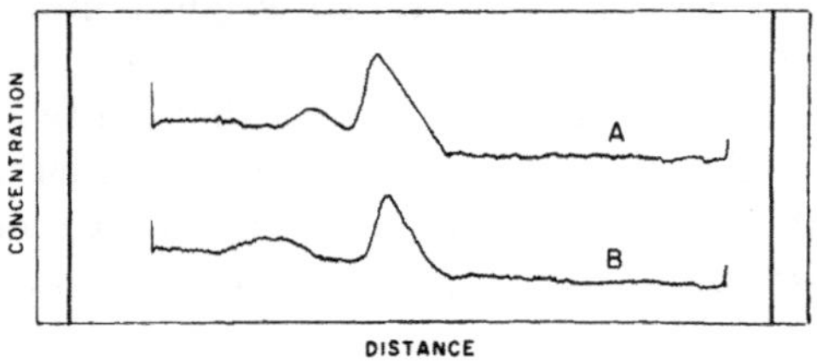

FIG. 4.—Effect of sonication of polyoma DNA. (*A*) untreated; (*B*) sonicated 1.5 min. Lamella: 15 μl, 25 μg/ml DNA. Bulk solution: 1.0 *M* NaCl, 0.01 *M* $NaHPO_4$, pH 8.0. Photographed 64 min after reaching full speed. 35,600 rpm, 20°.

(5) *Effects of heating at 100°C:* Component III is selectively melted out and becomes polydisperse in neutral sedimentation velocity experiments performed after heating solutions of polyoma DNA in 0.4 *M* NaCl 1–2 min at 100° followed by rapid cooling. The skewness disappears from the main band in buoyant density experiments at neutral pH. Component III forms a band of denatured DNA with a buoyant density 1.721 gm cm^{-3}, 0.003 gm cm^{-3} less than the buoyant density of denatured polyoma DNA. In these experiments the relative amounts and the sedimentation velocities of I and II were unchanged. Longer heating leads to gradual conversion of I into a 16 S DNA.

When fresh polyoma DNA at a concentration of 1–2 μg/ml was heated at 100° 1–2 min in buffers of low ionic strength followed by rapid cooling, I renatured nearly completely, while an amount corresponding to II and III remained denatured as judged by buoyant density. This result indicates that intact cyclic coils renature readily. Upon further heating, increasing amounts of denatured DNA were formed. Heated materials analyzed in alkali contained decreasing amounts of 53 S DNA and increasing amounts of 15 S DNA. Depurination which occurs on heating and leads to chain scission in alkali[16–18] is in part responsible for the conversion of double-stranded cyclic coils to random coils.

Electron microscopy: The striking and important result is that cyclic molecules are observed (see *Appendix*).

Discussion.—The major fraction of polyoma DNA denatures in alkali and remains double-stranded. A smaller fraction denatures in alkali to form single strands with about half the molecular weight of the double-stranded form. Comparison of the sedimentation coefficients in alkali of these two denatured DNA's with each other and with the sedimentation coefficients in alkali of other DNA single strands of known molecular weight leads to the conclusion that the denatured double-stranded polyoma DNA has an abnormally low frictional coefficient. A simple model for the native DNA which can account for the properties mentioned is a cyclic helix with separately continuous strands (Fig. 3).

The requirement that the approximately 500 turns originally present in the helical structure be conserved as long as the strands remain continuous leads necessarily to a configuration for the denatured molecule which is more compact than the random-coil configuration. Any spatial rearrangement of the phosphodiester chains upon denaturation gives rise to supercoiling. This new type of DNA coil, which we refer to as the *double-stranded cyclic coil*, has an abnormally high sedimentation coefficient and a higher buoyant density in alkaline cesium chloride than the random coil.

The introduction of one discontinuity in one of the strands in the native or denatured double-stranded molecule results in the formation in alkali of single strands, one linear and one cyclic. The finding that the molecular weight is always first reduced by a factor of 2 upon degradation shows that the strands in polyoma DNA are separately continuous. None of the foregoing excludes the additional presence of a cross-link labile in alkali. Further evidence for a cyclic structure for component I is the relatively high resistance to shearing by sonication, and the presence in the electron micrographs of about 90 per cent cyclic molecules.

We now examine the evidence that II is an open or linear form of the cyclic helix I. The buoyant density of II is essentially the same as I. Upon thermal denaturation of II, a buoyant density shift characteristic for double-stranded DNA is observed. The alkaline buoyant density shift is also normal for a linear helix. We conclude therefore that II is a linear helix. I is converted into 16 S helical DNA by the action of concentrated CsCl or heating at 100° in 0.4 *M* NaCl. The sedimentation coefficient of II is slightly lower than I, which is reasonable for the proposed relation between I and II. That II has the same molecular weight as I is indicated by the formation from II in alkali of 15 S single strands of one half the molecular weight of I. These arguments all suggest that II is an open or linear

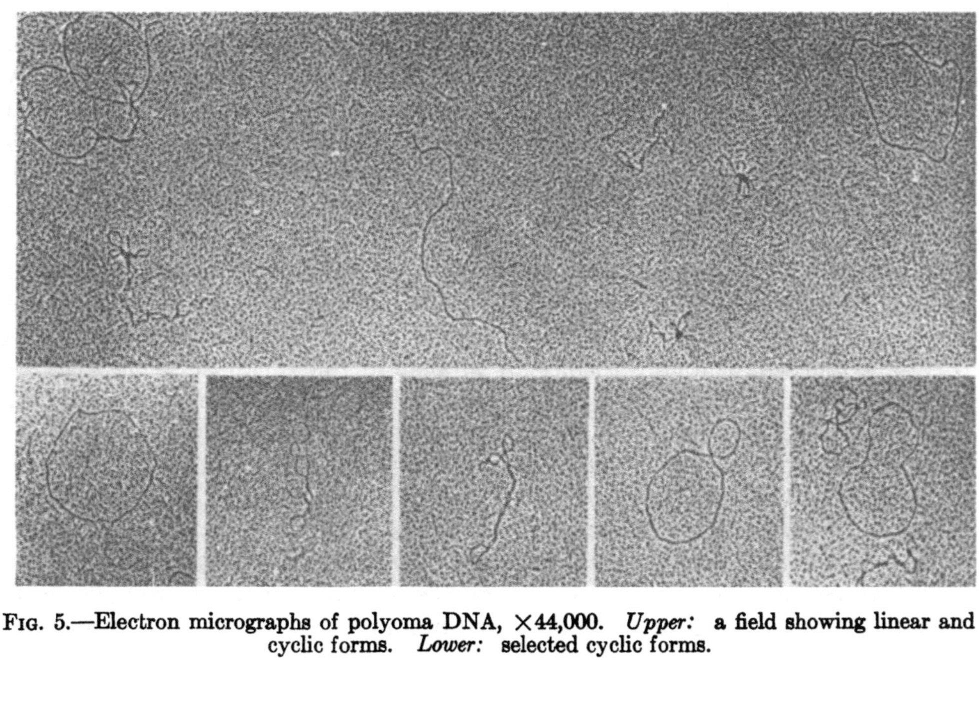

FIG. 5.—Electron micrographs of polyoma DNA, ×44,000. *Upper:* a field showing linear and cyclic forms. *Lower:* selected cyclic forms.

form of the cyclic helix I. Consistent with the above conclusion is the greater sensitivity of II to sonication, and the appearance in the electron micrographs of linear and cyclic molecules with the same mean length.

The origin of III, which was present in small amounts in all preparations, is not known. Its homogeneity with respect to sedimentation velocity and buoyant behavior rules out an adventitious contamination by host cell DNA. Component III was not found by Dulbecco and Vogt[3] in radioactively labeled polyoma DNA preparations, or by Crawford (cf. ref. 10) with boundary-sedimentation velocity experiments. Either the methods used by these investigators were not sufficiently sensitive or III represents a well-defined host cell DNA molecule which was synthesized prior to viral infection and thus remained unlabeled.

The thermal stability[1] of the infective titer of polyoma DNA upon heating at 100° can now be interpreted. The initial threefold increase in infective titer is the result of a more efficient attachment of the DNA to the mouse cells in the assay for infectivity.[1] The essentially unchanged titer during the next 10–20 min is attributed to a steady state between a thermal activation and an inactivation process. In the activation process the cyclic double-stranded molecule is opened to produce a molecule of higher infectivity. If a cleavage occurs at a phosphodiester bond, a pair of linear and cyclic single strands is formed. These separate unless held together by a cross link which we have not ruled out. If a cleavage at the site of a linker connecting four chain ends occurs, either single- or double-stranded linear molecules are formed. When all cyclic double-stranded molecules have been opened, the reservoir of potential infectivity is exhausted. Further heating leads to an exponential decrease in the infective titer.

The opening of the ring at high pH or as a result of heating may proceed by well-known reactions such as hydrolysis or depurination followed by hydrolysis; and these may occur at any site along the chains. The slow cleavage of the ring in concentrated neutral CsCl solutions is a surprising result which suggests that polyoma DNA contains a labile site in the chain not present in other biologically active DNA's which have been routinely studied in and recovered from equilibrium density gradient experiments in concentrated CsCl solutions.

APPENDIX BY WALTHER STOECKENIUS

Three different configurations of polyoma viral DNA are visible in the electron micrographs (Fig. 5): coiled molecules without visible free ends, fully extended cyclic molecules, and linear molecules. If a coiled form contained not more than two twists, its length could be accurately measured. In a given typical area, 83 coiled, 8 extended, and 9 linear molecules were observed. The mean lengths of 66 cyclic and 68 linear molecules were 1.58 $\mu \pm 0.16$ S.D. and 1.53 $\mu \pm 0.16$ S.D., respectively. These lengths correspond to a molecular weight of 3.0×10^6, if we assume that the mass per unit length of DNA molecules in the protein film is the same as in the B-form in fibers of NaDNA.[19–21]

Electron microscopy thus confirms that polyoma DNA molecules exist in linear and cyclic forms of equal molecular weight. The proportion of I and II by electron microscopy corresponds to that found by sedimentation analysis, as described below.

A striking feature of the electron micrographs is the high proportion of tightly coiled molecules. In otherwise identical preparations of T-7 DNA, linear molecules and a very few tightly coiled molecules were found. The latter form is probably due mainly to intramolecular cross-linking by protein.[21] The absence of free ends in the tightly coiled polyoma DNA molecules indicates that they are cyclic; the high proportion of such forms suggests that cyclic molecules are more readily cross-linked by protein. Component III could not be identified in the electron micrographs.

A sample containing 2 μg/ml polyoma DNA in 0.006 *M* KCl, 0.01 *M* Tris, pH 7.5 was shipped in dry ice, thawed, and frozen three times during these studies, and the remainder returned in dry ice. Re-examination (by R. W. and J. V.) at neutral pH in 1.0 *M* NaCl and 3.5 *M* CsCl showed an unchanged composition, 85% I and about equal amounts of II and III. In alkaline NaCl only slow polydisperse strands were found.

Grids were prepared by the method of Kleinschmidt and Zahn,[22] except that diisopropylphosphoryl trypsin (Worthington Biochemical Corp., no. TRL-DIP104) was used instead of cytochrome C and the solution was delivered to the surface via a glass rod.[23] One ml of solution (0.24 μg DNA, 100 μg DIP-trypsin in 1.0 *M* NH_4 acetate, pH 8.0) was delivered onto 0.015 *M* NH_4 acetate, pH 8.0. The resulting surface film was compressed to 1.0 dyne/cm at about 450 cm^2. Randomly selected portions of the surface film were picked up on platinum grids and shadowed with uranium at an angle of 7° while the specimen was rotating.

Summary.—The results obtained in this study show that polyoma viral DNA contains linear and cyclic helical molecules. The double-stranded cyclic helix denatures in alkaline solutions to form a double-stranded cyclic coil, a new type of coiled molecule in which all of the turns originally present in the helix are conserved.

It is a pleasure to thank R. W. Watson for expert technical assistance, R. D. Bruner for helpful discussions during the preparation of the manuscript, R. Dulbecco for providing the polyoma virus, R. L. Sinsheimer and P. F. Davison for gifts of materials, and R. Dulbecco and M. Vogt for a copy of their unpublished manuscript. This work was supported in part by grants HE 03394 and GM 08812 from the U.S. Public Health Service.

[1] Weil, R., these PROCEEDINGS, **49**, 480 (1963).

[2] Dulbecco, R., in *Monograph of the 17th Annual Symposium on Fundamental Cancer Research*, M. D. Anderson Hospital, Houston, Texas, Feb. 20, 1963 (Baltimore: Williams and Wilkins), in press.

[3] Dulbecco, R., and M. Vogt, these PROCEEDINGS, **50**, 236 (1963).

[4] Winocour, E., *Virology*, **19**, 158 (1963).

[5] Vinograd, J., R. Bruner, R. Kent, and J. Weigle, these PROCEEDINGS, **49**, 902 (1963).

[6] Vinograd, J., and J. Hearst, *Prog. in Chem. Org. Nat. Prods.*, **20**, 372 (1962).

[7] Vinograd, J., J. Morris, N. Davidson, and W. F. Dove, Jr., these PROCEEDINGS, **49**, 12 (1963).

[7a] These solutions were not free of carbon dioxide.

[8] Freifelder, D., and P. F. Davison, *Biophys. J.*, **3**, 49 (1963).

[9] Schildkraut, C. L., J. Marmur, and P. Doty, *J. Mol. Biol.*, **4**, 430 (1962).

[10] This value was obtained from the relative values of sedimentation coefficients of I and III in CsCl, $\rho = 1.35$, and the value for $s^{\circ}_{20,w}$ for I in 1.0 *M* NaCl. L. V. Crawford in *Virology*, **19**, 279 (1963), obtained 21 S and 14 S for two components in polyoma DNA.

[11] Similar results were obtained independently by Dulbecco and Vogt.[3]

[12] Sinsheimer, R. L., *J. Mol. Biol.*, **1**, 43 (1959).

[13] Davison, P. F., and D. Freifelder, *J. Mol. Biol.*, **5**, 643 (1962).

[14] Fiers, W., and R. L. Sinsheimer, *J. Mol. Biol.*, **5**, 424 (1962).

[15] Rosenkranz, H. S., and A. Bendich, *J. Am. Chem. Soc.*, **82**, 3198 (1960).

[16] Greer, S., and S. Zamenhof, *J. Mol. Biol.*, **4**, 123 (1962).

[17] Tamm, C., H. S. Shapiro, R. Lipshitz, and E. Chargaff, *J. Biol. Chem.* **203**, 673 (1953).

[18] Fiers, W., and R. L. Sinsheimer, *J. Mol. Biol.*, **5**, 420 (1962).
[19] Wilkins, M. H. F., G. Zubay, and H. R. Wilson, *J. Mol. Biol.* **1**, 179 (1959).
[20] Zubay, G., and M. H. F. Wilkins, *J. Mol. Biol.*, **4**, 444 (1962).
[21] Stoeckenius, W., *J. Biophys. Biochem. Cytol.*, **11**, 297 (1961).
[22] Kleinschmidt, A. K., and R. K. Zahn, *Z. Naturforsch.*, **14b**, 770 (1959).
[23] Trurnit, H. J., *J. Colloid Sci.*, **15**, 1 (1960).

THE TWISTED CIRCULAR FORM OF POLYOMA VIRAL DNA

J. Vinograd, J. Lebowitz, R. Radloff, R. Watson, and P. Laipis

The major part of the DNA from polyoma virus has been shown to consist of circular base-paired duplex molecules without chain ends.[1-3] The intertwined circular form accounts for the ease of renaturation[4] of this DNA and the failure of the strands to separate in strand-separating solvents.[1-3]

In previous studies[1-3] a minor component, II, observed in variable amounts in sedimentation analyses of preparations of polyoma DNA at neutral pH, was regarded to be a linear form of the viral DNA. Both the major component I (20S) and II (16S) were infective.[1, 5] In our further investigations of the minor component the following results, which are reported below, have been obtained: (1) The minor component is a ring-shaped duplex molecule. (2) It is generated by introducing one single-chain scission in component I by the action of pancreatic DNAase or chemical-reducing agents. (3) The sedimentation coefficient of II is insensitive to several single-strand scissions. (4) The conversion products, when not excessively attacked, are infective.

The foregoing results raised a new problem. Why does the viral DNA, an intact duplex ring, sediment 20 per cent faster than a similar duplex ring containing one or more single-strand scissions? Experiments bearing on this problem, presented below, indicate the presence of a *twisted circular structure* in polyoma DNA I. A mechanism for the formation of this locked-in twisted structure is proposed.

Methods.—Isolation and purification of the virus and extraction of the DNA: Two methods[6, 7] for purification of the virus were used. The DNA was isolated by Weil's method[4] except that the phenol was freshly distilled under argon.

Ultracentrifugation: Sedimentation analyses were performed in a Spinco model E ultracentrifuge by band centrifugation.[8] Some of the results were recorded with the photoelectric scanning attachment.[9, 10] Sucrose density gradient experiments were performed at 4°, 30,000 rpm, and 9 hr. The 3% and 20% sucrose solutions contained SSC (0.15 *M* NaCl and 0.015 *M* Na citrate) and 0.05 *M* Tris chloride pH 8.0.

Enzymes: Pancreatic DNAase, 1 × crystallized, was obtained from Worthington Biochemicals Corp. *E. coli* endonuclease I, 1000 units/ml,[11] and *E. coli* phosphodiesterase, 2000 units/ml,[12] were gifts from Professor I. R. Lehman. BSA, 30% bovine albumin solution, sterile, was obtained from Armour Pharmaceutical Co. The endonuclease I, 0.12 units/μg DNA, converted 60% of I into linear molecules in 8 min at 20° in the incubation mixture described by Lehman.[11]

Sedimentation velocity-pH titration: Fifteen μl, 40 μg/ml DNA in SSC/10, flowed from the sample well of the type III[13] band-forming centerpiece onto an alkaline CsCl bulk-solution. This solution was prepared by titrating 10 ml (Harshaw Chemical Co.) optical grade CsCl, $\rho = 1.35$, with 1 *M* KOH in CsCl, $\rho = 1.35$, at 20° under argon, and was transferred to the cell assembly under argon. Usually four samples in a pH series were analyzed simultaneously. A Beckman research model pH meter, a general purpose probe glass electrode, and a calomel reference electrode modified with a ground glass junction[14] were used.

Plaque assay: Infectivity of polyoma DNA was measured as described by Weil.[4]

Electron microscopy: Specimens were prepared by the method of Kleinschmidt and Zahn.[15]

Results.—Preparation of polyoma DNA II: Polyoma II can be prepared from I by treatment with several mild chemical-reducing agents (Table 1). These reagents

TABLE 1

ACTION OF REDUCING AGENTS ON POLYOMA DNA

Reagent	Ma	pH	Time (min)	Conversion[b] of I to II (%)
Hydroquinone[c]	0.0002	8.5	30	100
$FeCl_2$	0.001	8.6	30	90
Na_2SO_3 (1×10^{-3} M Cu_2SO_4, 0.1 M NH_4OH)[d]	0.01	10.8	60	45
Thiols[e]	0.01–0.02	3.8–4.2	60	80–90

a Final concentration of reducing agents which were diluted fivefold with SSC/10, 0.01 M Tris pH 8.5 containing 40 μg/ml DNA. The thiols were first dissolved in 0.4 M acetic acid before addition to the DNA solution.

b Sedimentation analysis at neutral pH.

c The reaction product was assayed for infectivity: 0.5×10^3 pfu/μg DNA compared with 1.5×10^3 obtained for untreated DNA in the same assay.

d Cu^{++} acts as a catalyst.[16] In the absence of Cu^{++} or NH_4OH 5% conversion was observed. In the absence of SO_3^- no conversion occurred.

e Mercaptoethanol, cysteine, and glutathione. In experiments with acetic acid without thiols, 10% conversion occurred.

were suggested by the observation that rigorous exclusion of impurities from the phenol, used in the isolation of the DNA, diminished the amount of the minor component II in the final DNA preparation. A simple conversion of I to II without intermediates and without detectable degradation products was observed in sedimentation analyses at pH 8.0 (Table 1). Based on the earlier assignment[1, 2] of a linear form to component II, the reactions with reducing agents and the infective nature of the products indicated a specific duplex cleavage. Dulbecco and Vogt[1] reported a similar conversion of I to II with low concentrations of pancreatic DNAase. These authors postulated that a bond opposite the single-strand scission introduced by the enzyme hydrolyzed under the influence of ring strain. The foregoing puzzling results are clarified by the experiments below.

Structure of Polyoma II.—The products from the action of pancreatic DNAase and of the reducing agents (Table 1) were examined in the electron microscope. They were found to be in the circular form. Figure 1*b* is typical of electron micrographs of materials obtained by treatment of I with pancreatic DNAase or with the several reducing agents. The possibility that linear molecules were selectively excluded in the preparation of the grids was eliminated by the results of a reconstruction experiment. A synthetic mixture of 10 per cent II and 90 per cent linear polyoma III (cf. below) showed the expected proportions of linear molecules.

The circular form for II is compatible with the proposal that the reducing agents and pancreatic DNAase introduce single-strand scissions into polyoma DNA I. It may be calculated that the circular form of the molecule should be retained until on the average about 50 single-strand scissions per molecule have been introduced.[17] The material shown in Figure 1*b* contained, on the average, about three breaks per molecule as calculated from the Poisson distribution.

A still milder treatment with pancreatic DNAase should give rise to single-stranded rings and uniform single-stranded linear molecules in strand-separating solvents, such as alkaline NaCl or CsCl.[2, 18] An example of such a result with 0.6 breaks per molecule is given in Figure 2*b*. These two components, $s^\circ_{w,20}$ = 18.4S ± 0.4, 15.7S ± 0.3 for the alkaline Na DNA, have been identified as single-stranded rings (18S) and single-stranded linear molecules (16S), respectively, by the following variation of an experiment originally performed with φX DNA by Fiers and Sinsheimer.[19] An aliquot of the product of the pancreatic DNAase digestion

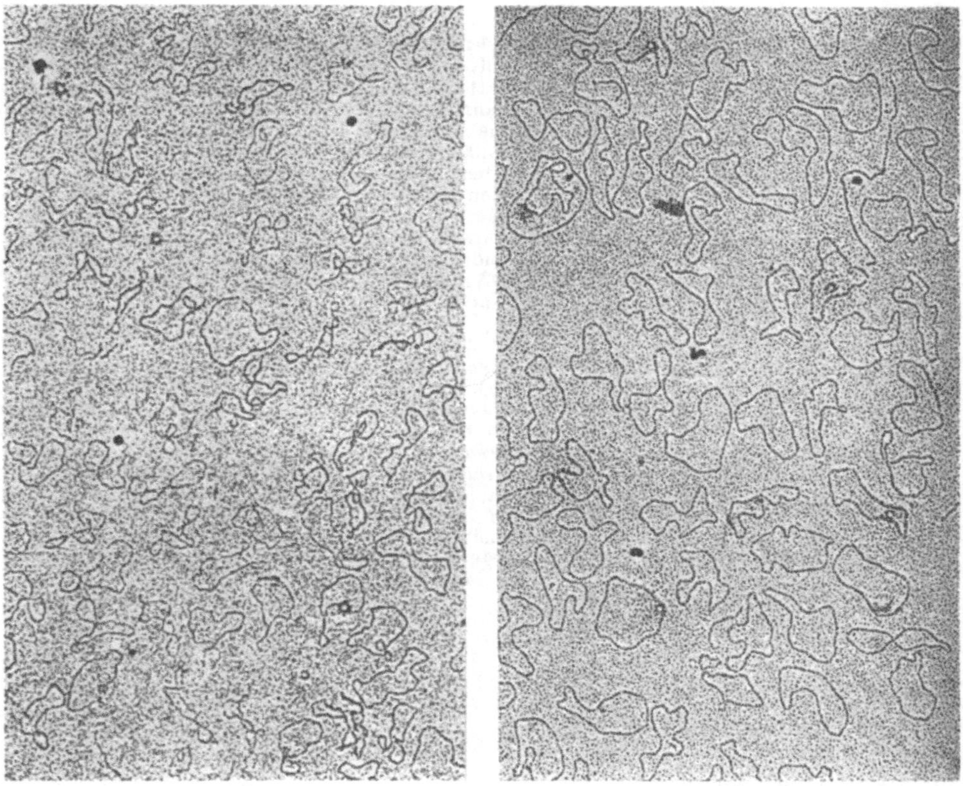

(a) (b)

FIG. 1.—Electron micrographs of polyoma DNA × 21,000. The materials in (a) and (b) were prepared by treatment of polyoma I with pancreatic DNAase, as described under Fig. 3. (a) was withdrawn from the reaction mixture after 5% conversion of I to II; (b) after 95% conversion.

(40% conversion) was heat-denatured and treated with *E. coli* phosphodiesterase. This enzyme attacks single-stranded DNA with a free 3′OH group. It is seen that the amount of the 16S component was substantially diminished, while the 18S component was resistant to the exonuclease (Fig. 2*c*). Thus polyoma DNA II can contain a wholly intact circular strand. At this level of digestion the second strand in the molecule will contain only one or two breaks if the attack is statistical.

We now examine the possibility that one single-strand scission in the duplex is adequate to convert polyoma ring I to ring II. If only one-chain scission is necessary, the rate of conversion of I to II should be the same as the rate of conversion of the 53S component (intact, denatured, double-stranded, cyclic molecules) to the slower moving single-stranded molecules in alkali. If more than one break were necessary to convert I to II, a faster rate of conversion would be seen in alkali. Dulbecco and Vogt[1] have already reported that the two rates are alike. In view of the importance of the result, we have repeated this experiment with the analyses performed in the analytical ultracentrifuge.

In Figure 3*a* it is seen that the alkaline analyses and the neutral analyses give, within the experimental error, the same extent of conversion, a result which confirms the Dulbecco and Vogt finding. Therefore, the conversion of I to II occurs

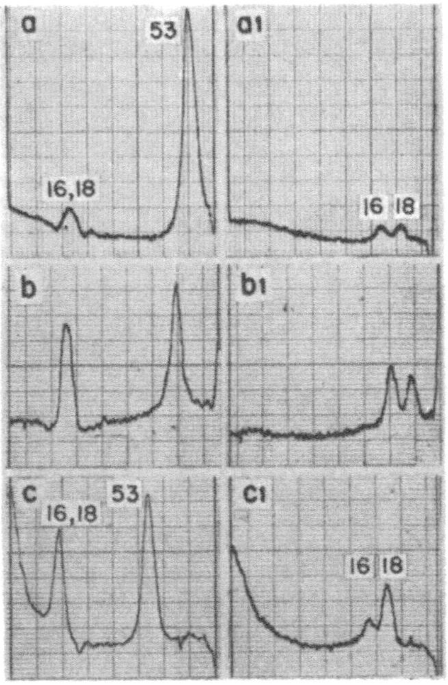

FIG. 2.—Sedimentation velocity patterns of polyoma DNA in alkaline CsCl. The *left* and *right* patterns are scans at about 30 min and 90 min after sedimentation begins. The field is directed toward the right. CsCl, ρ = 1.35 gm cm^{-3}, pH 12.5, 44,000 rpm. (*a*, *a1*) *Control*: Component I isolated in a sucrose gradient experiment treated identically as in (*b*) and (*c*) except for the absence of enzymes and BSA. Separate experiments showed the BSA to be free of DNAase activity under the conditions used. (*b*, *b1*) *Pancreatic DNAase treatment*: 106 μg/ml pure I in 0.048 *M* NaCl, 0.0075 *M* $MgCl_2$, 0.01 *M* Tris pH 8.0, 40 μg/ml BSA, and 2.7 $\times$ 10^{-4} μg/ml enzyme, 20 min at 20°. Reaction stopped by $^1/_{15}$ volume 1 *M* glycine buffer, pH 9.8. The leading band in (*a*), (*b*), and (*c*) is the 53S component. The resolved slower bands in (*a1*), (*b1*), and (*c1*) are the 16S and 18S components. (*c*, *c1*) *Effect of heat denaturation followed by E. coli phosphodiesterase treatment*: Product of (*b*) heated 5 min 100°, cooled rapidly. 70 μg/ml DNA in 0.03 *M* NaCl, 0.005 *M* $MgCl_2$, 0.007 *M* Tris, 0.067 *M* glycine pH 9.8, 0.90 mg/ml BSA, 71 units/ml *E. coli* phosphodiesterase, 90 min at 37°. Reaction stopped by $^1/_{10}$ volume 0.1 *M* EDTA.

whenever the first single-strand scission is introduced. The conversion appears to be first order. While the infectivity (Fig. 3*b*) declines at a slower rate than the conversion of I to II, the scatter in the data precludes any conclusions regarding the kinetics of inactivation. It is clear, however, that the first single-strand scission in this duplex DNA is not lethal.

More extensive treatment with pancreatic DNAase or the chemical-reducing agents so as to completely convert I to II (>4 average breaks per molecule) caused no detectable change in the sedimentation coefficient of II.

Preparation of Linear Polyoma DNA with E. coli Endonuclease I.—Polyoma I was partially converted into the linear form with *E. coli* endonuclease I, which is known to cleave duplex DNA.[11] The sedimentation velocity of the homogeneous linear molecules was 14.5S

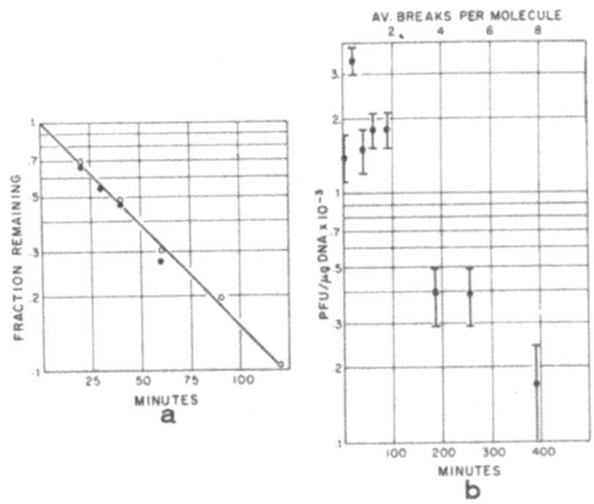

FIG. 3.—Chemical and biological effects of pancreatic DNAase treatment. (*a*) Analyses for single-stranded and double-stranded DNA. Extent of conversion was determined by band-sedimentation velocity experiments with photoelectric scanner. ⊙, (I)/(I + II) in neutral CsCl bulk solutions. ●, (53S)/(total) in alkaline CsCl pH 12.3. Areas under bands were corrected for radial dilution. Incubation mixture and conditions were the same as those given in legend to Fig. 2*b*, except for enzyme concentration, 2.0 $\times$ 10^{-4} μg/ml. 20-μl samples were withdrawn at the indicated times and added to 4 μl 0.1 *M* EDTA pH 8.5. The samples were frozen prior to analyses. (*b*) Infectivity of samples withdrawn from incubation in (*a*). The time for a unit average number of hits was obtained from (*a*) at 63% conversion. The error bars give the standard deviations from 16 replicate plates.

at pH 8.0, the same as previously reported[2] for the minor component III, and 16S in alkali. Polyoma II was not produced in detectable amounts in the above conversion of I to linear molecules. Electron micrographs confirmed the assignment of a linear form to the enzymatic product and also to the minor component III isolated by sucrose gradient sedimentation.

Structure of Component I.—The high sedimentation coefficient of I relative to II indicates that the viral component is either more compact or larger in mass than the circular conversion product. In the extreme case of no increase in friction, a 20 per cent reduction in mass is required to account for the change in *s*. An equal amount of mass would have to have been lost as a result of the action of pancreatic DNAase and the variety of reducing agents used. An excision of viral DNA would have been detected by Dulbecco and Vogt,[1] who were unable to find small fragments of labeled DNA after preparative band sedimentation of polyoma DNA treated with pancreatic DNAase. The identical buoyant densities of I and II[2] make it unlikely that a nonlabeled, non-DNA mass is removed.

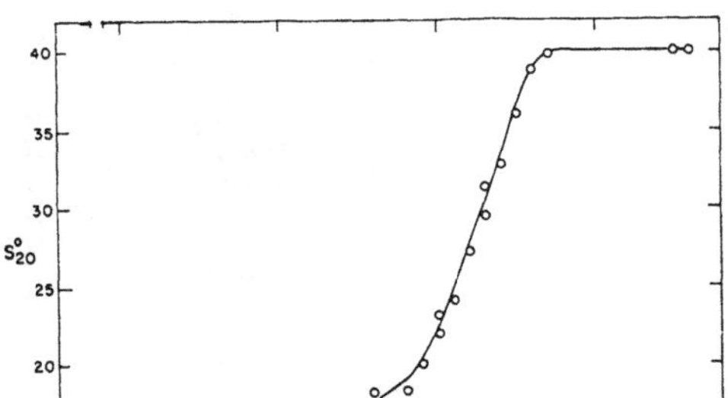

FIG. 4.—Sedimentation velocity-pH titration of the three components in polyoma DNA. I, ○; II, ●; III, △; ◑ only band present, cf. text; mixture of unresolved single strands in alkali, x; 29,500 rpm. Single linear, □, and single circular, ○, strands in alkali, 44,770 rpm. Sedimentation coefficients at 20° in CsCl, ρ = 1.35 gm cm^{-3}, are not corrected for solvent viscosity, η_r = 0.925, or buoyancy effects. The values at pH 8.0 and 12.4 are the means of 12 and 7 determinations, respectively.

Three kinds of experiments suggest that a particular kind of compact structure—*a twisted circular form*—is responsible for the high sedimentation coefficient of polyoma DNA. (1) The electron micrographs of the grids prepared from polyoma I contained[1] twisted circles to a variable extent (Fig. 1*a*). Grids prepared from polyoma II contained extended circles and no twisted configurations (Fig. 1*b*). The DNA samples used to prepare the grids for Figures 1*a* and *b* were identical except for the time of incubation with the enzyme. The spreading forces acting on the macromolecules in the monolayer appear to remove loops and crossovers that are not locked in the structure of the DNA.

(2) A study of the sedimentation velocity in 3 *M* CsCl of a mixture of the three components of polyoma DNA as a function of pH from 8 to 12.5 revealed a complicated pH-melting curve for component I (Fig. 4). Component II behaved normally [18, 20] and moved faster as denaturation increased until strand separation occurred with an attendant sudden drop in sedimentation velocity at pH 11.8. Component I, like II, was at first insensitive to pH. At pH 11.5, however, the sedimentation coefficient first dropped, and then in the pH range 11.6 to 11.8 was the same as for polyoma II. Only one moving band was observed in this pH range. The sedimentation coefficient of I then increased to the very high value characteristic of the *double-stranded cyclic coil* previously reported.[2] Essentially the same results were obtained in 1.0 *M* KCl solutions. The dip in the sedimentation velocity-

pH curve was initially unexplainable. If, however, polyoma DNA I contains left-handed tertiary turns, such a dip in the pH-melting profile would be required. In the early stages of denaturation some of the duplex turns, which are known to be right-handed, unwind. The unwinding of the duplex must be accompanied by a right-handed twisting of the remainder of the molecule. If the tertiary turns were originally left-handed (Fig. 5), progressive unwinding would cause the molecule to pass through configuration I′ characterized by the absence of tertiary turns. The extended configuration I′ is similar to that in polyoma II (Fig. 5) and both I′ and II would have similar sedimentation velocities. Further unwinding of duplex I′ is accompanied by continued right-hand twisting of the whole molecule until finally the double-stranded cyclic coil[2] configuration develops.

(3) The twisted circular structure provides a satisfactory explanation for the configurational change that occurs when one single-strand scission is introduced into the molecule. Such a scission generates a site for the rotation of the helix in the complementary strand opposite the break. The swivel relieves the topological restraint responsible for the twisted configuration.

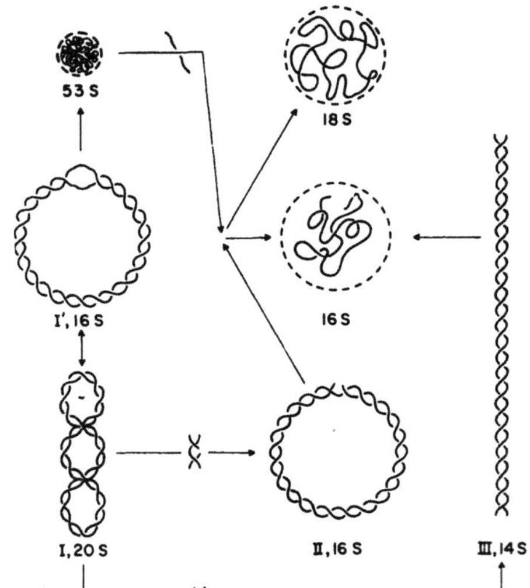

FIG. 5.—Diagrammatic representation of the several forms of polyoma DNA. The duplex segments shown contain 12 turns, about one fortieth of the total number. The twisted circular duplex shown contains one left-hand tertiary turn. 8% of the right-hand duplex turns in the model are unwound to form *I′*. The dashed circles around the denatured forms indicate the relative hydrodynamic diameters. The sedimentation coefficients were measured in neutral and alkaline NaCl solutions.

Discussion.—A mechanism for the formation of the twisted circular structures suggested by the above analysis of the pH-melting curve. According to this mechanism the last closure of chain ends occurs before all of the winding of the two DNA strands into the Watson-Crick structure is completed. The closing leaves the duplex in the configuration I′ (Fig. 5) restrained from converting to I by an as yet unknown factor participating in the DNA synthesis. Removal of the restraint then allows I′ to wind spontaneously into a complete Watson-Crick structure and form the twisted circular structure, I, with no change in winding number.

An alternative proposal is that the molecule in form II, which contains a swivel, is twisted by some organizer, e.g., the virus protein. The last covalent backbone bond is then made while the DNA remains twisted under the constraint of the organizer. This alternative is unlikely in view of Dulbecco's[21] finding that the polyoma DNA made before virus production begins has the sedimentation velocity of component I.

It is not possible at the present time to estimate reliably the number of tertiary turns. A turn is defined as a 360° rotation of the helix. The electron micrographs of I usually contain some molecules that are completely extended; these may have suffered a single break during grid preparation or may have been in state I′ due to denaturation induced by the spreading forces. This latter action would result in the unwinding of the twisted circles to form extended circles. The maximum number of crossovers that can be distinguished is 8, which corresponds to 4 turns or the unwinding of 40 base pairs. This limit may be low because it is difficult to count crossovers in tightly coiled forms. That the number of tertiary turns in the molecule is not large is suggested by the fact that the transition of I to I′ occurs before substantial melting of II takes place, as indicated by the *s* versus pH plot.

With the new assignment of structure to the three components of polyoma DNA, it is found that satisfactory agreement obtains between the observed sedimentation coefficient of the linear form, 14.5S, and the 15.3 ± 0.5S predicted by Studier's relation[18] for a molecular weight[2] of $3.0 \pm 0.3 \times 10^6$. The values, 18.4S and 15.7S for the alkaline single-stranded circular and linear forms similarly agree with the predicted values of 17.4S and 15.6S, respectively. The effect of ring closure of III to form component II is to increase S by 10 per cent. An effect of similar magnitude has been reported[22] for the cyclization of λ DNA.

The twisted circular structure observed here for polyoma DNA may be a common characteristic of covalently closed, circular duplex DNA. A part of the DNA from rabbit papilloma virus,[23] SV40 virus,[3] and the replicating form of ϕX DNA[24–26] have all been shown to be circular duplex molecules which do not strand-separate in alkali or after heating in formaldehyde. Two sedimentation velocity components differing by 20–30 per cent have been reported for the above DNA's.[27] Crawford and Black[3] observed sedimentation velocity-denaturation curves that are similar to our pH-melting curve upon heating SV40 DNA and polyoma DNA in formaldehyde solutions to various temperatures. No explanation was offered for this behavior, which we interpret as indicating the presence of a left-handed, twisted circular structure.[28]

Burton and Sinsheimer[29] have shown that the slow component II in RF-ϕX DNA dissociates in alkali to form linear and circular single-stranded molecules and have concluded that both of the undenatured forms of the DNA are circular. While this communication was in preparation, Jansz and Pouwels[30] reported that the pancreatic DNAase-induced conversion of I to II in RF-ϕX DNA represents a conversion between circular duplex molecules. No explanation for the change in the sedimentation coefficient was offered. In view of the results described here, it is likely that RF-ϕX DNA is in the twisted circular form. A common mechanism for the incorporation of the tertiary turns during replication is a strong possibility, and allows us to predict that the tertiary turns in the RF-ϕX DNA will be found to be left-handed.

Summary.—The results of this study show that circular duplex polyoma DNA may be converted to a less compact circular duplex by introducing a single-strand scission. The viral form contains tertiary turns which appear to have been locked in during replication.

It is a pleasure to thank I. R. Lehman for the generous gift of the *E. coli* enzymes, R. L. Sinsheimer, J. Petruska, and M. Fried for helpful discussions, R. Dulbecco for allowing us to quote his unpublished results, and T. Benjamin, L. Wenzel, and A. Drew for advice and assistance in the culture of the virus and the assay of the DNA. This work was supported in part by grants HE 03394 and CA 08014 from the U.S. Public Health Service.

This work was reported in part at the 9th Annual Meeting of the Biophysical Society, February 24, 1965. Contribution 3227 of Gates and Crellin Laboratories of Chemistry.

[1] Dulbecco, R., and M. Vogt, these PROCEEDINGS, **50**, 236 (1963).

[2] Weil, R., and J. Vinograd, these PROCEEDINGS, **50**, 730 (1963).

[3] Crawford, L. V., and P. H. Black, *Virology*, **24**, 388 (1964).

[4] Weil, R., these PROCEEDINGS, **49**, 480 (1963).

[5] Crawford, L. V., R. Dulbecco, M. Fried, L. Montagnier, and M. Stoker, these PROCEEDINGS, **52**, 148 (1964).

[6] Winocour, E., *Virology*, **19**, 158 (1963).

[7] Murikami, W., *Science*, **142**, 56 (1963).

[8] Vinograd, J., R. Bruner, R. Kent, and J. Weigle, these PROCEEDINGS, **49**, 902 (1963).

[9] Beckman Instruments Co., Spinco Division, Palo Alto, Calif.

[10] Hanlon, S., K. Lamers, G. Lauterbach, R. Johnson, and H. K. Schachman, *Arch. Biochem. Biophys.*, **99**, 157 (1962).

[11] Lehman, I. R., G. G. Roussos, and E. A. Pratt, *J. Biol. Chem.*, **237**, 819 (1962).

[12] Lehman, I. R., *J. Biol. Chem.*, **235**, 1479 (1960).

[13] Vinograd, J., R. Radloff, and R. Bruner, *Biopolymers*, in press.

[14] Lebowitz, J., and M. Laskowski, Jr., *Biochemistry*, **1**, 1044 (1962).

[15] Kleinschmidt, A. K., and R. K. Zahn, *Z. Naturforsch.*, **14b**, 770 (1959).

[16] Swan, J. M., *Nature*, **180**, 643 (1957).

[17] Thomas, C. A., Jr., *J. Am. Chem. Soc.*, **78**, 1861 (1956). The estimate of 50 was obtained for the case of a 10% lowering of the weight average molecular weight.

[18] Studier, F. Wm., *J. Mol. Biol.*, **11**, 373 (1965).

[19] Fiers, W., and R. L. Sinsheimer, *J. Mol. Biol.*, **5**, 408 (1962).

[20] Davidson, P. F., and D. Freifelder, *J. Mol. Biol.*, **5**, 643 (1962).

[21] Dulbecco, R., private communication.

[22] Hershey, A. D., E. Burgi, and L. Ingraham, these PROCEEDINGS, **49**, 748 (1963).

[23] Crawford, L. V., *J. Mol. Biol.*, **8**, 489 (1964).

[24] Kleinschmidt, A., A. Burton, and R. L. Sinsheimer, *Science*, **142**, 961 (1963).

[25] Burton, A., and R. L. Sinsheimer, *Science*, **142**, 962 (1963).

[26] Burton, A., and R. L. Sinsheimer, *Abstracts*, 8th Annual Meeting of the Biophysical Society, 1964.

[27] R. Weil has informed us that the slow component II in rabbit papilloma DNA is circular as seen in the electron microscope.

[28] Human papilloma DNA appears to be similar in configuration to SV40 DNA and polyoma DNA (Crawford, L. V., manuscript submitted for publication).

[29] Burton, A., and R. L. Sinsheimer, private communication.

[30] Jansz, H. S., and P. H. Pouwels, *Biochem. Biophys. Res. Commun.*, **18**, 589 (1965).

A DYE-BUOYANT-DENSITY METHOD FOR THE DETECTION AND ISOLATION OF CLOSED CIRCULAR DUPLEX DNA: THE CLOSED CIRCULAR DNA IN HELA CELLS

Roger Radloff, William Bauer, and Jerome Vinograd

Covalently closed circular duplex DNA's are now known to be widespread among living organisms. This DNA structure, originally identified in polyoma viral DNA,[1, 2] has been assigned to the mitochondrial DNA's in ox[3] and sheep heart,[4] in mouse and chicken liver,[3] and in unfertilized sea urchin egg.[5] The animal viral DNA's—polyoma, SV40,[6] rabbit[7] and human[8] papilloma—the intracellular forms of the bacterial viral DNA's—ϕX174,[9, 10] lambda,[11, 12] M13,[13] and P22[14]—and a bacterial plasmid DNA, the colicinogenic factor E_2,[15] have all been shown to exist as closed circular duplexes. Other mitochondrial DNA's[16, 17] and a portion of the DNA from boar sperm[18] have been reported to be circular, but as yet have not been shown to be covalently closed.

The physicochemical properties of closed circular DNA differ in several respects from those of linear DNA or of circular DNA containing one or more single-strand scissions.[19] The resistance to denaturation,[2, 20] the sedimentation velocity in neutral and alkaline solution, and the buoyant density in alkaline solution are all enhanced in the closed circular molecules. These three effects are a direct consequence of the topological requirement that the number of interstrand cross-overs must remain constant in the closed molecule.

The principal methods currently used for the detection and the isolation of closed circular DNA are based on the first two general properties. In this communication we describe a method based on the buoyant behavior of closed circular DNA in the presence of intercalating dyes.

The binding of intercalative dyes has recently been shown to cause a partial unwinding of the duplex structure in closed circular DNA.[22–24] In such molecules any unwinding of the duplex causes a change in the number of superhelical turns, so that the total number of turns in the molecule remains constant. A small and critical amount of dye-binding reduces the number of superhelical turns to zero. Further dye-binding results in the formation of superhelices of the opposite sign or handedness. The creation of these new superhelices introduces mechanical stresses into the duplex and a more ordered conformation into the molecule. These effects increase the free energy of formation of the DNA-dye complex. The maximum amount of dye that can be bound by the closed molecule is therefore *smaller* than by the linear or nicked circular molecule. Correspondingly, since the buoyant density of the DNA-dye complex[23, 25] is inversely related to the amount of dye bound, the buoyant density of the closed circular DNA-dye complex at saturation is *greater* than that of the linear or nicked circular DNA-dye complex.[23] Bauer and Vinograd have shown that the above effect results in a buoyant density difference of approximately 0.04 gm/ml in CsCl containing saturating amounts of ethidium bromide, an intercalating dye extensively studied by Waring[26] and Le Pecq.[27]

The method has been tested with known mixtures of nicked and closed circular viral DNA's, and has been used to isolate closed circular DNA from the mitochondrial fraction of HeLa cells and from extracts of whole HeLa cells.

Materials and Methods.—Preparation of HeLa cell extracts: HeLa S3 cells were grown on Petri dishes in Eagle's medium containing 10% calf serum. H^3-thymidine, 18 c/mmole, was obtained from the New England Nuclear Corporation. Ten μc per ml of medium were added to each plate 20 hr before the cells were harvested. After washing with isotonic buffer and decanting, the cells were treated by the method described by Hirt[28] for the separation of polyoma DNA from nuclear DNA. Approximately 2 ml of 0.6% sodium dodecylsulphate (SDS), in 0.01 *M* ethylenediaminetetraacetate (EDTA), 0.01 *M* tris, pH 8, were added to the plates. After 30 min at room temperature, the viscous extracts were gently scraped from the plates with a rubber policeman and transferred with a widemouth pipet to a centrifuge tube. Either 5 *M* NaCl or 7 *M* CsCl was then added with gentle mixing to a final salt concentration of 1 *M*. The resulting solution was cooled to 4° and centrifuged for 15 min at 17,000 $\times$ *g* in a Servall preparative ultracentrifuge. The supernatant solution was dialyzed at 4° overnight against two changes of 0.01 *M* EDTA, 0.01 *M* tris, pH 8 buffer in order to remove H^3-thymidine. The mitochondria from HeLa cells were isolated by differential centrifugation of an homogenate followed by banding in a sucrose gradient.

Preparation of viral DNA: Polyoma viral DNA was prepared as described previously.[19] The intracellular lambda DNA was kindly supplied by John Kiger and E. T. Young, II, of the Biology Division.

Chemicals: The ethidium bromide was obtained as a gift from Boots Pure Drug Co., Ltd., Nottingham, England. Harshaw optical grade cesium was used. The SDS was obtained from Matheson Company. All other chemicals were of reagent grade.

Preparative ultracentrifugation: The experiments were performed in SW50 rotors in a Beckman Spinco model L preparative ultracentrifuge at 20°C. The CsCl solutions, in either cellulose nitrate or polyallomer tubes, were overlaid with light mineral oil. After centrifugation, the tubes were fractionated with a drop-collection unit obtained from Buchler Instruments. The drops were collected in small vials or on Whatman glass-fiber GF/A filters. The dried filters were immersed in 10 ml of toluene-PPO-POPOP, and the samples were counted in a Packard Tri-Carb scintillation counter.

Electron microscopy: Specimens for electron microscopy were prepared by the method of Kleinschmidt and Zahn[29] and were examined in a Phillips EM200 electron microscope. All electron micrographs were made at a magnification of $\times$5054. The magnification factor was checked with a grating replica. Cytochrome *c* was obtained from the California Biochemical Corporation.

Fluorescence: Prior to drop collection, the centrifuge tubes were examined in a darkened room with 365 mμ light from a Mineral Light Lamp or preferably from a "Transilluminator" supplied by Ultraviolet Products, Inc., San Gabriel, California. The tubes were photographed on Polaroid type-146L film through a "contrast filter" from Ultraviolet Products, Inc. The fluorescence measurements were performed with a double-monochromator apparatus constructed in this laboratory by W. Galley and N. Davidson. The instrument was calibrated with solutions of ethidium bromide (5×10^{-3} to 1.0 μg/ml) in calf thymus DNA, 20 μg/ml, 1 *M* CsCl. The intensity of fluorescence was measured at 590 mμ with an exciting wavelength of 548 mμ.[30]

Results.—Selection of initial dye concentration, cesium chloride concentration, and centrifugation variables: The conditions for the experiments in the swinging-bucket rotor in the model L ultracentrifuge were selected with the object of obtaining separations similar to those obtained at equilibrium within a reasonable period of time. At equilibrium, the separation between components is approximately constant at initial dye concentrations between 50 and 100 μg/ml.[23] At these dye levels the buoyant densities are between 1.57 and 1.62 gm/ml. Figure 1 presents the dye and density distributions in CsCl solutions centrifuged for 24 and 48 hours. The initial dye concentration in both cases was found at a distance from the meniscus corresponding to four-tenths the length of the liquid column. In

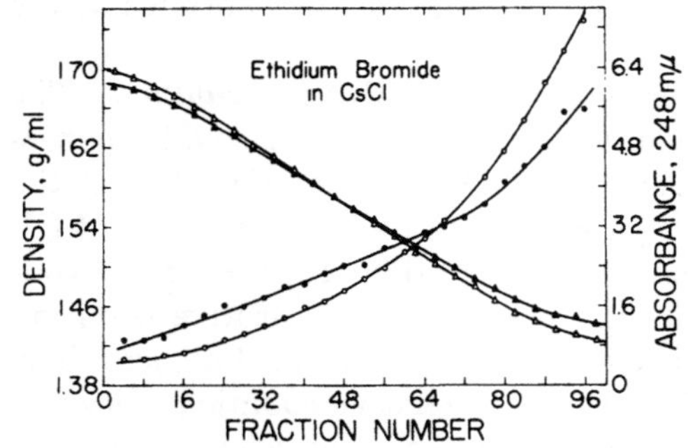

FIG. 1.—Density and dye distributions in CsCl density gradient columns, 3.00 ml 1.550 gm/ml CsCl, 100 μg/ml ethidium bromide, 24 (●, ▲) and 48 (O, △) hr at 43 krpm in a SW50 rotor at 20°.

this region of the cell, the time dependence of the dye concentration is minimal, and the constant density gradient is equal to approximately 0.010 gm/cm[4].

Results with purified mixtures of nicked and closed circular DNA: Figure 2 presents the results obtained with 1.5 μg of tritiated polyoma DNA. Substantially complete resolution was obtained with this DNA which has a molecular weight of three million daltons. Since the fractional amount of closed circular material corresponded to the fraction obtained in analytical sedimentation velocity analyses,[21] it is concluded that single-strand scissions did not occur during the course of the experiments. An experiment in which the DNA was contained in a thin lamella at the top of the liquid column was performed to test the effect of the initial DNA distribution. The results were substantially the same as those obtained when the DNA was uniformly distributed in the CsCl solution.

The results obtained with a mixture of approximately 60 μg of covalently closed and nicked circular lambda DNA, mol wt = 3×10^7, are presented in Figure 3.

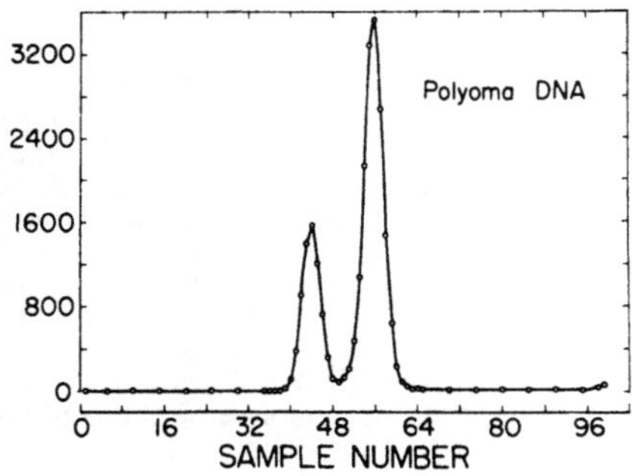

FIG. 2.—A mixture of purified polyoma DNA I and II 1.5 μg in buoyant CsCl, 3.00 ml 1.566 gm/ml CsCl, 100 μg/ml ethidium bromide, 24 hr at 43 krpm, 20°. The band maxima were separated by 12 fractions (four 7.5-μl drops per fraction). The buoyant densities of I and II are 1.588 and 1.553 gm/ml, respectively. The sample contains 30% I as indicated above compared with 32% by analytical band centrifugation.[21]

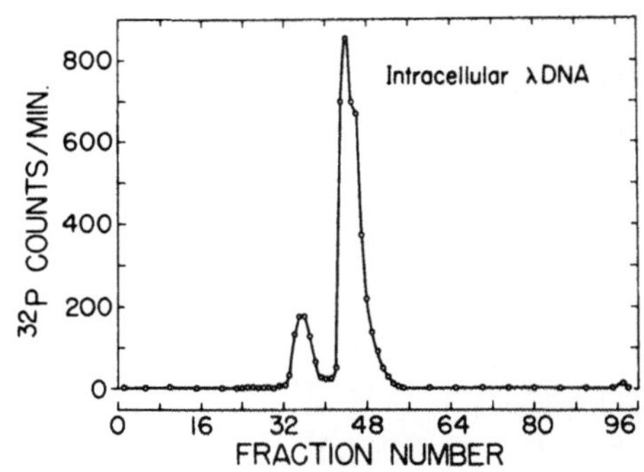

FIG. 3.—A mixture of purified intracellular lambda DNA I and II, 3.00 ml 1.55 gm/ml CsCl, 100 μg/ml ethidium bromide, 24 hr at 43 krpm, 20°. The centroids are separated by 9.5 fractions and 0.31 ml. Component I accounts for 16% of the total counts.

In this experiment, the nicked circular DNA formed a band of relatively high viscosity that may have distorted and broadened the light band during drop collection. An experiment to determine the effect of ethidium bromide on the protoplast assay for lambda DNA[32] was performed in collaboration with J. Kiger and E. T. Young, II. It was found that linear lambda had a normal titer when diluted by a factor of 1000 from a solution containing ethidium bromide and CsCl at the concentrations which occur at band center.

Detection of DNA by absorbance, fluorescence, and scintillation counting: The data in Figures 2 and 3 were obtained by scintillation counting of dried filter papers containing dye, cesium chloride, and labeled DNA. To examine the effect of the presence of the dye upon counting efficiency, a series of sixteen 50-μl samples of H^3-thymidine in a 1.55 gm/ml CsCl solution containing ethidium bromide in concentrations varying from 0 to 286 μg/ml were counted. The relative counting efficiency decreased linearly with a least-squares slope of 8.6×10^{-4} ml/μg. The dye gradient thus caused a difference of 1 per cent in relative counting efficiency between the two bands in Figure 2, while the dye depressed the relative counting efficiency by 8 per cent.

Red bands containing about 5 μg of DNA can be observed visually in the centrifuge tube. If an adequate amount of DNA is present, the fractionated gradient may be assayed spectrophotometrically at 260 mμ. At saturation, the increase in absorbance caused by dye-binding is about 40 per cent with linear DNA and about 20 per cent with closed circular DNA.

The photograph in Figure 4 shows the fluorescent emission from two DNA bands. The fluorescence from DNA in amounts as low as 0.5 μg per band are detectable visually. Since the separation between the two bands corresponds reliably to 0.30–0.36 ml, it was possible to use the fluorescent, less dense band as a marker to locate closed circular DNA present in amounts below the limit of detectability by spectroscopic or radioactive procedures. The method is thus especially suitable for examining nonradioactive preparations from tissues of higher animals.

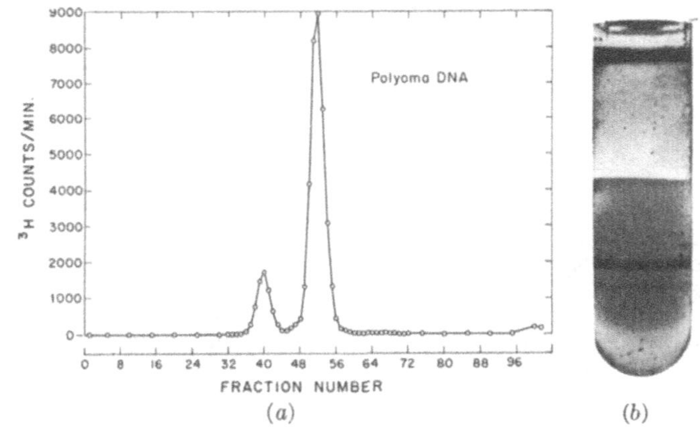

FIG. 4.—A mixture of purified polyoma DNA I and II, 3.00 ml, 1.558 gm/ml CsCl, 100 μg/ml ethidium bromide, 48 hr at 43 krpm, 20°. (*a*) The band maxima are separated by 12 fractions and 0.36 ml. The calculated buoyant densities are 1.592 and 1.556 gm/ml. (*b*) A photograph of the centrifuge tube prior to drop collection. The tube contains a total of 4 μg of DNA, and 0.64 μg of component I.

Removal of ethidium bromide from DNA solutions: It is often desirable to remove the dye quantitatively from a DNA sample. This was accomplished in a single passage of 1.0 ml of polyoma I DNA (40 μg/ml DNA, 100 μg/ml dye, 1 *M* CsCl) through a 0.8×4.5-cm column of analytical grade Dowex-50 resin. The fractions containing DNA were consolidated and were found to contain less than 1×10^{-2} μg dye in a fluorometric analysis.

Isolation of closed circular DNA from H^3-thymidine-labeled HeLa cells: Hela cell monolayers in Petri dishes containing 10^7 cells were treated with SDS by the procedure described by Hirt.[28] The results obtained when the extract was purposely sheared and when the shear was minimized (Figs. 5*a* and *b*) show that the relative amount of material in the light band was decreased when shear was minimized. The dense band in Figure 5*a* is clearly contaminated by material from the light band and reprocessing the dense band in a second CsCl-dye gradient would be necessary to obtain resolved bands.

Electron-microscope examination of the circular DNA in the dense band of HeLa cell extracts: Examination of electron micrographs of specimens prepared from fractions 33–36 in the dense band of HeLa cell extracts (Fig. 5*c*) showed less than 0.1

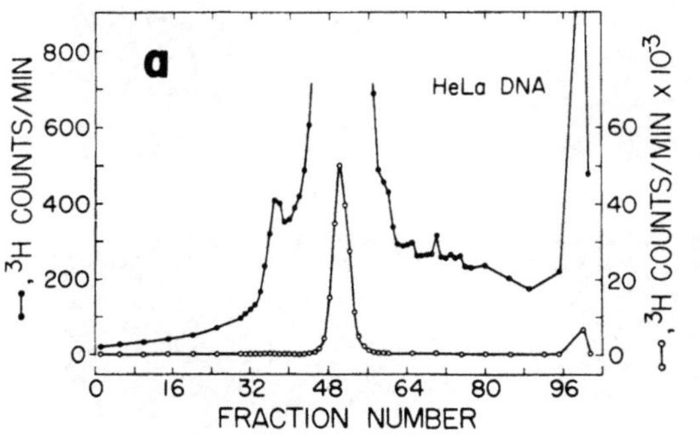

FIG. 5.—Buoyant-density profiles from extracts of whole HeLa cells, prepared as described in the text. Centrifuge conditions were the same as in Fig. 2. (*a*) DNA extract sheared during preparation. (*b*) DNA extract prepared with minimum shear. (*c*) Prepared with minimum shear for electron microscopy.

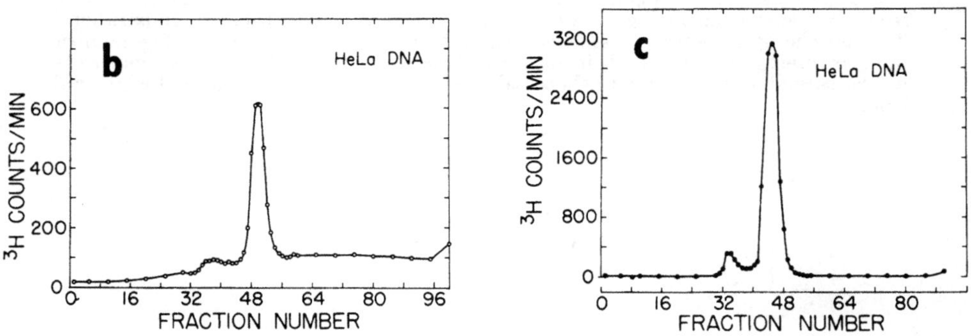

per cent of distinguishably linear DNA. The light band contained long linear DNA. The dense band contained an array of circles in three size groups: a homogeneous group of molecules with a mean length of 4.81 ± SE 0.24 microns; a heterogeneous population with lengths from 0.2 to 3.5 microns; and a paucidisperse set 2–4 times the length of the DNA in the homogeneous group. Figure 6 presents micrographs selected to illustrate the three size classes, and Figure 7 shows the frequency distribution among the first two groups. A survey of several hundred molecules on sparsely populated specimen grids revealed that the frequencies of the small and the large size classes were each about one-tenth the frequency of the homogeneous size class. In the large size class, the dimers were observed more frequently than the trimers or tetramers. The DNA from mitochondria isolated from HeLa cells consisted principally of molecules in the homogeneous size group. We conclude, therefore, that the homogeneous size class in Figure 7 is of mitochondrial origin. Small circular DNA's were not observed in significant amounts in the preparation of DNA from the isolated mitochondria.

Discussion.—The method described in this paper has proved to be a simple and direct procedure for obtaining closed circular DNA from extracts of whole cells and cell fractions. The method employs chemically mild conditions; single-strand scissions are not introduced by interaction with the dye, nor is there any permanent rearrangement of the DNA structure. All the steps of the separation

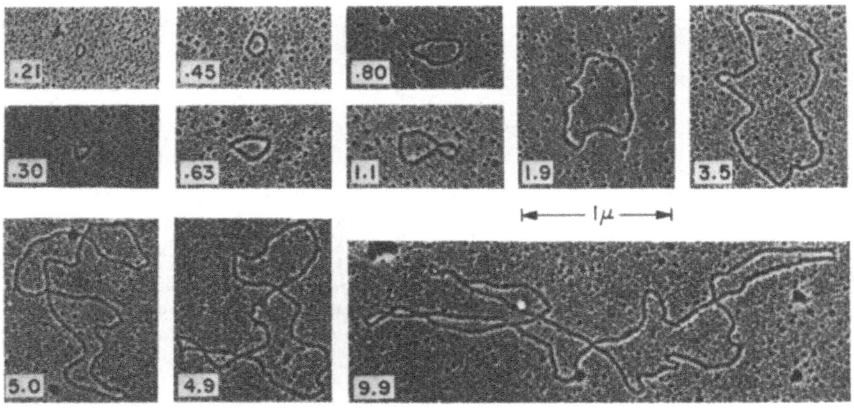

FIG. 6.—Electron micrographs of circular DNA from HeLa cells. Fractions 33–36 from the dense band in Fig. 5c were pooled and used in the specimen preparations. The top photographs present selected molecules of the small size range. The number in each insert gives the length in microns of the molecule. The first two molecules in the second row are of mitochondrial size; the third is twice the mitochondrial length.

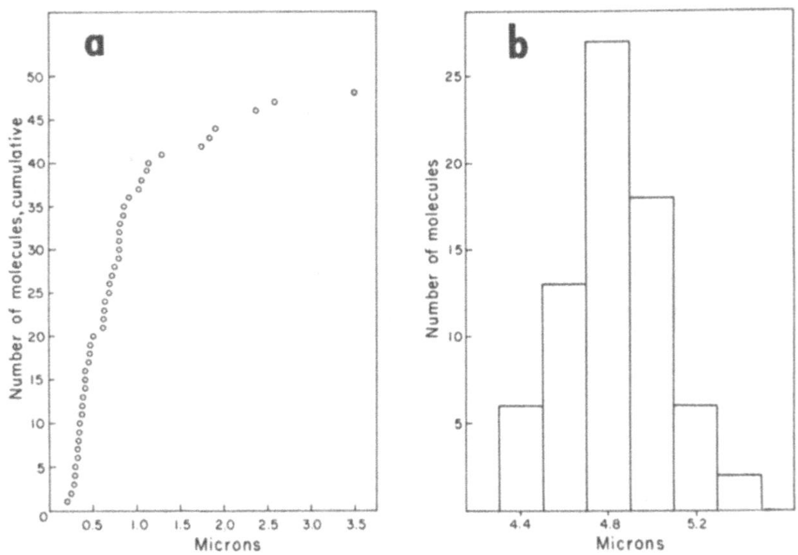

FIG. 7.—Frequency distribution of lengths of circular DNA isolated from the HeLa cell band referred to in the legend for Fig. 6. (*a*) A cumulative frequency distribution of lengths of molecules in the submitochondrial size class. (*b*) A histogram of the distribution of lengths of molecules in the mitochondrial size class.

are carried out at room temperature, or below, and at neutral pH. In the experiments performed so far, the maximum amount of DNA introduced in a single tube was 125 μg of polyoma DNA.

It has become clear, in the course of this initial study of the closed circular DNA in malignant human cells, that a portion of the mitochondrial DNA in HeLa cells is in the form of closed duplexes. These circular molecules are similar in form, size, and homogeneity to the mitochondrial DNA's from invertebrate and vertebrate species previously investigated.[3–5, 16, 17]

The heterogeneous group of small molecules, 0.2–3.5 microns, are to be compared with the size range, 0.5–16.8 microns, reported by Hotta and Bassel[18] to be present in preparations of unfractionated boar sperm DNA. The conclusion that the small circles, seen in our electron micrographs, represent DNA molecules rests on three observations. (1) The small circles were found at the same level in a CsCl-dye gradient as were the closed circular mitochondrial DNA molecules. (2) The grain pattern of the shadowed metal on the small circles viewed at a magnification of 1×10^5 was indistinguishable from the pattern on mitochondrial DNA molecules in the same field. (3) The small circles were seen only very infrequently in DNA preparations from isolated HeLa cell mitochondria. It is thus unlikely that the small circles arise from an artifact in the electron-microscope procedure. A more definitive characterization of these small DNA molecules, which can code for only 200–3500 amino acid residues, requires the preparation of larger quantities of material. The frequency of occurrence of the small molecules in the SDS extracts we have studied does not necessarily represent the frequency of occurrence in the HeLa cell. It is emphasized here that the dye-buoyant method segregates only *closed* circular duplexes. Molecules that contain even one single-strand scission, for whatever reason, find their way to the less dense band and, in whole cell extracts, intermingle with the large excess of linear DNA.

The larger-size circles, which were also seen at a frequency of about 10 per cent relative to mitochondrial DNA in the dense band, are clearly multiples of the mitochondrial length. The mean lengths from measurements of 43 double-length circles and a smaller number of larger multiples were 9.56 ± 0.42, 14.1 ± 0.4, and 19.8 ± 1.1 microns. We have so far not found a fully extended large circle. Nor have we found, in the several large circles which contain infrequent crossovers, any one which could not have arisen from a pairing of two or more mitochondrial molecules. An investigation of the significance of these multiple-size circles in HeLa DNA is in progress. Nass[17] has reported measurements of multiple lengths of mitochondrial molecules liberated from mitochondria by osmotic shock during specimen preparation. She attributed the multiples to the superposition of DNA molecules which originated from single mitochondria.

Summary.—A buoyant-density method for the isolation and detection of closed circular DNA is described. The method is based on the reduced binding of the intercalating dye, ethidium bromide, by closed circular DNA. In an application of this method we have found that HeLa cells contain, in addition to closed circular mitochondrial DNA of mean length, 4.81 microns, a heterogeneous group of smaller DNA molecules which vary in size from 0.2 to 3.5 microns and a paucidisperse group of multiples of the mitochondrial length.

Note added in proof: That double-length closed mitochondrial DNA molecules do occur has been shown by D. Clayton in this laboratory with preparations from leucocytes obtained from the blood of a donor with chronic granulocytic leukemia. Circular DNA molecules free of crossovers and of twice the mitochondrial length were observed in electron micrographs.

It is a pleasure to thank J. Huberman for providing us with the labeled HeLa cells; J. Kiger and E. T. Young, II, for their gift of closed circular lambda DNA and for allowing us to quote their unpublished analysis; G. Attardi for advice and assistance in the preparation of HeLa mitochondria; L. Wenzel and J. Eden for their assistance in the culture of the polyoma virus; R. Watson for his several technical contributions; and R. Kent for assistance in the preparation of the manuscript.

This work was supported in part by grants HE 03394 and CA 08014 from the U.S. Public Health Service and by fellowships from the U.S. Public Health Service and the National Science Foundation.

[1] Dulbecco, R., and M. Vogt, these PROCEEDINGS, **50**, 236 (1963).
[2] Weil, R., and J. Vinograd, these PROCEEDINGS, **50**, 730 (1963).
[3] Borst, P., and G. J. C. M. Ruttenburg, *Biochim. Biophys. Acta*, **114**, 645 (1965).
[4] Kroon, A. M., P. Borst, E. F. J. Van Bruggen, and G. J. C. M. Ruttenburg, these PROCEEDINGS, **56**, 1836 (1966).
[5] Pikó, L., A. Tyler, and J. Vinograd, *Biol. Bull.*, in press.
[6] Crawford, L. V., and P. H. Black, *Virology*, **24**, 388 (1964).
[7] Crawford, L. V., *J. Mol. Biol.*, **8**, 489 (1964).
[8] *Ibid.*, **13**, 362 (1965).
[9] Burton, A., and R. L. Sinsheimer, *J. Mol. Biol.*, **14**, 3276 (1965).
[10] Kleinschmidt, A. K., A. Burton, and R. L. Sinsheimer, *Science*, **142**, 1961 (1963).
[11] Young, E. T., II, and R. L. Sinsheimer, *J. Mol. Biol.*, **10**, 562 (1964).
[12] Bode, V. C., and A. D. Kaiser, *J. Mol. Biol.*, **14**, 399 (1965).
[13] Play, D. S., A. Preuss, and P. H. Hofschneider, *J. Mol. Biol.*, **21**, 485 (1966).
[14] Rhoades, M., and C. A. Thomas, private communication.
[15] Roth, T., and D. Helinski, private communication.
[16] David, I. B., these PROCEEDINGS, **56**, 269 (1966).
[17] Nass, M. M. K., these PROCEEDINGS, **56**, 1215 (1966).
[18] Hotta, Y., and A. Bassel, these PROCEEDINGS, **53**, 356 (1965).
[19] Vinograd, J., J. Lebowitz, R. Radloff, R. Watson, and P. Laipis, these PROCEEDINGS, **53**, 1104 (1965).
[20] Vinograd, J., and J. Lebowitz, *J. Gen. Physiol.*, **49**, 103 (1966).
[21] Vinograd, J., R. Bruner, R. Kent, and J. Weigle, these PROCEEDINGS, **49**, 902 (1963).
[22] M. Gellert, these PROCEEDINGS, **57**, 148 (1967).
[23] Bauer, W. R., and J. Vinograd, in preparation.
[24] Crawford, L. V., and M. J. Waring, *J. Mol. Biol.*, in press.
[25] Kersten, W., H. Kersten, and W. Szybalski, *Biochemistry*, **5**, 236 (1966).
[26] Waring, M. J., *Biochim. Biophys. Acta*, **114**, 234 (1966).
[27] Le Pecq, J., Dissertation, University of Paris (1965).
[28] Hirt, B., these PROCEEDINGS, **55**, 997 (1966).
[29] Kleinschmidt, A. K., and R. K. Zahn, *Z. Naturforsch.*, **14b**, 770 (1959).
[30] LePecq, J., and C. Paoletti, *Anal. Biochem.*, **17**, 100 (1966).
[31] Young, E. T., II, and R. L. Sinsheimer, private communication.

III
Integration State of Viral DNA in Transformed Cells

VIRAL DNA IN POLYOMA- AND SV40-TRANSFORMED CELL LINES

Heiner Westphal and Renato Dulbecco

It is clear that cells transformed by the small DNA-containing viruses, polyoma virus (Py) and SV40, harbor the viral DNA since they contain viral mRNA,[1] and some lines even yield infectious virus after fusion with cells able to support the multiplication of the virus.[2, 3] Very little is known, however, about the state of the viral DNA in these cells. In the past, several attempts were directed at determining the average number of copies of the viral DNA per cell by molecular hybridization of the DNA extracted from transformed cells with either viral DNA, or RNA made *in vitro* by copying the viral DNA. These efforts, however, were handicapped by two main technical difficulties. One is to prepare viral DNA uncontaminated by cellular DNA; the other difficulty is to have available a hybridization procedure capable of approaching the detection of one viral DNA molecule per cell, that is, one part of viral DNA among 2×10^6 parts of cellular DNA. In the experiments to be described below, both requirements have been met. Pure viral DNA was obtained by isolating the supercoiled component I[4–6] which is uncontaminated by cellular DNA. The hybridization procedure between cellular DNA and RNA made *in vitro* on purified component I DNA was that of Gillespie and Spiegelman,[7] with a number of small improvements which have increased its sensitivity and reproducibility. Fewer than five viral DNA molecules per cell can now be unambiguously identified.

In order to estimate the number of viral DNA copies per cell, it must be known whether complete molecules or only a part of them are present. This ambiguity does not exist in experiments employing the cell line SV3T3-47, from which infectious SV40 is recovered by fusion. For the other lines in which such recovery was impossible, the ambiguity persists. For this reason, we shall express the results in viral DNA equivalents per cell on the basis of reconstruction experiments.

Materials and Methods.—Viruses: Strains, purification, and DNA extraction of polyoma[8–10] and SV40[11] have been published elsewhere. The DNA extracted from the viruses was sedimented in a CsCl gradient (1.5 gm per ml 0.02 *M* Tris-HCl, pH 8.0, 0.002 *M* EDTA). The tube was centrifuged for 22 hr at 24,000 rpm (Spinco model L, SW25.1 head, 20°C); then fractions were collected from the bottom and the OD was determined. The fractions containing component I were pooled.

Cellular DNA: Mouse embryo DNA was isolated by a method previously used for solid tumors.[12] Nuclei and cytoplasm were prepared[13] from crowded tissue cultures, and the DNA was extracted with sodium dodecyl sulfate (SDS)-chloroform.[14] RNase treatment was carried out using 20 μg/ml of the enzyme for 30 min at 37°C. All DNA species were stored at 4°C with small amounts of chloroform added. The amounts of DNA were determined by the modified diphenylamine reaction.[15]

Enzymes: Purified DNA-dependent RNA polymerase (*E. coli* B) was the kind gift of Dr. Paul Berg. The enzyme was stored, without noticeable loss of activity during 6 months, in 50% glycerol at −20°C. The amounts of enzyme are given as amounts of protein of this preparation determined by Lowry's method.[16] DNase (RNase-free batch) was obtained from Worthington Biochemical, pancreatic RNase (5× crystallized) from Sigma Chemical; it was preheated at 80°C for 10 min at pH 5.0 to remove DNase traces and was then reneutralized. RNase T1 was obtained from Calbiochem.

In vitro synthesis of virus-complementary RNA (cRNA) was carried out according to Chamberlin and Berg.[17] The standard reaction mixture contained, in 400 μliters, 30 μg DNA-dependent RNA polymerase, 5 μg of SV40- or polyoma-DNA component I, 20 mμmoles each of tritiated cytidine 5′-triphosphate (CTP), 1.25 c/mmole; guanosine 5′-triphosphate (GTP), 1.0 c/mmole; uridine triphosphate (UTP), 2.3 c/mmole (all purchased from Schwarz BioResearch); and adenosine 5′-triphosphate (ATP), 8 c/mmole (purchased from the New England Nuclear Corp.). After an incubation time of 5 hr, the reaction was stopped by adding 20 μg DNase (15 min 37°C), and then SDS at 0.5% final concentration. One mg yeast-RNA (Worthington Biochemical) in 0.1 ml 0.1 × SSC was added. Then the mixture was fractionated on a Sephadex G-50 medium column (diameter 0.5 cm, bed volume 3 ml, flow rate 0.3 ml per min, equilibration buffer 0.1 × SSC, 0.1% SDS). The first peak to elute was collected, twice extracted with chloroform, dialyzed for 24 hr at 4°C against 0.1 × SSC, 0.01% SDS, and stored in small portions at −70°C. The calculation of the amount of cRNA formed was based on the assumption that the enzyme preparation does not contain nucleoside triphosphates, and that the rate of incorporation of each nucleoside triphosphate was equal. On this basis, the yield of a reaction mixture was, on the average, 20 μg cRNA, with a specific radioactivity of 5×10^6 cpm/μg (using a Beckman scintillation counter). The RNA had a sedimentation coefficient of *ca.* 4*S* (determined by sucrose gradient sedimentation together with marker HeLa ribosomal RNA) and contained 35–40% of the entire input of radioactivity.

DNA/RNA hybridization: The method of hybridizing RNA with DNA immobilized on nitrocellulose membranes[7] was used with the following modifications (unmodified steps are not described). Unless otherwise described, all DNA preparations were treated with DNase[1] to convert the component I of the viral DNA into a heat-denaturable form. DNA (100 μg) was treated with 8×10^{-4} μg DNase from the same batch, stabilized with 8 μg bovine serum albumin (fraction V) (Armour Pharmaceutical), in 1 ml of 0.01 *M* Tris-HCl, pH 8.0, containing 0.02 *M* $MgCl_2$ and 0.04 *M* NaCl, for 60 min at 30°C. Under these conditions, more than 90% of SV40 component I DNA was converted into component II, as shown by analytical ultracentrifugation. The DNase-treated DNA was subsequently poured into 2 vol boiling H_2O, boiled 15 min, chilled in ice, adjusted to 6 × SSC (final concentration), and slowly passed through Millipore membrane filters (HAWP 2400). Prior to use, these filters had been washed with boiling 6 × SSC in a Petri dish and then repeatedly with cold 6 × SSC on the filtration apparatus. The filters were incubated for 22 hr at 66°C in vials containing, in 1 ml 6 × SSC, tritiated cRNA, 1 mg yeast RNA, and SDS at a final concentration of 0.1%. The addition of SDS did not interfere with the specific RNA/DNA hybridization, but markedly reduced the background counts. The proportion of counts present in the input cRNA that remained on an empty filter under these conditions was 10^{-4}. During incubation, the vials were slowly shaken in a water bath. After the hybridization the filters were treated with 20 μg/ml pancreas RNase and 10 units/ml RNase T1 60 min 37°C.

After being counted in a toluene-scintillation mixture, the filters were assayed for DNA content by the diphenylamine reaction.[12] The amount of input cRNA was estimated in each case by placing 20 μliters of the hybridization mixture on a filter containing immobilized DNA, without filtering it through, but letting it dry in air. The dried filter was then counted. This procedure eliminated the need for a quenching correction.

Results.—Optimal conditions for RNA/DNA hybridization (Fig. 1): The conditions for hybridizing the viral RNA synthesized *in vitro* with cellular DNA were optimized in reconstruction experiments. These employed mixtures of viral DNA and a large excess of carrier DNA from nontransformed cells, in proportions similar to those to be expected in transformed cells. The filters

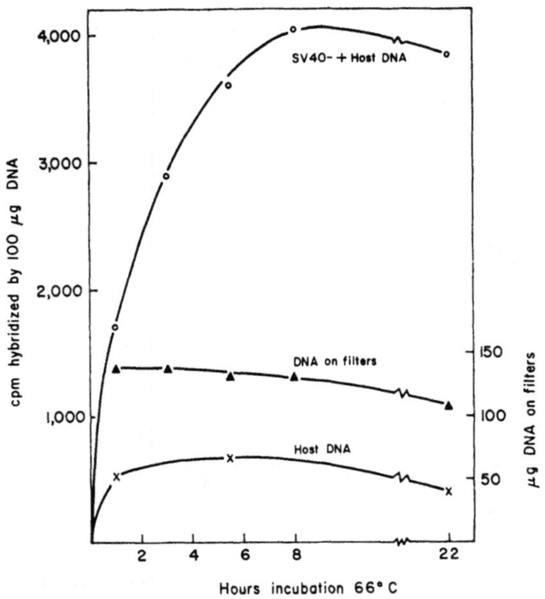

FIG. 1.—Kinetics of DNA/cRNA hybridization. Py3T3-6 DNA (which does not contain SV40-specific sequences) was mixed with SV40 DNA at a ratio corresponding to 100 viral DNA molecules per host cell, and hybridized with RNA complementary to SV40 component I DNA for the indicated periods of time. The amount of DNA immobilized per filter was 160 μg. The input of tritiated cRNA per filter was 2×10^5 cpm. The hybrid cpm/filter were normalized to an average of 100 μg DNA per filter (*see* legend, Table 3). Each measurement represents the average of five replicate filters. × — ×, cpm bound to filters with carrier DNA only; O — O, cpm bound to filters containing the virus/host DNA mixture; ▲ — ▲, amount of DNA on the filters, determined after counting.

contained 5×10^{-3} μg SV40 DNA per 100 μg of 3T3 DNA, corresponding to 100 viral DNA molecules per cell (Table 1). The number of counts immobilized on a filter reached a maximum after eight hours incubation at 66°C. A continuation of the incubation beyond this time caused a noticeable reduction of the amount of unspecific binding of cRNA to the carrier DNA. Although there was also a slight loss of DNA from the filters and of cRNA counts from filters containing viral DNA, an incubation time of 22 hours optimized the ratio of specific to nonspecific binding and was therefore adopted in all experiments. Under these conditions, the amount of specific hybrid counts increased linearly with the input of cRNA over a large range. Approximately 3×10^6 cpm cRNA were needed to saturate 10^{-4} μg of polyoma DNA mixed with 100 μg of carrier DNA (Fig. 2). About 1000 cpm were bound, corresponding to 2×10^{-4} μg of RNA. The precise quantitative significance of this result cannot be evaluated owing to some uncertainty of the true specific activity of the RNA. Anyway, the result shows that most sequences of the viral DNA have been copied *in vitro*.

Of the total DNA applied to a filter, a considerable proportion (up to 60%) was lost during immobilization and incubation at high temperature. Presum-

TABLE 1. *Quantitative data related to hybridization experiments.*

Weight of the DNA of a cell nucleus	10^{-5} μg
Weight of the DNA of a cell cytoplasm	10^{-8} μg
Weight of one viral DNA molecule	5×10^{-12} μg
Proportion of viral DNA in a cell containing a single molecule of DNA	5×10^{-7}
Amount of viral DNA, in 100 μg of total DNA, corresponding to 1 viral DNA equivalent per cell	5×10^{-5} μg
Tritiated complementary RNA: specific activity	5×10^{6} cpm/μg

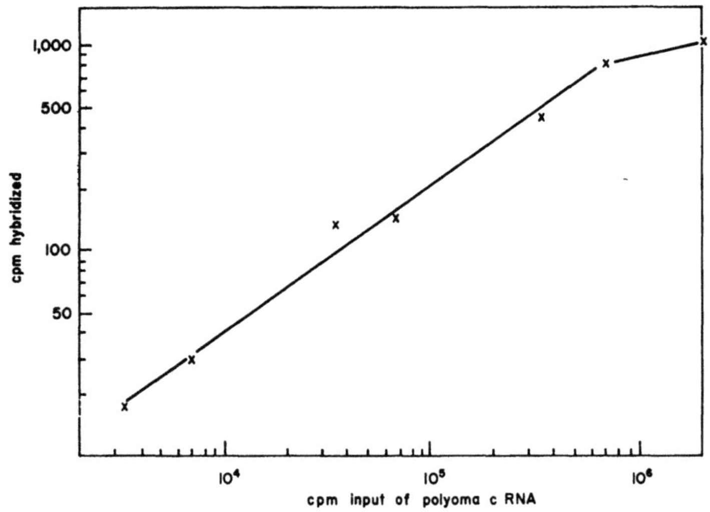

FIG. 2.—Saturation of polyoma DNA with homologous cRNA. Polyoma DNA was immobilized on nitrocellulose filters together with mouse embryo DNA (200 μg per filter) at a ratio of five viral DNA molecules per host cell (calculation, Table 1), and hybridized with given amounts of polyoma cRNA under standard conditions. The counts per minute specifically hybridized are obtained after subtracting the counts per minute bound to filters containing equal amounts of mouse embryo DNA alone. The DNA content of the counted filters was, on the average, 65 μg. Each measurement represents the average of three replicate filters (standard deviation $<10\%$).

ably this loss occurs whenever the amounts of DNA employed were near the capacity of the filters. Thus the question was whether there was any selective loss of viral DNA molecules owing to their small size. Reconstruction experiments, however, showed that the number of hybridized counts was a function of the ratio of viral to host DNA in the original hybridization mixture, irrespective of the amount of DNA that was lost during hybridization; therefore, loss is not selective.

Virus DNA equivalents in DNA extracted from nuclei and cytoplasm of various SV40- and/or polyoma-transformed cell lines: A description of the cell lines investigated in this work is given in Table 2. No infectious virus could be isolated from the pure cultures; by cell fusion with BSC-1 cells, infectious SV40 could be isolated only from two lines; no line yielded infectious plaque virus on fusion with mouse embryo cells (Watkins and Dulbecco[3]; Dulbecco, unpublished). During several years of maintenance in different laboratories, most of the lines have been transferred through many generations.

In order to determine whether viral DNA could be demonstrated, the DNA of each line was hybridized to viral cRNA in several replicate filters. The results were compared to controls in which an identical amount of DNA from the corresponding untransformed cells was hybridized with the same RNA, using the same number of filters. The determination of the number of viral DNA equivalents per cell is based on the comparison between the number of counts specifically

TABLE 2. *Cell lines investigated.*

Cell line	Reference and description	Parent cell line	Transf. agent	Virus Rescue: Direct methods	Virus Rescue: Cell fusion	T-Antigen
3T3	*Todaro, Green*[18] Normal established line	3T3	–	–	–	–
SV3T3-47	*Todaro, Green*[19] SV40-transformed derivative of 3T3	3T3	SV40	–	+(SV)	+
SV3T3-56	Clone of 3T3 cells transformed by nitrous acid-treated virus	3T3	SV40	–	–	+
Py3T3-6	*Benjamin*[1] Py-transformed derivative of 3T3, isolated in agar	3T3	SV40	–	–	+
SVPy3T3-11	*Todaro et al.*[20] SV40, Py doubly transformed	3T3	Py and SV40	–	+(SV)	+
BHK Cl.13	*Stoker, Abel*[21] Normal baby hamster kidney	BHK	–	Not tested	Not tested	–
Py8	Py transformed derivative of BHK Cl.13 isolated in agar	BHK	Py	–	–	+
H50	*Khera et al.*[22] Derivative from an SV40-induced hamster tumor	Hamster tumor	SV40	–	–	+
ME	Normal mouse embryos	–	–	Not tested	Not tested	–

hybridized, using the DNA of a transformed line, and a calibration curve obtained from the hybridization of mixtures of untransformed cell DNA with known amounts of viral DNA. Specifically hybridized counts are those fixed by the filters containing viral DNA after subtraction of counts fixed by filters containing untransformed DNA alone. There is an increase of specific hybridization as a function of the amount of transformed cell DNA bound to the filters.

Table 3 gives the results from different SV40- and/or polyoma-transformed lines and the calibration data. The average number of SV40 DNA equivalents detected in SV3T3-47 and in H50 cells are 20 and 60, respectively. For both experiments Py3T3-6 DNA was used as an additional negative control lacking SV40 sequences. The *t*-tests of the difference of the sample means and the resulting probabilities indicate the significance of the results. In order to determine the effect of culture age on the number of detectable viral DNA equivalents, cultures of SV3T3-47 cells that were actively growing and had not yet reached confluency were compared with well-crowded cultures (containing five times as many cells per culture). The number of SV40 DNA equivalents was equal in the two cases. The examination of many replicate DNA preparations per cell line, however, would be necessary to give an accurate calculation of the average number of viral DNA equivalents per transformed cell under different conditions.

Hybridization experiments were extended to several other lines transformed

TABLE 3. *SV40 DNA equivalents in SV40-transformed cell lines.*

DNA source	Cpm per 100 μg DNA	Specific Cpm*	Hybridization Probability	Viral DNA equiv./cell
Expt. 1				
3T3 (control)	1040 ± 87 (10)†	0	—	—
3T3 + 5 molecules SV40 DNA/cell	1390 ± 115 (9)	350	<0.002	8
3T3 + 16 molecules SV40 DNA/cell	1670 ± 164 (9)	630	<0.002	16
SV3T3-47	1840 ± 115 (10)	800	<0.002	20
Py3T3-6	1135 ± 214 (10)	95	>0.5	5
Expt. 2				
3T3 (control)	745 ± 108 (5)	0	—	—
3T3 + 20 molecules SV40 DNA/cell	1120 ± 113 (5)	375	<0.002	20
SV3T3-47	1135 ± 80 (5)	390	<0.002	21
Py3T3-6	695 ± 93 (5)	−50	>0.5	—
BHK Cl.13 (control)	600 ± 68 (9)	0	—	—
BHK Cl.13 + 10 molecules SV40 DNA/cell	805 ± 85 (9)	205	<0.002	10
H50	1800 ± 198 (9)	1200	<0.002	58
Expt. 3				
3T3 (control)	483 ± 53 (5)	0	—	—
SVPy3T3-11	1428 ± 35 (5)	945	<0.002	44
SV3T3-47	908 ± 54 (5)	425	<0.002	20
SV3T3-47Y‡	886 ± 133 (5)	403	<0.002	19
Expt. 4				
3T3 (control)	715 ± 99 (12)	0	—	—
SV3T3-47	1482 ± 70 (8)	767	<0.002	20
SV3T3-56	979 ± 100 (12)	264	<0.002	7

Using the standard procedure, tritiated RNA complementary to SV40 (component I) DNA was hybridized with the indicated DNA species. The H^3 input of cRNA per filter was: expt. 1, 4×10^5 cpm; expt. 2 and 4, 3×10^5 cpm; expt. 3, 2×10^5 cpm. 160 μg DNA was added per filter. The amount of DNA remaining on the filters after hybridization was determined and the hybrid cpm of each filter were normalized to an average of 100 μg DNA. SV40 DNA component I was used to calibrate the system.

* Cpm after subtracting the 3T3 control values from the transformed 3T3 DNA species; after subtracting the BHK Cl.13 control from H50.

† Cpm ± standard deviation and in parentheses, the number of filters.

‡ SV3T3-47Y DNA is derived from cells harvested when semiconfluent.

by SV40 and/or polyoma virus (Tables 3 and 4). It was found that the SVPy-3T3-11 line contains about 40 SV40 and 10 polyoma DNA equivalents per cell. The number of SV40 DNA equivalents in the line SV3T3-56 is only 7, i.e., one third of that found in SV3T3-47. Line Py3T3-6 and the Py8 line of polyoma-transformed hamster cells contain about 5 and 7 Py DNA equivalents, respectively. SV3T3-47 DNA and H50 DNA failed to give specific hybridization with polyoma cRNA.

When the distribution of the virus-specific DNA between nucleus and cytoplasm in transformed cells was examined, using lines Py3T3-6, SV3T3-47, and H50, all hybridizable counts were found in the nucleus (Table 5).

Discussion.—The results reported confirm the previous reports that viral DNA can be detected in cells transformed by either polyoma virus or SV40.[23, 24] Furthermore, the improved experimental technique allows for the first time a reliable estimation of the average numbers of virus DNA equivalents per cell based on reconstruction experiments. The main result is that the number of equivalents

TABLE 4. *Polyoma DNA equivalents in Py-transformed cell lines.*

DNA source	Cpm per 100 μg DNA	Specific Cpm	Hybridization Probability	Viral DNA equiv./cell
Expt. 1				
3T3 (control)	1400 ± 140 (10)	0	—	—
3T3 + 10 molecules Py DNA/cell	2100 ± 150 (9)	700	<0.002	10
3T3 + 20 molecules Py DNA/cell	2950 ± 320 (9)	1550	<0.002	20
Py3T3-6	1750 ± 130 (10)	350	<0.002	5
SV3T3-47	1400 ± 95 (10)	0	>0.5	—
Expt. 2				
3T3 (control)	891 ± 24 (6)	0	—	—
Py3T3-6	1100 ± 57 (6)	209	<0.002	5
SVPy3T3-11	1318 ± 68 (6)	427	<0.002	10
Expt. 3				
BHK Cl.13 (control)	530 ± 46 (9)	0	—	—
BHK Cl.13 + 7 molecules Py DNA/cell	747 ± 45 (9)	217	<0.002	7
Py8	745 ± 39 (9)	215	<0.002	7

Tritiated RNA complementary to polyoma (component I) DNA was hybridized with the indicated DNA species. The H^3 input per filter was: expt. 1, 5×10^5 cpm; expts. 2 and 3, 2×10^5 cpm. Polyoma DNA component I was used to calibrate the system. Further details are given in the legend of Table 3.

per cell varies in different lines from about 60 (H50 line) to a number close to the limit of significant detection (Py3T3-6), and they are localized in the nucleus. In many lines, the number of equivalents is much higher than the ploidy of the cells employed.

As to the biological significance of these differences, we shall remark that lines with a higher average number of viral DNA equivalents were not "more" transformed, as judged from the morphological and growth properties of the cells in culture. An especially interesting line is SV3T3-56, which was derived from a clone transformed by nitrous acid-treated virus and does not yield infectious virus on fusion with BSC-1 cells with the help of UV Sendai virus (unpublished observations), in contrast to SV3T3-47 and SVPy3T3-11.[3] Thus it was probably transformed by a mutant virus unable to cause a productive infection. The low number of equivalents present in this line, compared with other SV40-transformed derivatives of line 3T3, suggests that the expression of a viral gene function may be required for building up a large number of DNA equivalents per cell.

There is no correlation between numbers of viral DNA equivalents per cell

TABLE 5. *Distribution of the virus-specific DNA between nucleus and cytoplasm in transformed cells.*

Cell line	No. of Viral DNA Equivalents per Nucleus	Cytoplasm
SV3T3-47	20	$\leqslant 0.03$
Py3T3-6	5	$\leqslant 0.005$
H50	60	$\leqslant 0.04$

Cytoplasmic and nuclear DNA were separately hybridized with the appropriate complementary RNA (2×10^5 cpm/filter). After counting, 8 μg DNA were, on the average, bound per filter. Calibrations and controls were like those given in Table 3.

and ability of the cells to yield infectious virus after fusion with virus-susceptible cells. For instance, line H50, with 60 equivalents, does not yield virus on fusion. It may have a mutated viral DNA. There may be a correlation between the number of equivalents and the amount of *T*-antigen formed because minimal amounts of the antigen are formed in line SV3T3-56.

An interesting question concerns the origin of the many copies of viral DNA present in the cells of certain lines.

For understanding this problem the physical characteristics of the viral DNA molecules present in the cells must be determined. Work in this direction is in progress.

Summary.—The amounts of viral DNA present in polyoma- and SV40-transformed cell lines have been determined by hybridizing the cellular DNA with *in vitro* RNA complementary to viral DNA. The numbers of virus DNA equivalents per cell varies from 5 to 60 in different lines. The viral DNA is located in the nuclei of the cells. The biological significance of these findings is discussed.

This work was carried out with the aid of research grant CA-07952 from the National Cancer Institute.

[1] Benjamin, T., *J. Mol. Biol.*, **16**, 359 (1966).
[2] Koprowski, H., F. C. Jensen, and Z. Steplewski, these PROCEEDINGS, **58**, 127 (1967).
[3] Watkins, J. F., and R. Dulbecco, these PROCEEDINGS, **58**, 1396 (1967).
[4] Dulbecco, R., and M. Vogt, these PROCEEDINGS, **50**, 236 (1963).
[5] Weil, R., and J. Vinograd, these PROCEEDINGS, **50**, 730 (1963).
[6] Crawford, L. V., and P. H. Black, *Virology*, **24**, 388 (1964).
[7] Gillespie, D., and S. Spiegelman, *J. Mol. Biol.*, **12**, 829 (1965).
[8] Vogt, M., and R. Dulbecco, *Virology*, **16**, 41 (1962).
[9] Winocour, E., *Virology*, **19**, 158 (1963).
[10] Weil, R., *Virology*, **14**, 46 (1961).
[11] Hatanaka, M., and R. Dulbecco, these PROCEEDINGS, **56**, 736 (1966).
[12] Winocour, E., *Virology*, **25**, 276 (1965).
[13] Penman, S., K. Scherrer, Y. Becker, and J. E. Darnell, these PROCEEDINGS, **49**, 654 (1963).
[14] Marmur, J., *J. Mol. Biol.*, **3**, 208 (1961).
[15] Burton, K., *Biochem. J.*, **62**, 315 (1956).
[16] Lowry, O. H., A. L. Rosebrough, A. L. Farr, and R. J. Randall, *J. Biol. Chem.*, **193**, 265 (1951).
[17] Chamberlin, M., and P. Berg, these PROCEEDINGS, **48**, 81 (1962).
[18] Todaro, E. J., and H. Green, *J. Cell Biol.*, **17**, 299 (1963).
[19] Todaro, E. J., and H. Green, *Science*, **147**, 513 (1965).
[20] Todaro, E. J., K. Habel, and H. Green, *Virology*, **27**, 179 (1965).
[21] Stoker, M., and P. Abel, in *Cold Spring Harbor Symposia on Quantitative Biology*, vol. 27 (1962), p. 375.
[22] Khera, K. S., A. Ashkenazi, F. Rapp, and J. L. Melnick, *J. Immunol.*, **91**, 604 (1963).
[23] Axelrod, D., K. Habel, and E. T. Bolton, *Science*, **146**, 1466 (1964).
[24] Reich, P. R., P. H. Black, and S. M. Weissman, these PROCEEDINGS, **56**, 78 (1966).

THE INTEGRATED STATE OF VIRAL DNA IN SV40-TRANSFORMED CELLS

J. SAMBROOK, H. WESTPHAL, P. R. SRINIVASAN, AND R. DULBECCO

It has been shown by a modified DNA-RNA hybridization technique that the nuclei of cells transformed by either polyoma virus or SV40 contain viral DNA.[1] This method employs RNA synthesized *in vitro* with form I viral DNA[2-4] and *E. coli* RNA polymerase. The number of viral DNA equivalents per cell varies from 5 to 60, depending on the cell line. In this communication, we will report on the physical state of the viral DNA in SV3T3 cells, an SV40-transformed cell line that contains 20 SV40 DNA equivalents per cell.

Materials and Methods.—Viruses: The strains of polyoma virus and SV40 used and the methods of virus purification have already been published.[5-8]

Cells: The origins and the methods of cultivation of cell lines SV3T3-47 and Py3T3-6 have been described.[1]

DNA extraction: DNA was obtained from purified polyoma virus or SV40 by two extractions with phenol saturated with 0.5 *M* tris(hydroxymethyl)aminomethane (Tris), pH 7.8, and one with chloroform-isoamylalcohol (24:1). Forms I and II of the viral DNA were purified by sedimentation through 5–20% neutral sucrose gradients[9] containing 0.01 *M* Tris, pH 8.0, 0.001 *M* ethylenediaminetetraacetate (EDTA), and $^1/_{10}$ solution of 0.15 *M* NaCl, 0.015 *M* Na citrate, pH 7 (SSC/10). The gradients were centrifuged for 18 hr at 23,000 rpm in a Spinco SW 25.1 rotor at 20°.

DNA was prepared from isolated nuclei[10] by the method of Marmur,[11] except that the ribonuclease treatment was omitted.

High-molecular-weight DNA was isolated as follows: 2 ml of 0.5 *M* NaOH containing 0.1 *M* EDTA was layered on the top of a 54-ml alkaline sucrose gradient (10–30% sucrose in 0.3 *N* NaOH, 0.01 *M* EDTA, and 0.5 *M* NaCl) in polyallomer tubes. Low-optical-density sucrose (Harshaw Chemical Co., Cleveland, Ohio) was used in these experiments because sucrose from other sources caused high backgrounds in the hybridization experiments. From 2 to 4 $\times$ 10^6 cells in 0.3 ml of NaCl, 0.8%; KCl, 0.038%; Na_2HPO_4, 0.01%; Sigma 7-9, 0.3% (TD) were carefully layered into the alkaline layer on the top of the gradient. The gradients were stored at 4° for at least 12 hr to allow complete liberation of the DNA and then centrifuged in a Spinco SW 25.2 rotor at 25,000 rpm for 5 hr at 4°. The high viscosity of the DNA tends to create considerable turbulence in the tube during collection. To avoid this difficulty, fractions were collected from the top by pumping 40% sucrose into the bottom of the tube. The tube contents flowed through a nick in the side of the tube and were guided by a glass rod. This type of alkaline sucrose gradient has been used previously to isolate high-molecular-weight DNA from *E. coli*[12] and mammalian cells.[13]

DNA was also isolated from nuclei by the method of Hirt.[14] Freshly prepared nuclei from 7 $\times$ 10^7 cells were distributed in a thin film on the wall of a polyethylene tube. Twenty ml of 0.01 *M* Tris, pH 7.4, containing 0.01 *M* EDTA and 0.6% SDS were slowly added. After 15 min at room temperature, the cell lysate was adjusted to 1 *M* by the addition of 5 *M* NaCl. During the addition and subsequent mixing, care was taken to minimize the amount of shear of the DNA. After storage at 4° for 16 hr, the mixture was centrifuged at 35,000 $\times$ g in an angle head at 4°. The supernatant was carefully poured off; the pellet was washed with 5 ml of 0.01 *M* Tris, pH 7.4, and 0.01 *M* EDTA; and the washings were combined with the supernatant. DNA was extracted separately from the supernatant and the pellet by repeated extractions with chloroform-isoamylalcohol mixture until no protein could be observed at the interphase, followed by dialysis against SSC/100 containing 0.001 *M* EDTA.

Chromosomes: Monolayers of cells were treated with colcemid (0.5 μg per ml) in calcium-free medium for 18 hr. The cells arrested in metaphase (about 40%) were selectively detached from the glass by shaking. Chromosomes were isolated by the method of Maio and Schildkraut.[15] DNA was extracted from the purified chromosomes by the method of Marmur.[11]

Equilibrium centrifugation was performed two ways: (1) with ethidium bromide[9, 16] in CsCl gradients in a Spinco 50 angle head rotor at 43,000 rpm for 48 hr at 20°; fractions were collected by puncturing the bottom of the tubes with a 13-gauge needle, and ethidium bromide was removed by passing the fractions through a column of Dowex-50 (Na form); (2) with alkaline CsCl gradients in a Spinco SW 50L rotor at 30,000 rpm for 60 hr at 20°; the DNA-CsCl solution was adjusted to pH 13 with 1 *N* NaOH, and the initial density of the solution was 1.74 gm/ml.

Labeling of virus and cells: P^{32}-labeled polyoma virus and SV40 were prepared by infecting cells in Eagle's medium containing low phosphate (10^{-5} *M*) and 10% dialyzed bovine fetal serum. Twenty-four hours after infection, carrier-free P^{32} as orthophosphate (100 μc/ml) was added. Five days later when the cytopathic effect was maximal, the virus was harvested, purified, and the DNA was immediately extracted. This DNA had a specific activity equal to or greater than 5×10^5 cpm per μg.

Cellular DNA was labeled by incubating exponentially growing cells for 8 hr in medium containing 0.2 μc of 2-C^{14}-thymidine per ml.

Cellular protein was labeled by incubating exponentially growing cells for 8 hr in Eagle's medium lacking leucine and lysine, supplemented with 10% dialyzed bovine fetal serum, and containing 1 μc each of H^3-lysine and H^3-leucine per ml.

DNA-RNA hybridizations: The method of preparation of H^3 complementary RNA (cRNA) has already been published, as have the details of the modified hybridization technique used in this laboratory.[1] The amount of DNA on each filter was determined after hybridization and counting by Burton's method.[17]

Radioactivity: Samples were precipitated with 5% trichloracetic acid onto Whatman GF/C paper and were counted in a Beckman scintillation counter using a toluene-based scintillation mixture.

Chemicals: Ethidium bromide was purchased from Boots Pure Drug Co., Nottingham England. Colcemid was a product of Ciba Pharmaceutical Co., New Jersey. Thymidine-2-C^{14} (52.8 mc/mmole) was obtained from Schwarz BioResearch, Inc., Orangeburg, New Jersey. L-leucine-4,5-H^3 (4.6 mc/μmole) and DL-lysine-4,5-H^3 (5 mc/μmole) were purchased from New England Nuclear Corp., Boston, Massachusetts.

Results.—(1) *Evidence that transformed cells do not contain supercoiled forms of viral DNA:* Centrifugation to equilibrium in neutral CsCl containing ethidium bromide has been shown to resolve unnicked circular DNA from the main mass of linear cell DNA.[9] The cyclic DNA binds less dye and bands at a higher density.[16] This method was used to find out whether transformed cells contain any form I viral DNA. The separation obtained when P^{32}-labeled SV40 DNA was mixed with H^3-thymidine-labeled cell DNA and centrifuged to equilibrium in CsCl-ethidium bromide is shown in Figure 1. All the unnicked circular viral form I is separated from the cellular DNA; viral form II behaves as linear DNA in these gradients and remains with the cell DNA. DNA's from 3T3 cells transformed by SV40 (i.e., SV3T3) and 3T3 cells transformed by polyoma virus (i.e., Py3T3) were centrifuged in a similar way. Then the gradients were divided into two fractions: one containing the cell DNA, and the other the rest of the heavy side of the gradient. Both fractions were first dialyzed against 0.001 *M* Tris, pH 8.0, 0.001 *M* EDTA, SSC/100 to remove CsCl and then

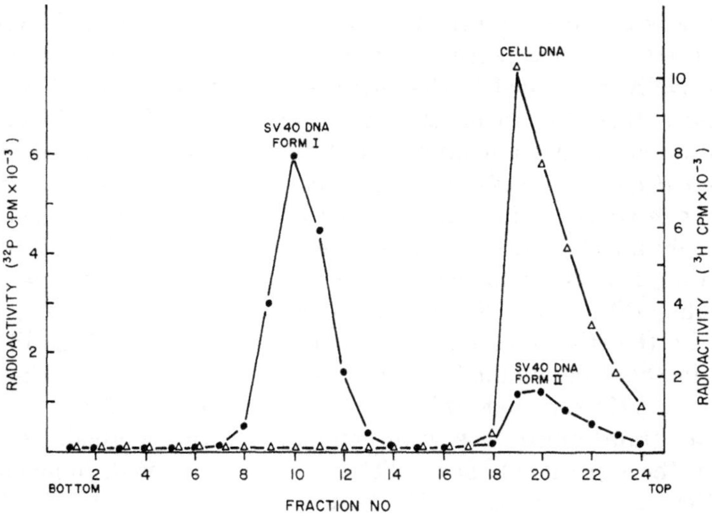

FIG. 1.—Equilibrium centrifugation of a mixture of 0.2 μg purified SV40 P^{32}-DNA (I and II) and 100 μg cell H^3-DNA in 5.00 ml CsCl, density 1.57 gm/ml, 100 μg/ml ethidium bromide for 36 hr at 43 krpm, 20°C in a Spinco 50 angle head. Fractions were collected and the DNA was precipitated and counted as described.

passed through Dowex-50 to remove ethidium bromide. The DNA was heat-denatured and then immobilized on Millipore filters and hybridized with H^3-cRNA prepared *in vitro* by using SV40 form I DNA as template. The results are shown in Table 1 where the counts hybridized with cellular DNA are in column *C* and those hybridized with the denser part of the gradient are in column *V*. There is no difference in counts in column *V* between SV3T3 and Py3T3 cells. If SV3T3 cells had contained free SV40 viral DNA form I, it would have been detected as shown by a reconstruction experiment whose results are given in line 3. In this experiment, SV40 DNA containing 80 per cent of form I and 20 per cent of form II was mixed with Py3T3 DNA and centrifuged. The results indicate that form I was entirely recovered in the *V* fraction and form II in the *C* fraction.

A highly significant difference between SV3T3 and Py3T3 cells is present in column *C*. All the virus-specific hybridizable counts are found in the band containing the cellular DNA. The question arises whether this result could be

TABLE 1. *Hybridization of DNA from transformed cells after centrifugation in CsCl/ethidium bromide gradients.*

Cell type	Fraction from Gradient (cpm) *V*	*C*
SV3T3	69 ± 5	358 ± 2
Py3T3	64 ± 7	129 ± 12
Py3T3 + 20 SV40 genomes per cell	385 ± 11	198 ± 5

H^3-RNA complementary to SV40 (form I) DNA was hybridized with the indicated fractions. The amount of DNA remaining on the filters after hybridization was determined and the figures in column *C* are the hybrid cpm normalized to 50 μg DNA. The figures in column *V* are uncorrected hybrid cpm. **The H^3-cRNA input per filter was 1.2 × 10^6 cpm in this experiment.**

produced by selective loss of form I DNA or its conversion to form II during the extraction of DNA. This possibility was excluded by experiments in which labeled viral DNA was added to the nuclei before extraction of DNA. Under these conditions, there was no loss of viral form I DNA. We therefore conclude that there is no substantial amount of form I DNA in the transformed cells.

(2) *Evidence that transformed cells do not contain free viral DNA of a size similar to that present in virions:* The Hirt method of DNA extraction when applied to cells lytically infected with polyoma virus separates viral DNA irrespective of its form (i.e., I or II) from cell DNA.[14] The efficiency of the method was tested by adding P^{32}-labeled SV40 DNA to freshly isolated nuclei and extracting the mixture by the Hirt procedure. From 90 to 95 per cent of the P^{32} counts were found in the supernatant fraction. Less than 2 per cent of the cellular DNA is present in the supernatant if care is taken to minimize shear.

Nuclei from transformed cells were extracted by this method. DNA was isolated from the supernatant and pellet fractions, denatured, immobilized on Millipore filters, and hybridized. The results are shown in Table 2. There is

TABLE 2. *Hybridization of DNA extracted from transformed cells by Hirt's method.*

Cell type	Fraction	
	Pellet	Supernatant
SV3T3	694 ± 31	143 ± 20
Py3T3	287 ± 23	114 ± 20
Py3T3 + 140 SV40 genomes per cell	670 ± 18	3550 ± 360

H^3-RNA complementary to SV40 (form I) DNA was hybridized with the indicated DNA. The amount of DNA remaining on the filters after hybridization was determined, and the hybrid cpm given in the column under "Pellet" have been normalized to 50 μg DNA. The hybrid cpm for the supernatant fractions are uncorrected values. The H^3-cRNA input in this experiment was 1.7×10^5 cpm/filter.

no significant difference between the SV3T3 cells and the control, Py3T3, cells in the number of counts hybridized to DNA isolated from the supernatant fractions. However, marked differences are found in the hybridization to DNA obtained from the pellet fractions. Reconstruction experiments in which 140 SV40 DNA molecules per cell were added before extraction show that 90 per cent of the specifically hybridizable counts can be removed in the supernatant fraction. The viral DNA in transformed cells therefore is not in a free form of size comparable to that of DNA present in virions.

(3) *Evidence that the viral DNA is integrated in transformed cells:* (a) *Hybridizable material in isolated chromosomes:* Chromosomes were isolated from SV3T3 cells, the DNA was extracted, heat-denatured, immobilized on Millipore filters, and hybridized with SV40-cRNA. Control filters contained SV3T3 DNA isolated from cells that remained attached to the glass after the metaphase cells had been harvested, and also DNA's extracted from nuclei of Py3T3 and SV3T3 interphase cells. The results are presented in Table 3. The DNA isolated from chromosomes hybridized to the same extent as DNA from SV3T3 interphase cells and to a significantly greater extent than DNA from Py3T3 interphase cells. Nuclear contamination of the chromosome preparation could account for no more than 10 per cent of the DNA in the chromosome DNA preparation. The SV40 DNA in SV3T3 cells is therefore associated with chromosomes during the metaphase.

TABLE 3. *Hybridization of DNA extracted from chromosomes and nuclei.*

Source of DNA	Cpm hybridized
SV3T3 chromosomes	680 ± 63
SV3T3 nuclei*	690 ± 87
Py3T3 nuclei	290 ± 32
SV3T3 nuclei	740 ± 47

H^3-RNA complementary to SV40 (form I) DNA was hybridized with the indicated DNA. The amount of DNA remaining on the filters after hybridization was determined and the hybrid cpm have been normalized to 50 μg DNA. The H^3-cRNA input in this experiment was 2.1×10^5 cpm/filter.

* DNA extracted from nuclei of SV3T3 cells treated with colcemid.

(b) *Hybridizable material in high-molecular-weight DNA:* High-molecular-weight DNA was isolated by alkali treatment of both SV3T3 and Py3T3 cells at the top of an alkaline sucrose gradient followed by centrifugation. The denatured DNA had a sedimentation coefficient of about 110*S* (Fig. 2). The cellular DNA was well separated from viral forms I and II. Moreover, it was free of protein as evidenced by the absence of counts in the cellular DNA peak when the cells were labeled with a mixture of H^3-leucine and H^3-lysine (Fig. 3). Under these conditions, linear DNA of molecular weight of 6×10^7 equivalent to 20 viral genomes will have a sedimentation coefficient of about 50*S*.[18] The high-molecular-weight DNA obtained from the gradients was neutralized with 1.1 *N* HCl containing 0.2 *M* Tris, dialyzed against large volumes of SSC/100 containing 0.001 *M* EDTA, concentrated, redialyzed against the same buffer, adjusted to 6 × SSC, immobilized on Millipore filters, and hybridized against SV40 cRNA. The results are given in Table 4. The high-molecular-weight DNA isolated from SV3T3 cells hybridizes to the same extent as DNA isolated from the nuclei of SV3T3 cells. However, the counts hybridized are markedly higher than those

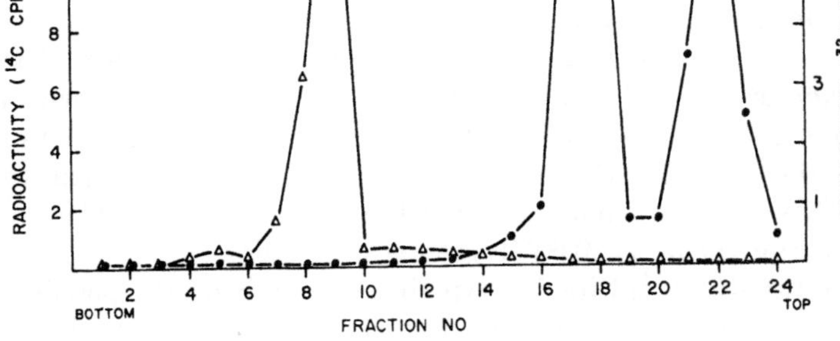

FIG. 2.—Velocity sedimentation of a mixture of polyoma P^{32}-DNA, I and II, and cell C^{14}-DNA in alkaline sucrose gradients. For details, see *Materials and Methods.*

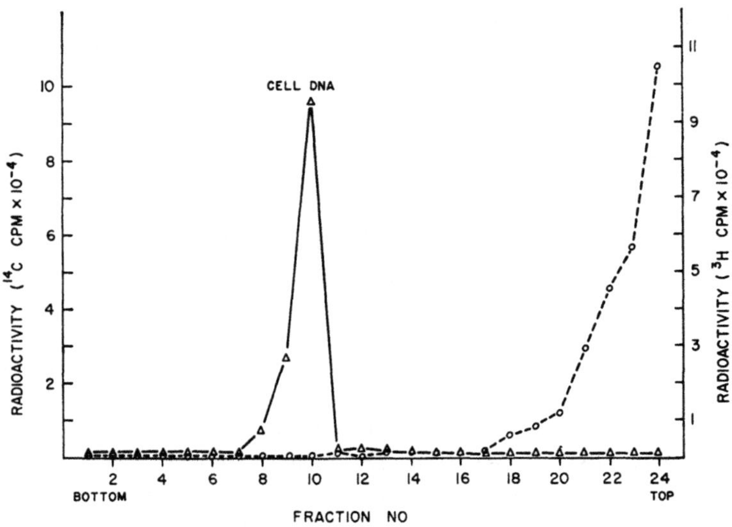

FIG. 3.—Velocity sedimentation of lysed cells labeled with C^{14}-thymidine and H^3-leucine and H^3-lysine. For details, see *Materials and Methods.*

obtained with either high-molecular-weight Py3T3 DNA or Py3T3 DNA isolated from nuclei. The hybridizable counts found with SV3T3 are comparable to the increase seen when 20 genomes of SV40 per cell are added either to high-molecular-weight Py3T3 DNA or to Py3T3 DNA isolated from nuclei.

In a further experiment, the high-molecular-weight DNA obtained from alkaline sucrose gradients was neutralized, dialyzed, concentrated, and centrifuged to equilibrium in alkaline CsCl gradients. At the end of the run, fractions were collected from the gradients, the refractive index of each fraction was measured, and the density calculated. The fractions of density 1.74–1.81 gm/ml were pooled, neutralized, and dialyzed. The DNA from these fractions was hybridized in the normal manner. The results are given in Table 4, experiment 2. SV3T3 DNA again hybridizes to a greater extent than Py3T3 DNA, and there is little difference in cpm hybridized between DNA's obtained directly from alkaline sucrose gradients and DNA's that were subsequently centrifuged to equilibrium in alkaline CsCl.

Discussion.—In these experiments, DNA's isolated by a variety of methods from SV3T3 and Py3T3 cells have been hybridized with RNA made *in vitro* with SV40 DNA as template. In every case, it was possible to isolate a DNA fraction from the SV3T3 cells that hybridized to a greater extent than the comparable DNA fraction from the control cell line, Py3T3. This increased hybridization is equivalent to 20 SV40 genomes per cell.

The results of the ethidium bromide experiments eliminate the possibility that viral DNA in SV3T3 cells is in the form of supercoiled circles of any size; and the fact that no hybridizable counts were present in the supernatant of the Hirt extraction eliminates the possibility that the viral DNA is in a free form of any configuration of molecular weight comparable to the DNA of the virion.

TABLE 4. *Comparison of hybridization of DNA obtained from transformed cells by alkaline sucrose gradients and by Marmur's method.*

	Source of DNA	Cpm hybridized
Expt. 1:	Alkaline sucrose gradients	
	SV3T3	354 ± 22
	Py3T3	144 ± 16
	Py3T3 + 20 genomes of SV40/cell	394 ± 36
	Nuclei extracted by Marmur's method	
	SV3T3	342 ± 15
	Py3T3	186 ± 13
	Py3T3 + 20 genomes of SV40/cell	386 ± 22
Expt. 2:	Alkaline sucrose gradients	
	SV3T3	593 ± 30
	Py3T3	210 ± 14
	Py3T3 + 25 genomes of SV40/cell	780 ± 30
	Alkaline sucrose gradients, followed by equilibrium centrifugation in alkaline CsCl	
	SV3T3	620 ± 12
	Py3T3	248 ± 2

H^3-RNA complementary to SV40 (form I) DNA was hybridized with the indicated DNA. The amount of DNA remaining on the filters after hybridization was determined, and the hybrid cpm normalized to 50 μg DNA. The H^3-cRNA input per filter was 1×10^5 cpm in expt. 1 and 1.5×10^5 cpm in expt. 2.

There are two further possible states of the viral DNA: (1) free form II molecules of a size larger than normal viral DNA, derived by tandem duplication; (2) covalently bound to cellular DNA or bound to other structural components of the nuclei.

The maximum size of a molecule described in the first possibility would depend on the average number of viral DNA equivalents per cell and on their distribution between individual cells. In the case of SV3T3 cells, there are about 20 viral genome equivalents per cell, and three lines of evidence suggest that they are uniformly distributed throughout the cell population. (*a*) The SV3T3 cells used in this study (SV3T3-47) are a clonal line; (*b*) all SV3T3-47 cells show a similar amount of T antigen in their nuclei as judged by immunofluorescence; and (*c*) the average number of SV40 DNA equivalents per cell does not differ in populations of cells grown under very different physiological conditions. So it is unlikely that any SV3T3-47 cell contains more than 60×10^6 daltons of viral DNA. If this DNA were present as a single molecule, it would have a sedimentation coefficient of about 50*S* in alkali,[18] and no hybridization would have been detected in the 110*S* DNA peak collected from the alkaline sucrose gradients.

Clearly the data support the alternative hypothesis that the SV40 DNA is attached to a larger molecule. It is unlikely that this is a structural component of the cell since the high-molecular-weight-DNA fraction that hybridizes with RNA complementary to SV40 has the same density as denatured DNA in alkaline CsCl (i.e., 1.78 gm/ml).

Thus the results presented here provide evidence that the viral DNA molecules in SV3T3 cells are integrated with cell DNA by alkali-stable covalent linkages. The viral DNA may be linked to the cellular DNA as one large piece at a single site, or it may be connected at multiple sites after many individual insertions. This problem is currently under investigation.

We wish to acknowledge useful suggestions from Drs. John Lett, Paul Berg, J. Vinograd, and R. Sinsheimer.

This work was carried out with the aid of research grant no. CA-07592 from the National Cancer Institute.

[1] Westphal, H., and R. Dulbecco, these PROCEEDINGS, **59,** 1158 (1968).
[2] Dulbecco, R., and M. Vogt, these PROCEEDINGS, **50,** 236 (1963).
[3] Weil, R., and J. Vinograd, these PROCEEDINGS, **50,** 730 (1963).
[4] Crawford, L. V., and P. H. Black, *Virology,* **24,** 388 (1964).
[5] Vogt, M., and R. Dulbecco, *Virology,* **16,** 41 (1962).
[6] Winocour, E., *Virology,* **19,** 158 (1963).
[7] Weil, R., *Virology,* **14,** 46 (1961).
[8] Hatanaka, M., and R. Dulbecco, these PROCEEDINGS, **56,** 736 (1966).
[9] Radloff, R., dissertation, California Institute of Technology (1966).
[10] Penman, S., K. Scherrer, Y. Becker, and J. E. Darnell, these PROCEEDINGS, **49,** 654 (1963).
[11] Marmur, J., *J. Mol. Biol.,* **3,** 208 (1961).
[12] McGrath, R. A., and R. W. Williams, *Nature,* **212,** 534 (1966).
[13] Lett, J. T., I. Caldwell, C. J. Dean, and P. Alexander, *Nature,* **214,** 790 (1967).
[14] Hirt, B., *J. Mol. Biol.,* **26,** 365 (1967).
[15] Maio, J. J., and C. L. Schildkraut, *J. Mol. Biol.,* **24,** 29 (1967).
[16] Bauer, W., and J. Vinograd, *J. Mol. Biol.,* **33,** 141 (1968).
[17] Burton, K., *Biochem. J.,* **62,** 315 (1956).
[18] Studier, F. W., *J. Mol. Biol.,* **11,** 373 (1965).

IV
Detection of Virus-Specific (T) Antigen

A SPECIFIC COMPLEMENT-FIXING ANTIGEN PRESENT IN SV40 TUMOR AND TRANSFORMED CELLS

Paul H. Black, Wallace P. Rowe, Horace C. Turner, and Robert J. Huebner

Many experimental tumors, both carcinogen-induced[1] and virus-induced,[2-5] contain new cellular antigens, generally demonstrable by transplant rejection procedures. Huebner *et al.*[6] first demonstrated the presence of new, noninfectious, complement-fixing (CF) antigens clearly under viral genetic control, in adenovirus-induced tumors in hamsters and rats.

A new transplantation antigen(s) has been found in SV40-induced hamster tumors,[7-9] but it is not established whether its synthesis is under viral or host cell genetic control. This paper presents evidence for a new CF antigen in SV40 tumors and transformed tissue culture cells, formed from information contained within the viral genome. Preliminary results of these studies were presented by Huebner *et al.*[6]

Materials and Methods.—The CF procedure was identical with that used by Huebner *et al.*;[6] this is a Bengtson procedure done in microplates using overnight fixation at 4°C, with two exact units of complement. Tumor extracts consisted of 10% suspensions in Eagle's basal medium, clarified by centrifugation at 2500 rpm for 30 min, and stored at −60°C. The extracts were tested for antigens only if the undiluted extract was not anticomplementary (AC). The primary SV40-induced hamster tumors used in these studies are from experiments described in detail elsewhere.[10] Tumor extracts used as standard CF antigens were selected for having high titer reactivity with sera from tumorous hamsters.

Suspensions of both normal and transformed tissue culture cells of various species,[11] as well as tissue culture cells from a variety of hamster tumors, were prepared in the following manner. Cells grown in 32-oz Blake bottles were washed with phosphate-buffered saline (PBS) (pH 7.2), scraped off the glass with a rubber policeman, centrifuged at 150 *g* for 8 min, and resuspended in 9 volumes of PBS. These suspensions were stored at −60°C before use.

"Viral antigen" was prepared by inoculating SV40 strain 776[12] into a continuous tissue culture cell line (strain BSC-1) of African green monkey kidney (AGMK) at a multiplicity of about 10^{-4}, and harvesting the cells and fluid together when cytopathogenicity was maximal. The cell suspension was stored at −20°C, thawed, and used without further processing.[13]

Four types of sera were used as standard reagents; to avoid undue heterogeneity of antibodies, sera of individual animals generally were used: (1) serum from tumorous hamsters, selected for having high CF antibody titer against SV40 tumor extracts, and no reaction with viral antigen; (2) serum from similar hamsters, but having high CF antibody titers for both tumor extracts and viral antigen; (3) convalescent serum (#2159) from a monkey inoculated with SV40 virus;[14] and (4) serum of weanling hamsters hyperimmunized with SV40 virus grown in AGMK cultures.

Tests for CF antigens in tumor extracts were done by testing against hamster serum diluted (generally 1:10 or 1:20) to contain 4–16 units of antitumor antibody, and against the monkey immune serum at a dilution of 1:20, which corresponded to 1–2 units of antibody reactive with tumor extracts. The same monkey serum was used at a dilution of 1:80 to test for viral antigen in AGMK harvests; this dilution of the serum contained 8 units of antiviral antibody (*vide infra*). Hamster sera were screened for CF antibody at a dilution of 1:5 or 1:10, using 8 units of tumor antigen or 4 units of viral antigen.

The methods used for virus assay and neutralization tests have been described elsewhere.[10]

Ultracentrifugal fractionation of antigens was done by the following procedure. Frozen material was thawed and reclarified by centrifugation at 2,500 rpm for 15 min. The supernatant fluid was centrifuged at 30,000 rpm (59,310 *g* av) for 2 hr in a Spinco no. 40 rotor at 4°C; the

TABLE 1

REPRESENTATIVE PATTERNS OF CF ANTIBODY RESPONSES TO TUMOR AND VIRAL ANTIGEN

Antigen	Reciprocals of Antibody Titer Versus Undiluted Antigen and Antigen Titer Versus 1:20 Serum: Serum of hamster with primary tumor; #5299		Serum of hamster with transformed cell tumor from clone 3-3*; #5246		Serum of hamster with transformed cell tumor from clone 3-3; #5074		SV40 immune monkey serum #2159		Hamster hyperimmune anti-SV40 serum #6141		Virus titer ($TCID_{50}$/0.1 ml)
	Antibody titer	Antigen titer	Antibody titer	Antigen titer	Antibody titer	Antigen titer	Antibody titer	Antigen titer	Antibody titer	Antigen titer	
Primary SV-40 hamster tumor #1369	≥80	8	≥80	16	≥80	≥16	40	8	0†	0	$<10^0$
1st passage transplant of primary SV40 hamster tumor #2007	≥80	8	≥80	16	≥80	≥16	20	8	0	0	$<10^0$
Tumor from transformed hamster kidney cell clone 3-3; #1918	≥80	8	≥80	8	≥80	8	(20)‡	1	0	0	$<10^0$
SV40 virus (BSC-1 TC§)	0	0	10	≥ 4	0	0	640	4	20	4	$10^{8.0}$
Control (BSC-1 TC)			0	0			0	0	0	0	
Polyoma-induced hamster sarcoma	0	0	0	0	0	0	0	0	0	0	

* Clone 3-3 = clonal line derived from THK-1 line of SV40 transformed hamster kidney cells.[17]
† 0 = <20 serum antibody titer or <1 antigen titer.
‡ (20) = partial reaction at 1:20.
§ TC = tissue culture.

upper and lower halves of the supernatant fluid were saved separately. The pellet was rinsed with Eagle's basal medium, and resuspended to one-tenth volume in PBS.

Results.—Antigen in SV40 hamster tumors: Of 35 extracts of primary SV40-induced sarcomas, 32 (91%) reacted in CF with late sera from hamsters with primary tumors. Three transplanted primary tumors and seven tumors induced by inoculation of transformed hamster kidney or hamster embryo tissue culture cells all reacted with these sera. The majority of primary, transplanted, and transformed cell-induced tumors also reacted with the 1:20 dilution of the SV40 monkey immune serum. The quantity of antigen reacting with the hamster antiserum was not related

TABLE 2

PRESENCE OF "TUMOR" ANTIGEN IN CELLS OF VARIOUS SPECIES INFECTED WITH SV40

Antigen		CF Antigen Titer Versus Indicated Serum				
Species	Material	Primary SV40 hamster tumor serum #5586 1:10 (16 antibody units)	SV40 immune monkey serum #2159 1:20	Primary adeno 12 hamster tumor serum 1:10	Polyoma hamster tumor serum 1:10	SV40 negative African green monkey serum 1:20
Hamster	SV40 primary tumor	8	8	0[a]	0	
	TC[b] cells from SV40 primary tumor	2	0	0	0	0
	SV40 transformed TC cells-THK-1[c] (P-59)[d]	8	≥4	0	0	0
	SV40 transformed TC cells clone 3-3[e] (P-34)	≥8	2	0	0	0
	SV40 transformed TC cells clone 4-1[e] (P-37)	≥8	≥4	0	0	0
	SV40 transformed TC cells clone 4-2[e] (P-36)	≥8	≥4	0	0	0
	SV40 transformed TC cells[f] #3775	≥8	≥4	0	0	0
	Controls[h]	All 0	All 0			
Rabbit	SV 40 "transformed" kidney TC (P-12)	≥8	2	0	0	0
	Normal kidney TC (P-1)	0	0			
Swiss mouse	SV 40 "transformed" kidney TC (P-16)	≥8	≥4	0	0	0
	Normal kidney TC (P-1)	0	0			
C3H mouse	SV40 "transformed" kidney TC (P-4)	≥8	≥4	0	0	0
	Normal kidney TC (P-1)	0	0			
Pig	SV40 "transformed" kidney TC (P-2)	≥8	≥4	0	0	0
	Normal kidney TC (P-2)	0	0			
Cow	SV40 infected kidney TC (P-2)	2	0			
	Normal kidney TC (P-2)	0	0			
	Bovine papilloma transformed cells (P-23)[g]	0	0			
Guinea pig	Normal kidney suspension	0	1			

[a] 0 = no reaction with undiluted antigen.
[b] TC = tissue culture.
[c] THK-1 = continuous cell line of SV40 transformed hamster kidney cells.[16]
[d] P = passage level.
[e] Clonal lines derived from THK-1 line.[17]
[f] Tissue culture line of cells from tumors induced with SV40 transformed hamster embryo cells; these tumors were kindly supplied by Dr. Joseph Melnick.
[g] Continuous cell line of bovine conjunctiva transformed by bovine papilloma virus.[20]
[h] Normal carcass; normal kidney TC; primary polyoma tumor; polyoma primary tumor TC cells (Stanton); polyoma-transformed hamster embryo cells (Dulbecco); adenovirus 12 tumor; adenovirus 12 tumor TC cells P-26 (Schlesinger); adenovirus 18 tumor; Rous tumor.

TABLE 3

EFFECT OF HEAT ON CF ANTIGEN TITERS OF TUMOR EXTRACTS AND VIRUS

Antigen	CF Antigen Titer versus Indicated Serum: Primary hamster tumor serum #5586 1:20 (8 antibody units)	Primary hamster tumor serum #2004 1:40 (≥4 antibody units)	SV40 immune monkey serum #2159 1:20	SV40 immune monkey serum #2159 1:80
1st passage transplant of primary SV40 hamster tumor, #2007				
Unheated	16	≥64	8	
Heated 56° 30′	0	0	0	
Transformed hamster cells, #3775*				
Unheated	32	32	8	
Heated 56° 30′	0	0	0	
Transformed hamster cells, #3797*				
Unheated	32		≥16	
Heated 50° 30′	(1)†		0	
Heated 56° 30′	0		0	
SV40 Virus				
Unheated	0	4		8
Heated 50° 30′	0	≥ 2		8
Heated 56° 30′	0	≥ 2		8

* #3775 and #3797 are tissue culture cell lines of SV40 transformed hamster embryo cultures (Melnick). See Table 2.

† Partial reaction undiluted.

to size of tumor; primary tumors less than 1 cm in diameter were as often positive, and in as high a titer, as medium-sized and large tumors. Likewise, there was no apparent relationship to presence or titer of extractable infectivity in the tumor.

Table 1 shows representative titration results with the various antisera. The three SV40 tumors showed complete reciprocal cross-reactions, indicating that the same antigen(s) was present in primary, transplanted, and clone-induced tumors. None of these tumors contained demonstrable infectivity. Although the tumor extracts reacted with the monkey immune serum, the antibody titer obtained was far lower than the titer of antiviral antibodies, and there was no reaction between these extracts and hyperimmune hamster SV40 antiserum.

Presence of antigen in transformed cells of various species origin: SV40 infection induces a proliferative response accompanied by alterations in morphology and growth pattern of tissue culture cells derived from a number of mammalian species.[11] The continuous cell lines of transformed cells derived from these cultures of various species origin provided a unique opportunity to approach the question of whether the new antigen detected in hamster tumors was formed in response to information contained in the host or in the viral genome. As seen in Table 2, "transformed" cells of hamster, rabbit, Swiss mouse, C3H mouse, and porcine origin reacted with the SV40 hamster tumor and monkey immune sera in a fashion similar to tumor extracts. The antigen was also present, in low titer, in a suspension of second passage cells from a fetal bovine kidney culture chronically infected with SV40, but showing no evidence of transformation; the supernatant culture fluid at this passage level contained only trace amounts of virus.

A variety of control antigens gave no significant reaction with these sera; it is particularly noteworthy that the positive hamster serum did not react with Forssmann antigen in guinea pig kidney, although the monkey serum gave a slight reaction. Most of the positive and control hamster antigens reacted with a low dilution of rabbit antisheep erythrocyte serum, indicating the presence of the Forssmann antigen in many of these cell lines;[15] however, all antigens of the other species had negative to partial reactions with this serum. The positively reacting antigens did not react with sera of hamsters with other virus-induced tumors or with a monkey serum negative for SV40 antibodies.

Comparison of some properties of tumor and viral antigens: Although the reaction of tumor extracts and transformed cells with the antiviral monkey serum suggested that the tumors contained antigens in common with the viral antigen preparation, the properties of the two types of antigen differed in at least two important respects. Firstly, antigens in tumor extracts were rendered nonreactive by heating at 50 or 56°C for 30 min, whereas the viral antigen was unaffected (Table 3). Secondly, the tumor antigens were chiefly "soluble," remaining unsedimented after centrifugation, whereas the viral antigen was completely sedimented (Table 4).

Characteristics of antibody responses to viral and tumor antigens in tumor-bearing hamsters: In a previous report,[10] we described the patterns of neutralizing and viral

TABLE 4

ULTRACENTRIFUGATION OF CF ANTIGENS IN TUMOR EXTRACTS AND VIRAL HARVEST

Material tested[c]	Primary Tumor #2007[a]		Transformed Cells #3775[b]		SV40 Virus (BSC-1 TC)	
	Primary hamster tumor serum #5586 1:10 (16 antibody units)	SV40 immune monkey serum #2159 1:20	Primary hamster tumor serum #5586 1:10 (16 antibody units)	SV40 immune monkey serum #2159 1:20	Primary hamster tumor serum #5586 1:10	SV40 immune monkey serum #2159 1:80
Starting suspension	16[d]	8	16	4	0	2
Centrifugal fractions						
Upper supernatant	8	2	8	4	0	0
Lower supernatant	16	4	16	8	0	0
Pellet (10×)	16	(AC)[e]	1	0	0	8

[a] This tumor extract was not reclarified prior to ultracentrifugation. Precipitates formed as a result of storage in the frozen state may have been responsible for the appearance of antigen in the pellet fraction.
[b] For description see Table 2.
[c] For centrifugation procedure see *Materials and Methods* section.
[d] CF antigen titer.
[e] The low titer reaction observed could not be evaluated because of anticomplementary activity at a 1:2 dilution.

TABLE 5

RELATIONSHIP OF "ANTITUMOR" CF ANTIBODY TITER TO SIZE OF TUMOR

Type of tumor	Size of tumor*	CF Antibody Titer to Tumor Antigen <5	5–20	40–>160	No. sera pos†: no. tested	
Primary	Pretumor	35‡	0	0	0/35	
	Small	5	11	5	18/23	(78%)
	Medium	8	10	14	29/37	(78%)
	Large	2	3	6	10/12	(83%)
Transplanted THK-1 cells§	Pretumor	11	0	0	0/11	
	Small	6	2	2	4/10	(40%)
	Medium	3	4	27	31/34	(91%)
	Large	1	0	8	8/9	(89%)

* Small = 0.1–1.0 cm mean tumor diameter. Medium = 1.1–3.5 cm mean tumor diameter. Large = >3.5 cm mean tumor diameter.
† Includes 8 sera which were positive in screening test but on which titers were not determined.
‡ Number of sera with indicated titer.
§ For description of THK-1, see Table 2.

TABLE 6

COMPARISON OF ANTITUMOR AND ANTIVIRAL ANTIBODIES IN HAMSTER SERA

"Antitumor" CF antibody titer	Number of Sera with Titers in Range Indicated: Primary Tumors				Transformed Cell-Induced Tumors			
	Antiviral CF antibody titer							
	0	5–10	20–40	≥80	0	5–10	20–40	≥80
0	9	3	1	0	20	0	0	0
5–10	17	3	0	0	1	0	0	0
20–40	12	2	0	0	9	0	0	0
≥80	6	5	3	1	32	2*	1*	0
	Neutralizing antibody titer							
	0	5–10	20–40	≥80	0	5–10	20–40	≥80
0	1	3	4	0	12	2	0	0
5–10	5	6	0	0	0	0	0	0
20–40	2	7	1	0	9	0	0	0
≥80	0	0	4	4	23	1*	2*	3*

* In the tests of sera from hamsters with tumors induced with transformed cells, the 3 sera positive in CF and the 6 sera positive in neutralization tests were all from 2 animals.

CF antibody responses in hamsters developing tumor after inoculation with virus when 2 days old. Tumor development in these animals was accompanied by development of rising titers of neutralizing antibody, the titer tending to parallel the size of tumor. CF antibody to viral antigen also developed, but usually much later and in lower titer than the neutralizing antibody.

One hundred and seven sera from 34 hamsters with primary SV40-induced tumors were tested for CF antibody reactive with tumor extracts. These sera are serial samples from the animals described in reference 10. The upper portion of Table 5 summarizes the relationship of this CF antibody to size of tumor at the time of bleeding. No antibody was detected prior to tumor development, while antibody appeared in moderate titer in the majority of hamsters with early tumors, and tended to rise in titer as tumors enlarged. In all, 31 of the 34 animals developed antitumor CF antibody.

The great majority of hamsters with tumors produced by transplantation of the THK-1 line[16] of *in vitro* transformed hamster kidney cells, or of the clones derived therefrom,[17] also developed antibodies to tumor extracts (Table 5, bottom), and in generally higher titers than in the animals with primary tumors.

Many of these serial serum specimens were also tested for CF and neutralizing antibody to SV40; the relationships of titers of antiviral and antitumor antibodies are shown in Table 6. Hamsters with primary tumors showed a slight correlation between tumor CF antibody and neutralizing or viral CF antibody titers, in that animals with high titer tumor CF antibody tended to have the highest antiviral titers. In view of the correlation of all three titers with tumor size, the lack of clear correlation here is all the more striking. Tumors induced with transformed cells rarely induced antiviral antibodies; the viral CF and neutralizing antibodies which were detected were confined to sera of two animals with large tumors and high antitumor CF antibody. Neither of these two tumors contained demonstrable infectivity, as judged by tests of three serial biopsy specimens from one, and an autopsy specimen from the other. Although these data suggest some relationships between viral and tumor antigens, they also provide clear evidence that they are not identical.

Discussion.—The data presented here have established that cells which have become tumorous or which have undergone transformation subsequent to infection with SV40 consistently elaborate a specific CF antigen(s) not found in normal tissue or other tumors. The predominant antigen is clearly distinct from the SV40 antigen(s) detected in viral suspensions by the routine CF and neutralization procedures, both in serologic reactivity, as evidenced by the reactions with hamster antisera, and physically, as shown by the differences in thermal inactivation and sedimentation characteristics. However, present data do not exclude the possibility that this "tumor" antigen may be present in subdetectable amounts in viral harvests, and do not necessarily imply that its presence is restricted to transformed cells or cells with an integrated cell-virus relationship.

Although the predominant antigen in tumor extracts is distinct from the presently known "viral" antigen(s), it is not clear whether the latter is also present in the neoplastic cells in immunologically significant amounts. It appears that SV40 tumors[18, 19] and transformed cells[16, 17] contain the viral genome in a highly integrated state, and may occasionally liberate small amounts of infectious virus. However, as judged by the relation between infectivity and CF antigen titers in AGMK harvests, the quantity of virus produced by tumor cells is far too small to be detected by the CF test, and is probably inadequate even to induce antibody response. Previous studies of antibody responses of hamsters with primary SV40-induced tumors suggested that significant amounts of viral antigens were indeed elaborated, since neutralizing antibody titers rose proportionally to size of tumor, even in animals with noninfectious tumors.[10] However, a few animals with large tumors failed to develop neutralizing or viral CF antibody. As reported here, tumors induced with transformed tissue culture cells rarely stimulated antiviral antibody response, suggesting that there may be a difference between primary and transformed cell-induced tumors with regard to content of viral antigen. This discrepancy may be due to differences in degree of expression of the viral genome, or to the booster effect of small amounts of viral antigen in primary tumors on the minimal antibody response produced by the initial virus inoculum. In this regard it is of interest that two hamsters with apparently noninfectious tumors induced by transplantation of transformed cells developed rising titers of neutralizing and viral CF antibody similar to the responses observed in primary tumors.

The consistent reactivity of hamster tumor and transformed cell suspensions with the SV40 immune monkey serum could also indicate presence of viral CF antigens. However, the reacting antigen showed the heat instability and low sedimentation rate which were characteristic of the "tumor-specific" antigen; also, the quantitative aspects of the reactivity differed markedly from the reaction of this serum with viral antigen. It seems more likely that the monkey serum contains two distinct types of antibody, one against virus and one against "tumor-specific" antigen formed during the course of a subclinical and presumably nononcogenic infection.

The relationship of the new cellular antigen described here to that responsible for rejection of SV40 tumor transplants by hamsters immunized with SV40[7-9] is not known. The only available evidence suggests that they are distinct, since animals hyperimmunized with SV40 did not develop CF antibody to tumor antigens (Table 1) and these animals presumably would be resistant to tumor challenge.

The presence of the tumor CF antigen in all species transformed cells tested provides strong evidence that it is formed from information present in the viral genome. The alternative, that the information for formation of the antigen is of host cell origin, made functional by the viral infection, seems incompatible with the specificity of the antigen and with its presence in five genera of mammals. The presence of the antigen, coupled with the difficulty of detection of infectious virus in transformed rabbit and mouse cells,[11] suggests that the genetic information of SV40 is present in these cells in an integrated state similar to that in hamster cells. It is noteworthy that this apparently integrated virus-cell relationship has not led to overtly malignant cell behavior in the case of the C3H mouse cells, which do not transplant in isologous hosts;[11] also, in the case of the bovine cell cultures, production of the antigen was not accompanied by noticeable alteration in cellular morphology or growth pattern.

The findings reported herein are similar in many respects to the findings of Heubner *et al.*[6] in virus-free hamster tumors induced by adenovirus types 12 and 18; however, in the adenovirus tumors, noninfectious viral antigens are easily demonstrated. There is also evidence for tumor specific antigens not found in viral harvests, but their identification as wholly new antigens is not clearly resolved.

Summary.—Suspensions of hamster tumors and various species cells transformed by SV40 contain a specific antigen(s) demonstrable by CF tests with sera of hamsters bearing SV40 tumors. This antigen has several properties which are distinct from the SV40 viral antigen(s); it is soluble, heat labile, and has a different serologic reactivity. These results, taken together with other evidence, support the hypothesis that the new antigen is synthesized by information from SV40 viral genome integrated in the tumor and *in vitro* transformed cells.

Antibody to the "new" cellular antigen appears in the serum of hamsters soon after the development of tumor, and rises in titer as the tumors increase in size. Response to this antigen is largely independent of antiviral antibody response.

[1] Klein, G., and E. Klein, in *Basic Mechanisms in Animal Virus Biology*, Cold Spring Harbor Symposia on Quantitative Biology, vol. 27 (1962), p. 463.

[2] Habel, K., *Proc. Soc. Exptl. Biol. Med.*, **106,** 722 (1961).

[3] Habel, K., *J. Exptl. Med.*, **115,** 181 (1962).

[4] Sjögren, H. O., I. Hellstrom, and G. Klein, *Cancer Res.*, **21,** 329 (1961).

[5] Sjögren, H. O., *Virology*, **15,** 214 (1961).

[6] Huebner, R. J., W. Rowe, H. Turner, and W. Lane, these Proceedings, **50,** 379 (1963).

[7] Habel, K., and B. E. Eddy, *Proc. Soc. Exptl. Biol. Med.*, **113,** 1 (1963).

[8] Koch, M., and A. B. Sabin, *Proc. Soc. Exptl. Biol. Med.*, **113,** 4 (1963).

[9] Defendi, V., *Proc. Soc. Exptl. Biol. Med.*, **113,** 12 (1963).

[10] Black, P. H., and W. P. Rowe, *J. Natl. Cancer Inst.*, in press.

[11] Rabson, A. S., and R. L. Kirschstein, *Proc. Exptl. Biol. Med.*, **111,** 323 (1962); Black, P. H., and W. P. Rowe, *Virology*, **19,** 107 (1963); Shein, H. M., J. F. Enders, J. D. Levinthal, and A. E. Burket, these Proceedings, **49,** 28 (1963); Koprowski, H., J. A. Pontén, F. Jensen, R. G. Ravdin, P. Moorhead, and E. Saksela, *J. Cellular Comp. Physiol.*, **59,** 281 (1962); Shein, H. M., and J. F. Enders, these Proceedings, **48,** 1164 (1962); Black, P. H., and W. P. Rowe, *Proc. Soc. Exptl. Biol. Med.*, in press.

[12] Meyer, H. M., Jr., H. E. Hopps, H. G. Rogers, B. E. Brooks, B. C. Bernheim, W. P. Jones, A. Nisalak, and R. D. Douglas, *J. Immunol.*, **88,** 796 (1962).

[13] This antigen was prepared by Dr. Klaus Schell of Microbiological Associates, Inc., under contract PHS-43-62-200.

[14] The monkey serum was kindly supplied by Dr. Joseph L. Melnick.
[15] Fogel, M., and L. Sachs, *J. Natl. Cancer Inst.*, **29**, 239 (1962).
[16] Black, P. H., and W. P. Rowe, these PROCEEDINGS, **50**, 606 (1963).
[17] Black, P. H., W. P. Rowe, and H. L. Cooper, these PROCEEDINGS, **50**, 847 (1963).
[18] Sabin, A. B., and M. A. Koch, these PROCEEDINGS, **49**, 304 (1963).
[19] Gerber, P., *Science*, **140**, 889 (1963).
[20] Black, P. H., J. W. Hartley, W. P. Rowe, and R. J. Huebner, *Nature*, **199**, 1016 (1963).

DETECTION OF SPECIFIC ANTIGEN IN SV40-TRANSFORMED CELLS BY IMMUNOFLUORESCENCE

JOHN H. POPE AND WALLACE P. ROWE

Immunofluorescent techniques have been used by several investigators to study the replication of SV40 virus (1–3); in all cases, antiviral immune serum was used, which would detect antigens on the surface of the viral particle. In African green monkey kidney cell cultures (AGMK) infected with SV40 virus, viral antigen appeared in the nucleus at 18 hours and reached a peak at 38 hours, after which nuclear antigen decreased and a cytoplasmic stage occurred (1). Nuclear fluorescence was also observed 10 days after infection of cells of grivet monkey kidney, rhesus monkey kidney, human embryo kidney, and Syrian hamster kidney (2). The same techniques were applied to cells transformed by SV40 virus in a search for evidence of viral antigen. Faint nuclear fluorescence, interpreted as non-specific, was observed in transformed Syrian hamster kidney cells (2). In cultures of transformed human embryo kidney cells, which were producing virus, less than 0.1 per cent of cells showed nuclear fluorescence (3) and it seems probable that these were the cells responsible for virus replication.

It has recently been demonstrated that SV40 tumors and transformed cells contain a specific complement-fixing (CF) antigen, distinct from that of the virus particle, which reacts with sera of hamsters bearing SV40 tumors (4). Based on this information, an indirect immunofluorescent test for SV40-transformed cells was developed using the serum of hamsters bearing SV40 tumors. It was hoped that such a technique would prove effective in detection of transformed cells and that information would be gained on the location of the antigen and the percentage of cells elaborating antigen.

Materials and Methods

Tissue Cultures.—All immunofluorescent studies were done with cells grown in tissue culture, on 11 x 22 mm coverslips in 60 x 15 mm plastic Petri dishes. The growth medium was

Eagle's basal medium 2 (BME) containing 5 or 10 per cent heated (56°C/30 minutes) newborn agamma calf serum (Hyland Laboratories, Los Angeles), 100 units of penicillin and 100 μg of streptomycin per ml. Plates were incubated at 37°C in an atmosphere of 5 per cent CO_2.

SV40 transformed cell lines included the transformed hamster kidney line THK-1 (5); transformed kidney cells of C3H mouse, rabbit, and porcine origin (6); and a transformed hamster embryo cell line (H50-1) (7) received from Dr. J. L. Melnick. Control cell lines included lines from 2 hamster tumors induced by adenovirus type 12, polyoma-transformed hamster embryo cells (received from Dr. R. Dulbecco), and the normal hamster cell BHK-21 (8).

Primary cultures of hamster tumors induced by SV40, polyoma, or adenovirus 12 and 18 were established by the trypsinization procedure. When possible, coverslips of cultures of tumors and cell lines were harvested before a complete cell sheet was formed.

Secondary cultures of AGMK were inoculated with $10^{6.7}$ ID_{50} of SV 40 virus strain 777 and harvested at 53 hours. BSC-1 cells (9) were similarly infected, and harvested at 38 hours. Maintenance was on 2 per cent heated agamma calf serum in BME in an atmosphere of 5 per cent CO_2.

Antisera.—H50-1 cell-induced tumors were used for production of hamster antitumor serum. This tumor line was maintained by serial transplantation in weanling hamsters by subcutaneous inoculation of tumor fragments. Serum was collected as late as possible after transplantation, and tested for CF antibody to tumor antigen. Two pools of CF-positive sera were prepared. Pools 1 and 2 titered 1:40 and 1:10, respectively, against SV40 tumor antigen, were negative for antiviral CF antibody at 1:10 dilution, and had no virus-neutralizing activity at 1:5 dilution. Control sera were pooled sera from normal hamsters and from hamsters bearing tumors induced by adenovirus type 12 (Ad. 12); the latter serum pool contained CF antibodies to both Ad. 12 tumor and viral antigen (10).

Conjugates.—For the indirect test, fluorescein-labeled goat antiserum to hamster serum globulin was used; this conjugate was kindly supplied by Dr. Roger Wilsnack. The conjugate was diluted 1:10 in a normal hamster brain extract (20 per cent in phosphate-buffered saline pH 7.2 (PBS) clarified at 2000 RPM/15 minutes), held at 4°C for 1 hour, and then centrifuged at 15,000 RPM for 15 minutes in a Spinco 40 rotor. The direct test was conducted with the conjugated serum of a Cercopithecus monkey which had been inoculated with SV40 virus; this serum had a CF titer of 1:320 to virus, a neutralizing antibody titer of 1:1280, and $<1:10$ CF titer to SV40 tumor antigen.

Immunofluorescent Tests.—Coverslips were washed twice in PBS, fixed for 10 minutes in acetone at −60°C, air-dried, and stored in stoppered vials at 4°C. In the indirect test, cells were treated with a 1:5 dilution of unheated hamster serum at 37°C for 30 minutes in a humid atmosphere, washed twice in PBS for 5 minutes, and rinsed in distilled water. Anti-hamster conjugate was then added for 30 minutes at 37°C, the washing process was repeated, and the coverslip mounted in buffered glycerol (pH 8.5). Similar incubation and washing procedures were used with the direct method. The specificity of reactions in the indirect test was checked repeatedly by testing uninfected cells treated with the SV40 hamster antiserum, and infected cells treated with control serum. The direct test was controlled by testing the conjugated antiserum against uninfected cells.

RESULTS

Table I summarizes the results obtained with various combinations of antisera and cells. Specific immunofluorescence was readily demonstrated when the hamster tumor sera were tested with SV40 tumor and transformed cells. In all instances, the fluorescence of cells in interphase was confined to the nucleus,

TABLE I

Immunofluorescent Staining Reactions of SV40 Antitumor and Antiviral Sera

	Tissue culture passage level*	Hamster tumor serum (indirect test)			Anti-SV40 monkey serum (direct test)
		Anti-SV40		Anti-Ad. 12	
		Pool 1	Pool 2		
		per cent	*per cent*		*per cent*
SV40-transformed cell lines					
Hamster (H50-1)	>10	++++‡ 100		–	–
Hamster (THK-1)	76	++++ 100	++++ 100	–	–
C3H mouse kidney (TCMK-1)	21	++++ 100	++++ 100	–	–
Rabbit kidney (TRK-1)	20	++++ 100	++++ 100	–	–
Porcine kidney (TPK-1)	11	++++ 100	++++ 100	–	++ 0.01
Primary SV-40-induced hamster tumor	1		++++ 30	–	–
Ad. 12-induced hamster tumors					
Prototype (Huie)	1	–	–	++++	–
Field strain 10534	1	–		++++	
Ad. 12-induced mouse tumor	1 and 2	–	–	++++	
Ad. 18-induced hamster tumor	1	–			
Polyoma hamster tumor	1	–			
BHK-21 hamster cell line	37	–			
Normal hamster embryo	1	–	–		
SV40-infected					
AGMK (53 hrs.)			++++ 95		++ 90
BSC-1 (38 hrs.)		++++ 95			++ 90
Controls, uninfected					
AGMK		–	–		–
BSC-1		–	–		±

* Number of tissue culture passages since removal from hamsters. The H50-1 and polyoma cells were from tumors produced by inoculation of transformed tissue culture cells; the other cell lines had not been passed in animals.

‡ The – to ++++ scale refers to the intensity in individual cells, and the percentage refers to the estimated proportion of cells showing specific fluorescence.

the nuclei appearing finely granular and of uniform brightness throughout, except for small unstained areas the size and shape of nucleoli. However, cells in mitosis showed uniform cytoplasmic staining, the chromosomal material being completely negative (Fig. 1).

In the continuous transformed cell lines of all four host species, essentially every cell demonstrated strong fluorescence. Multinucleated giant cells were

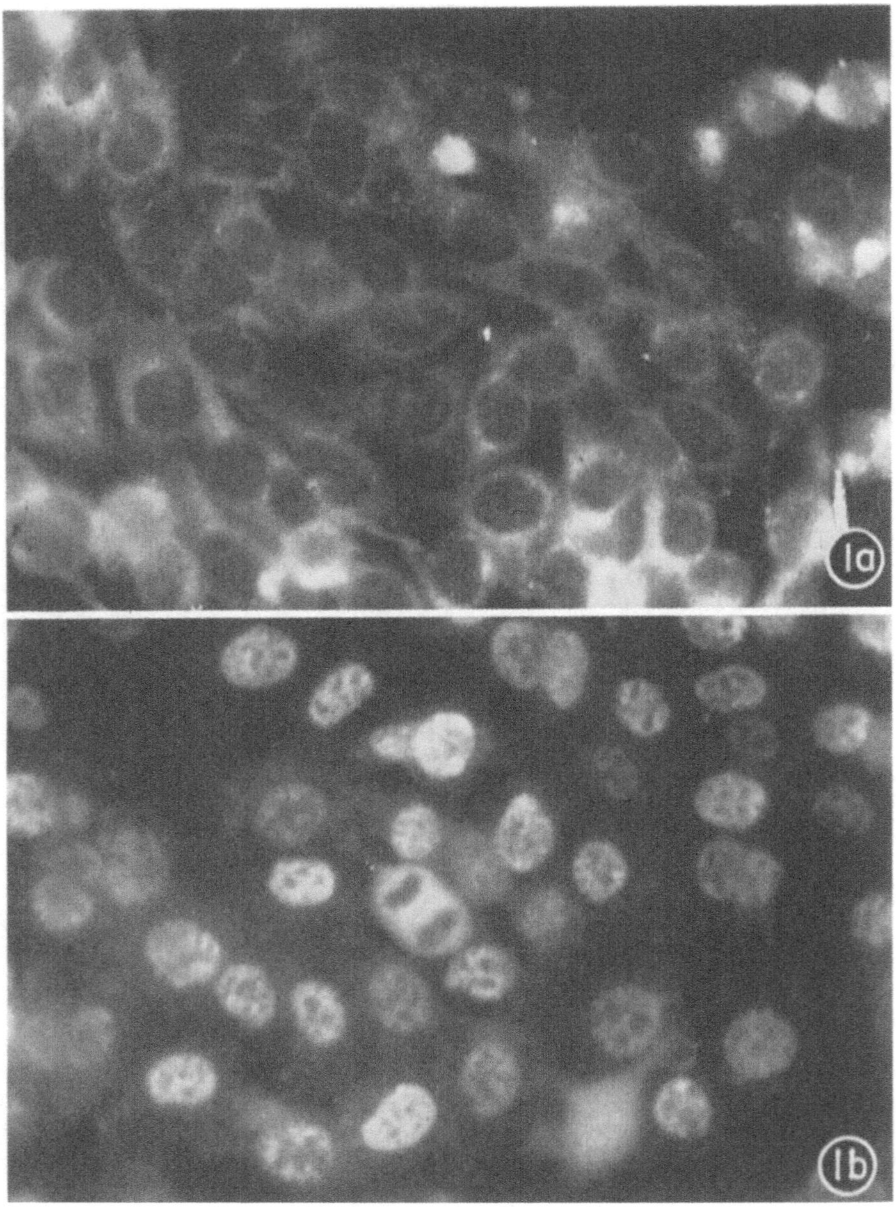

FIGS. 1 *a* and 1 *b*. Fluorescent staining of H50-1 tumor cells, in first tissue culture passage, with sera of tumorous hamsters (indirect test). × 450.

FIG. 1 *a*. Stained with negative serum (from hamster with Ad. 12 tumor). This picture was overexposed relative to Fig. 1 *b* to show cellular outlines. Nuclei are completely unstained.

FIG. 1 *b*. Stained with SV40 hamster tumor serum pool 1. Every nucleus is stained. A cell in mitosis, in center of figure, shows diffuse staining of cytoplasm.

evident in the H 50-1 cell line; in these cells, every nucleus was brightly fluorescent (Fig. 2). First passage tissue cultures from tumors showed fluorescence in a smaller proportion of cells, presumably because of admixture of tumor cells with normal cells; large fibroblast cells, which were clearly distinct from tumor cells, were invariably negative.

In contrast, the antiviral serum gave no fluorescence with the SV40-transformed cells of hamster, rabbit, and mouse origin; these cells do not contain extractable virus, although the THK-1 cells yield virus when overlaid on AGMK monolayers (5).

AGMK and BSC-1 cell cultures acutely infected with SV40 virus reacted with both the SV40 antiviral and antitumor sera. With both sera, the fluorescence was confined to the nucleus.

The various control preparations, listed in Table I, gave no staining, with the exception of faint nuclear fluorescence of uninfected BSC-1 cells treated with the monkey antiviral serum.

Hamster tumor serum pool 2, with CF antibody titer of 1:10, was found to produce nuclear fluorescence of full intensity up to a dilution of 1:20. Nuclear fluorescence was markedly reduced, both in incidence of fluorescent nuclei and in intensity of staining, by absorption of serum pool 2, at a final dilution of 1:10, with SV40 tumor extract, while absorption with Ad. 12 tumor extract or normal hamster brain had no apparent effect.

An attempt was made to compare reactivity of individual hamster sera when tested in CF against SV40 tumor antigen and in the fluorescent antibody test against SV40-transformed cells of the THK-1 line. Nine hamsters were inoculated subcutaneously with THK-1 cells, and serum was obtained 10 weeks later, at which time seven bore tumors. The two animals without tumors, and one tumorous animal, were negative in both the CF and fluorescent antibody tests. Five tumorous animals were negative for CF antibody, but gave faint (1 to 2 plus) nuclear fluorescence; the other tumorous hamster had a CF antibody titer of 1:40 and gave 4 plus nuclear staining. Thus, there appeared to be a parallel between antitumor CF antibody and ability to stain transformed cells, with the latter test somewhat more sensitive.

DISCUSSION

The results of the immunofluorescent studies confirm the findings of Black *et al.* (4, 5, 11) in several important respects. First, they confirm that cells of various species altered by SV40 virus continue to elaborate a specific antigen(s) which reacts with sera of hamsters carrying SV40 tumors. Secondly, they confirm that the antigen is serologically distinct from that reactive in neutralization and antiviral CF tests. Most important, they indicate that continued production of the antigen is an integral property of essentially every transformed cell. The studies of Sabin and Koch (12) and of Black *et al.* (11) indicated that a high proportion of transformed hamster cells carry the viral genome. The present studies indicate that this proportion is of the order of 100 per cent, and also that the same holds true for transformed rabbit, mouse, and porcine cells.

The validity of these conclusions is entirely dependent on the specificity of the fluorescent reactions observed. Three lines of evidence indicate that the reactions are specific for antigen(s) formed from viral genetic information. First, the lack of reaction in the various control preparations, which included tumors induced by other viruses, strongly indicates that the antigen occurs only in response to SV40 infection. Secondly, as pointed out by Black *et al.* (4), the presence of the antigen in each of the various species transformed cells indicates that the antigen is under viral genetic control. And thirdly, the formation of the antigen in acutely infected AGMK and BSC-1 cultures undergoing cytolytic response indicates its relationship to the virus; in this connection it should be stressed that hamster tumor sera reacted with the acutely infected cells even though devoid of virus-neutralizing and antiviral CF antibodies. This finding confirms the results of Hoggan, Black, and Rowe (in preparation) which showed the development during the cytolytic infection of AGMK cells of a CF antigen reacting with sera of hamsters bearing SV40 tumors.

It is not possible at present to say that the fluorescent antigen observed in infected AGMK is identical with the tumor CF antigen, although this seems

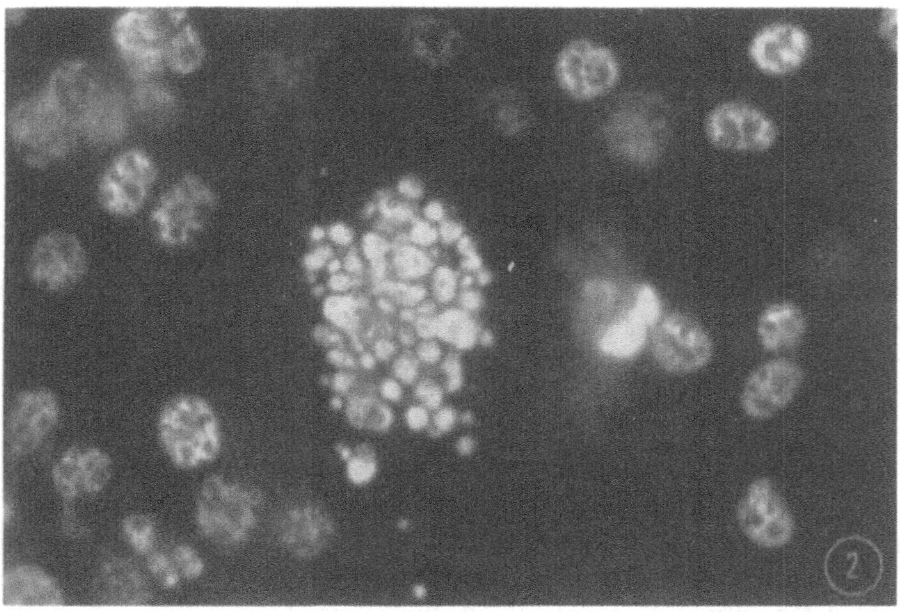

FIG. 2. Fluorescent staining of SV40-transformed hamster kidney cells (THK-1 line) with serum pool 2. All nuclei but one are stained. Multinucleated giant cell (center) with every nucleus brightly fluorescent. × 450.

probable. It may be that in both infected AGMK cells and transformed cells a specific antigen associated with an early stage of viral synthesis but distinct from that of the intact virus, is produced in the nucleus. In infected AGMK and BSC-1 cells, but not in transformed cells, the synthetic processes continue further and normally result in production of intact virus.

Melnick *et al.* (13) have shown that in the presence of 5-fluorouracil, infected AGMK cells produce large amounts of viral protein which is not incorporated into complete viral particles, but it is detectable by fluorescent antibody tests with antiviral immune serum. While it is conceivable that their test was detecting the "tumor antigen," it is more likely that it was revealing the viral surface antigens. Thus, the arrest of viral protein production in the transformed hamster, rabbit, and mouse cells appears to be more complete than that produced by 5-fluorouracil.

The nuclear localization of the fluorescent-stainable tumor antigen suggests that it is distinct from the SV40-specific transplantation antigen (14–16), which is presumably located in the cell membrane.

SUMMARY

With an immunofluorescent technique involving the use of serum of hamsters with SV40 tumors, nuclear fluorescence was detected in each of five cell lines, derived from four mammalian species, transformed by SV40 virus. Essentially all nuclei, including those of multinuclear cells, were fluorescent-stainable. Serum of hamsters bearing SV40 tumors was also found to give nuclear fluorescence in susceptible cells (AGMK or BSC-1) acutely infected with SV40 virus.

These findings provide further evidence that cellular incorporation of the SV40 viral genome, with partial expression of the genome by synthesis of at least one virus-specific antigen, is an integral property of all SV40 transformed cells.

BIBLIOGRAPHY

1. Mayor, H. D., Stinebaugh, S. E., Jamison, R. M., Jordan, L. E., and Melnick, J. L., Immunofluorescent, cytochemical, and microcytological studies on the growth of the simian vacuolating virus (SV-40) in tissue culture, *Exp. Mol. Path.*, 1962, **1,** 397.
2. Levinthal, J. D., and Shein, H. M., Morphology and distribution of SV40-infected cells in human, simian and hamster renal cell cultures as detected by immunofluorescence, *Proc. Soc. Exp. Biol. and Med.*, 1963, **112,** 405.
3. Shein, H. M., Enders, J. R., and Levinthal, J. D., Transformation induced by simian virus 40 in human renal cell cultures. II. Cell-virus relationships, *Proc. Nat. Acad. Sc.*, 1962, **48,** 1350.
4. Black, P. H., Rowe, W. P., Turner, H. C., and Huebner, R. J., A specific complement-fixing antigen present in SV40 tumor and transformed cells, *Proc. Nat. Acad. Sc.*, 1963, **50,** 1148.

5. Black, P. H., and Rowe, W. P., An analysis of SV40-induced transformation of hamster kidney tissue *in vitro*. I. General characteristics, *Proc. Nat. Acad. Sc.*, 1963, **50,** 606.
6. Black, P. H., and Rowe, W. P., SV-40 induced proliferation of tissue culture cells of rabbit, mouse, and porcine origin, *Proc. Soc. Exp. Biol. and Med.*, 1963, **114,** 721.
7. Ashkenazi, A., and Melnick, J. L., Tumorigenicity of simian papovavirus SV40 and of virus-transformed cells, *J. Nat. Cancer Inst.*, 1963, **30,** 1227.
8. MacPherson, I., and Stoker, M., Polyoma transformation of hamster cell clones-an investigation of genetic factors affecting cell competence, *Virology*, 1962, **16,** 147.
9. Hopps, H. E., Bernheim, B., Nisalak, A., and Smadel, J. E., Biological characteristics of a serially cultivated kidney cell culture derived from the African green monkey *Cercopithecus aethiops*, *Fed. Proc.*, 1962, **21,** 454.
10. Huebner, R. J., Rowe, W. P., Turner, H. C., and Lane, W. T., Specific adenovirus complement-fixing antigens in virus-free hamster and rat tumors, *Proc. Nat. Acad. Sc.*, 1963, **50,** 379.
11. Black, P. H., Rowe, W. P., and Cooper, H. L., An analysis of SV40-induced transformation of hamster kidney tissue *in vitro*. II. Studies of three clones derived from a continuous line of transformed cells, *Proc. Nat. Acad. Sc.*, 1963, **50,** 847.
12. Sabin, A. B., and Koch, M. A., Evidence of continuous transmission of noninfectious SV40 viral genome in most or all SV40 hamster tumor cells, *Proc. Nat. Acad. Sc.*, 1963, **49,** 304.
13. Melnick, J. L., Stinebaugh, S. E., and Rapp, F., Incomplete simian papovavirus SV40. Formation of non-infectious viral antigens in the presence of fluorouracil, *J. Exp. Med.*, 1964, **119,** 313.
14. Habel, K., and Eddy, B. E., Specificity of resistance to tumor challenge of polyoma and SV40 virus-immune hamsters, *Proc. Soc. Exp. Biol. and Med.*, 1963, **113,** 1.
15. Koch, M., and Sabin, A. B., Specificity of virus-induced resistance to transplantation of polyoma and SV40 tumors in adult hamsters, *Proc. Soc. Exp. Biol. and Med.*, 1963, **113,** 4.
16. Defendi, V., Effect of SV40 virus immunization on growth of transplantable SV40 and polyoma virus tumors in hamsters, *Proc. Soc. Exp. Biol. and Med.*, 1963, **113,** 12.

SYNTHESIS OF SV40 TUMOR ANTIGEN DURING REPLICATION OF SIMIAN PAPOVAVIRUS (SV40)

F. RAPP, T. KITAHARA, J. S. BUTEL, AND J. L. MELNICK

Hamsters bearing tumors induced either by simian papovavirus 40 (SV40) or by cells tranformed *in vitro* by the virus develop antibodies capable of detecting a new complement-fixing antigen in the transformed cells.[1] Antibodies also develop for a new intranuclear antigen present in all cells transformed by SV40 and detected by the immunofluorescence technique.[2, 3] It appears that the new antigens detected by complement-fixation and by immunofluorescence are similar or identical immunologically.[4] The new tumor antigen is presumably under control of at least a portion of the virus genome, but some of the cell lines producing the antigen do not go on to synthesize infectious SV40.[5] The present experiments, and those reported in the accompanying paper,[6] were designed to determine whether cells susceptible to SV40 produced the tumor antigens in the course of the replication of the virus.

Materials and Methods.—The cells used in this study were primary green monkey kidney cells (GMK), a stable line of cells derived from the same species (BSC1),[7] hamster embryo fibroblasts (HEF), and embryonic human lung cells (HEL). The cells were grown in 16-oz bottles for complement-fixation tests and on 15-mm round cover glasses in 60-mm plastic Petri dishes for the immunofluorescence tests. Nutrient fluids consisted of Eagle's basal medium, concentrations of calf serum varying from 2 to 10 per cent, 100 units of penicillin, and 100 μg of streptomycin per ml. The Petri dishes were incubated under 5 per cent CO_2; all cultures were kept at 37°C. Following inoculation with approximately 10 plaque-forming units of SV40 per cell, the cultures were reincubated at 37°C and harvested at varying intervals.

For complement-fixation tests, the cells in 16-oz bottles were harvested by scraping with a rubber policeman, the cells disrupted by rapid freezing and thawing or by treatment in a sonic oscillator, the suspensions clarified by centrifugation at 2500 rpm for 10 min, and the supernates used as antigen. Uninfected cells were treated in the same manner to serve as antigen controls. All preparations were tested for anticomplementary activity and both known positive and negative sera were included in all tests. The semimicro complement-fixation tests were carried out in plastic dishes.

Cells growing on cover glasses were harvested and fixed for 3 min in acetone as previously described.[3] The cells were exposed to (*a*) serum from hamsters with or without SV40-induced but virus-free tumors, and antihamster globulin labeled with fluorescein isothiocyanate, or (*b*) labeled anti-SV40 monkey serum previously shown to be highly specific for the detection of SV40 viral antigens.[8] All tests included positive and negative controls for the detection of both SV40 virus antigens and SV40 tumor antigens.

Results.—A large number of experiments were carried out to demonstrate the

TABLE 1

DETECTION OF ANTIGENS IN MONKEY CELLS INFECTED WITH SV40 BY COMPLEMENT-FIXATION

		Complement-Fixation: Antigen Titers		
Source of antigen	Days post-inoc.	SV40-immune monkey serum*	Tumor hamster serum†	Nontumor hamster serum
Infected GMK	3	16‡	8	<2
	7	8	<2	<2
Infected BSCl	0.25	<2	<2	<2
	0.5	<2	4	<2
	1	4	8	<2
	2	32	32	<2
Infected BSCl	2	16	32	<2
	4	32	32	<2
	7	16	<2	<2
2X-10 cells§		<2	32	<2
SV40 virus		32	<2	<2

* Diluted 1:80 to contain 4 antibody units.
† Diluted 1:40 to contain 4 antibody units.
‡ Numbers represent highest dilution of antigen yielding less than 50% hemolysis.
§ A line of hamster cells transformed *in vitro* by SV40 and known to contain SV40-induced cellular antigen in every nucleus.[3]

possible presence of the complement-fixing SV40 tumor antigen in GMK and BSCl cells following inoculation of the cultures with SV40. Representative experiments, summarized in Table 1, reveal the presence of this antigen in both types of cells. Tumor antigen was detected 12 hr following inoculation of BSCl cultures at a time when SV40 viral antigens were not present in detectable quantity. The amount of tumor antigen declined with time following inoculation as seen by the decrease in complement-fixing titer of cultures 7 days after inoculation when the antigen from such cultures was reacted with the tumor hamster serum. Neither nontumor hamster serum nor normal monkey serum reacted with any of the antigens tested. The specificity of the sera used was further controlled by reacting them in the tests with SV40 viral antigens and with antigens prepared from the 2X-10 cells; the latter is a line of cells transformed by SV40 and known to contain both the new complement-fixing cellular antigen[4] as well as the new intranuclear cellular antigen.[3] The SV40 viral antigen reacted only with SV40-immune monkey serum, while the 2X-10 antigens reacted only with the tumor hamster serum.

Experiments carried out to characterize further the SV40 viral and SV40 tumor antigens have shown that the SV40 viral antigen can be sedimented by centrifugation at 30,000 rpm for 2 hr but that the tumor antigen remains in the supernatant fluid when treated in the same fashion. Heating the tumor antigen (whether from virus-infected GMK or BSCl cells, or from virus-free 2X-10 cells) for 30 min at 50°C completely destroys its complement-fixing ability, but this amount of heating does not alter the capability of the SV40 viral antigen to fix complement in the presence of specific antibody. Thus the present findings confirm those recently obtained with the SV40 viral and tumor antigens obtained from transformed cells.[1] They also demonstrate that the new antigen which appears early in the infected cell has the same properties as the antigen in the tumor cells.

Immunofluorescent detection of both SV40 tumor and SV40 viral antigens in GMK cells revealed early synthesis of the tumor antigen. Twenty-four hours following the inoculation of the cultures, approximately 50 per cent of the cells were

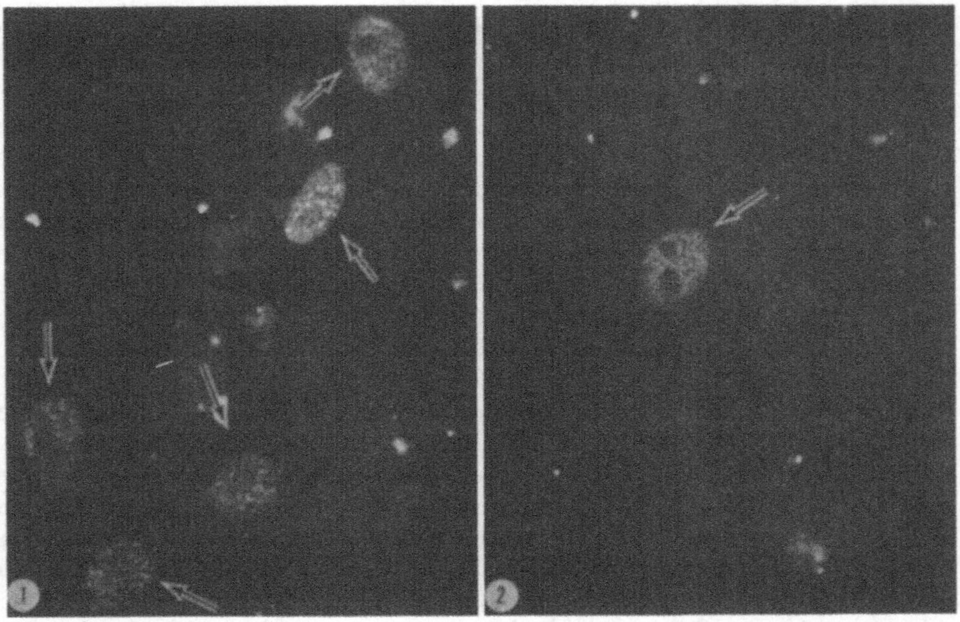

FIGS. 1 and 2.—Photomicrographs of immunofluorescent green monkey kidney cultures infected 24 hr earlier with SV40. (1) Reacted with anti-SV40 tumor antigen reagents. Arrows point to cells synthesizing intranuclear tumor antigen. ×425. (2) Reacted with anti-SV40 monkey serum. Arrow points to cell synthesizing SV40 viral antigen. ×425.

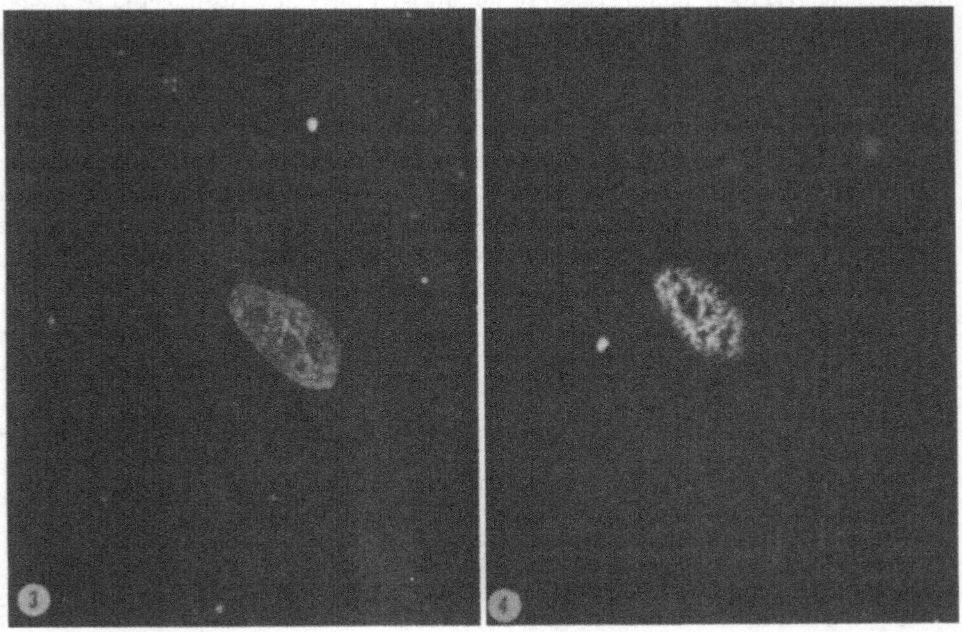

FIGS. 3 and 4.—Photomicrographs of immunofluorescent human embryonic lung cultures infected 72 hr earlier with SV40. (3) Reacted with anti-SV40 tumor antigen reagents. Note single cell with intranuclear tumor antigen. ×425. (4) Reacted with anti-SV40 monkey serum. Single cell synthesizing SV40 viral antigen in the nucleus. ×425.

synthesizing the intranuclear tumor antigen (Fig. 1), but only 1 per cent of the cells contained SV40 viral antigen (Fig. 2). With the progression of time, more cells contained viral antigen but then the tumor antigen could no longer be detected in many of the cells. The human HEL cultures synthesized both antigens (Figs. 3 and 4) but only a small proportion of the cells in the culture were able to do so, and it could not be determined whether the same cells were synthesizing both antigens. A rare cell in the hamster HEF cultures synthesized the tumor antigen but viral antigens could not be detected in any of these cells.

Discussion.—The data now accumulating in our laboratory as well as elsewhere[2, 6, 9] that monkey cells synthesize the SV40 tumor antigen during early stages of infection strongly support the hypothesis that information for the synthesis of this antigen in cells transformed by SV40 is directed by the viral genome. Failure to isolate infectious SV40 from transformed cultures[5] may be due to integration of only a portion of the viral genome with that of the cell genome. Such integration is implied by the finding that information for the synthesis of the antigen is transmitted to all daughter cells during mitosis.[3] While the tumor antigen does not appear to be incorporated into the virus capsid, it is unknown at the present time whether it is always extraviral, as it is in the transformed cells, or whether it is part of the internal structure of the virus particle.

Hamsters bearing SV40-induced, virus-free tumors simultaneously develop complement-fixing and immunofluorescent antibodies, suggesting that both antibodies are formed in response to the same or closely related antigens.[4] The high correlation found in the present study between the complement-fixation and immunofluorescence results in finding SV40 tumor antigen early in virus-infected cells again points to the probability that the two methods are detecting the same or similar antigens.

Summary.—Monkey cells infected with papovavirus SV40 regularly synthesize SV40 tumor antigen during early stages of the replication of the virus. This antigen is synthesized in the nucleus and can be detected by both immunofluorescence and complement-fixation. The antigen also develops in an occasional embryonic human lung and embryonic hamster fibroblast following inoculation of the cultures with SV40. The tumor antigens in infected monkey cells and in human and hamster cells transformed by SV40 are immunologically similar.

This investigation was aided by U.S. Public Health Service research grants CA-04600 from the National Cancer Institute; AI 05382 and 2 T1 AI 74 from the National Institute of Allergy and Infectious Diseases, NIH.

[1] Black, P. H., W. P. Rowe, H. C. Turner, and R. J. Huebner, these PROCEEDINGS **50**, 1148 (1963).

[2] Pope, J. H., and W. P. Rowe, *J. Exptl. Med.*, **120**, 121 (1964).

[3] Rapp, F., J. S. Butel, and J. L. Melnick, *Proc. Soc. Exptl. Biol. Med.*, in press.

[4] Kitahara, T., J. S. Butel, F. Rapp, and J. L. Melnick, *Nature*, in press.

[5] Melnick, J. L., K. S. Khera, and F. Rapp, *Virology*, **23**, 430 (1964).

[6] Sabin, A. B., and M. A. Koch, these PROCEEDINGS, **52**, 1131 (1964).

[7] Hopps, H. E., B. C. Bernheim, A. Nisalak, J. H. Tjio, and J. E. Smadel, *J. Immunol.*, **91**, 416 (1963).

[8] Melnick, J. L., S. E. Stinebaugh, and F. Rapp, *J. Exptl. Med.*, **119**, 313 (1964).

[9] Sabin, A. B., and M. A. Koch, these PROCEEDINGS, **50**, 407 (1963).

V
Activation of Infectious Tumor Virus SV40 in Transformed Cells

Studies on the Transfer of Subviral Infectivity from SV40-Induced Hamster Tumor Cells to Indicator Cells

PAUL GERBER

Some of the factors involved in the interaction between SV40-induced, virogenic hamster tumor cells and indicator cells were investigated. The number of detectable virogenic tumor cells was found to depend on their plating efficiency. Demonstration of virogenicity required viability of the tumor cells as well as direct contact with the indicator cells. This interaction could be enhanced by the presence of ultraviolet-irradiated Sendai virus. The tumor cells retained their virogenic character following prolonged passage in the presence of 10 μg/ml of cytosine arabinoside.

INTRODUCTION

During the course of studying the virus-tumor relationship in hamster ependymomas produced by simian virus 40 (SV40) we observed that cell-free tumor extracts failed to yield infectious virus. On the other hand, infectious virus could be consistently demonstrated by seeding the tumor cells directly onto sensitive indicator cells (Gerber and Kirschstein, 1962). We postulated then that the virus may persist in an "altered state." Evidence presented in subsequent reports (Gerber, 1963, 1964) suggested that the viral genome was permanently present in some of the tumor cells. This report summarizes the results of further studies on the interaction between virogenic tumor cells and indicator cells in culture.

MATERIALS AND METHODS

Cell cultures. The EPA cell line originally derived from a SV40-induced hamster ependymoma (Gerber and Kirschstein, 1962) has been stored at −70°. At periodic intervals several ampules were removed to reestablish a culture. Most of the present studies were carried out with cells of the 25th–70th passage, unless otherwise noted. The cells were grown in Eagle's basal medium containing 5% fetal calf serum and 100 μg/ml of neomycin.

Primary cultures of African green monkey kidney cells[1] (GMK) were grown in 0.5% lactalbumin hydrolyzate in Hanks' solution, 5% fetal serum and 100 μg/ml of neomycin. A continuous cell line, LLCMK$_2$[1] derived from rhesus monkey kidney cells (Hull *et al.*, 1962) was grown in the same medium as the EP cells. All cell cultures were incubated at 36° and maintained with Eagle's medium with 2% fetal calf serum.

Virus assay and identification. Cell-free lysates of test cultures were prepared by one cycle of freezing and thawing followed by centrifugation at 2500 rpm for 15 minutes. Each of six tube cultures of GMK cells was inoculated with 0.5 ml of a cell lysate. They were placed on a roller drum, incubated for 14 days, and observed for the appearance of specific cytopathic effects (CPE). The cells of tubes failing to show CPE were lysed by one cycle of freezing and thawing and the pooled lysates of each group were subcultured to fresh GMK cells which were maintained for 36–40 days. The identity of virus isolates was established by neutralization tests with rabbit anti-SV40 serum.

Sendai virus. An egg-adapted strain of Sendai virus was kindly provided by Dr. J. C. Parker, Microbiological Associates. This virus was isolated from a mouse colony and identified serologically.

Allantoic fluid virus was concentrated by centrifugation at 20,000 rpm for 1.5 hours. The pellet was resuspended in 1/10 of the original volume in modified Hanks' solution (Hanks' solution without glucose and bicarbonate, buffered to pH 7.6 with 0.01 *M* Tris). The virus concentrate contained 40,000–80,000 HA units/ml. It was diluted with modified Hanks' solution to a titer of 8000 HA units/ml and inactivated by exposure to ultraviolet radiation (UV) at a dose of 25 ergs/mm^2/sec for a period of 3.5 minutes. The loss of infectivity was ascertained by inoculation of GMK cells.

[1] Obtained from the Tissue Culture Section of the Division of Biologics Standards.

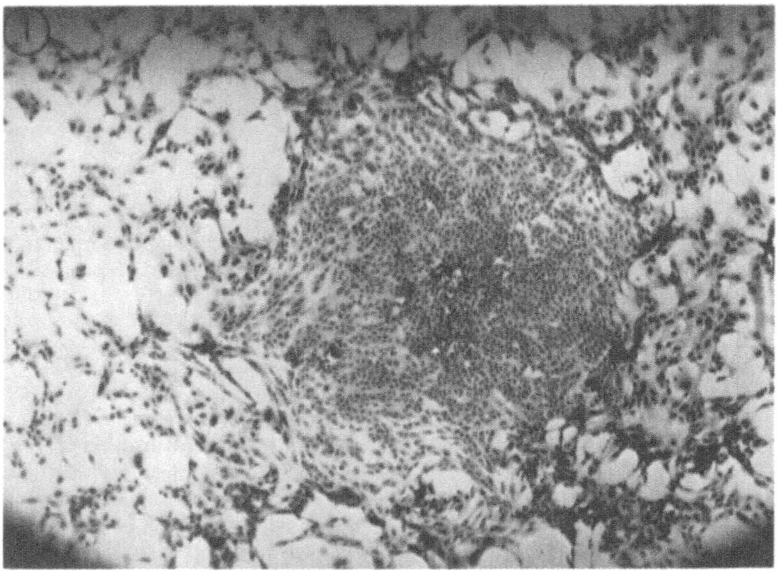

FIG. 1. Colony of EP cells in mixed culture with GMK cells. Hematoxylin and eosin stain. Magnification: × 90.

Immunofluorescence technique. The indirect method was used. Coverslip cultures were washed with Hanks' solution, air dried, and fixed in acetone for 3 minutes at room temperature. The cells were allowed to react with sera from hamsters bearing SV40 tumors and anti-hamster rabbit globulin conjugated with fluorescein-isothiocyanate. A Leitz fluorescence microscope equipped with an Osram HB-200 mercury arc vapor lamp was used.

Physical treatments. Suspensions of EP cells in growth medium, preparations of intact SV40 virus strain 777 or infectious viral DNA were subjected to each of the following treatments: X-irradiation was carried out with a Westinghouse Quadrocondex unit, operated at 200 kv, 15 ma, with a 0.25 mm Cu and 0.55 mm Al filter. The exposure was at a rate of 650 r per minute. A General Electric germicidal lamp was used for UV irradiation at a dose of 25 ergs/mm^2/sec. Sonic vibration was carried out with a Biosonik sonifier at a frequency of 20 kc/sec; the specimens were held at 0–4° and exposed to two treatments, each of 10 seconds' duration.

RESULTS

Number of Detectable Virogenic EP Cells in Uncloned Cultures

Our previous estimation (Gerber and Kirschstein, 1962) of the number of virus-yielding cells in primary SV40-induced ependymomas or in EP cell lines in early passage, indicated a ratio of 1:3000 to 1:30,000. Since the plating efficiencies of these cells was rather low, we decided to repeat these experiments using EP cell cultures at the 25th and 70th passage levels. The plating efficiency of these cells ranged from 15 to 60% in the absence of feeder layer cells. Cultures of EP cells during the logarithmic growth phase were dispersed with a mixture of 0.2% trypsin and 0.02% Versene. The resulting cell suspensions consisted mainly of single cells. The number of viable cells was determined by dye exclusion counts, and serial tenfold dilutions were prepared in growth medium, siliconized glassware being used. One milliliter of cell suspension was inoculated into four tube cultures of GMK cells and into four empty tubes (glass controls). After stationary incubation for 4 days, the cultures were refed and transferred to a roller drum. The cultures were observed at frequent intervals during a period of 20 days. Figure 1 shows the typical appearance of a colony of EP cells in mixed culture with GMK cells, 14 days after seeding. As early as 10–12 days after tumor cell seeding the CPE characteristics of SV40 could be seen in GMK cells adjacent to the well-defined colonies of the tumor cells (Fig. 2). All mixed cultures failing to show CPE and two of the glass controls were frozen and thawed once; the cell-free lysates of each group were pooled and subcultured to additional GMK tube cultures. These were observed for 4 weeks for the appearance of specific CPE. The results of representative experiments summarized in Table 1 suggest that the number of detectable virogenic cells depends on their plating efficiency. In all cases the isolated virus was identified as SV40 by serum neutralization tests. The cell-free lysates from the "glass controls" failed to show detectable virus.

Treatment of EP Cells with Subcellular Components of GMK Cells

The initial localization of the CPE in close proximity to the EP cells and the absence of demonstrable intra- or extracellular virus in EP cell cultures, suggested a cell-to-cell transfer of subviral material from the tumor cell to the indicator cell or, conversely, the transfer of subcellular components or metabolites, essential for virus synthesis from the indicator cell to the tumor cell. Accordingly, cells of bottle cultures of GMK cells were trypsinized and sedimented. The pellet containing 2×10^9 cells was resuspended in 4.5 ml maintenance medium and divided into 3 aliquots of 1.5 ml. The cells were lysed by multiple cycles of freezing and thawing, by ultrasonic treatment, or by combination of these methods; aliquots were stored at $-20°C$. Tube cultures of EP cells were then exposed to 0.1, 0.2, or 0.5 ml of the cell lysates for a period of 2 hours at 37°, followed by addition of maintenance medium. After 6 days of incubation at 35°, the cells were again treated with the respec-

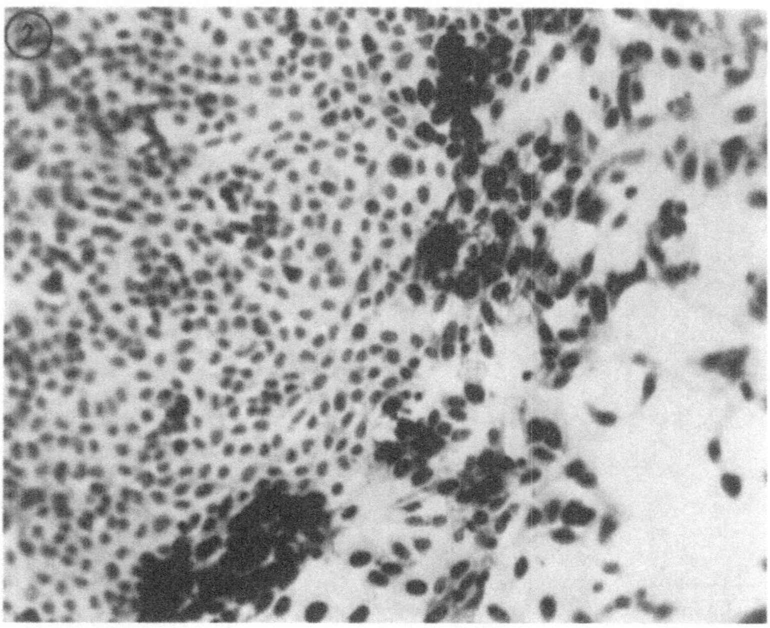

FIG. 2. Higher magnification (× 350) of Fig. 1. Note SV40 CPE in GMK cells adjacent to EP cells. Hematoxylin and eosin stain.

TABLE 1

NUMBER OF VIROGENIC CELLS DETECTABLE IN UNCLONED CULTURES OF EP CELLS

Number of EP cells seeded	EP cell passages seeded onto					
	GMK cells				Glass	
	25th		70th		25th	70th
	Cell growth	Virus isolation	Cell growth	Virus isolation	Virus isolation	
10^3	4/4[a]	+	4/4	+	−	−
10^2	4/4	+	4/4	+	−	−
10^1	2/4	+	3/4	+	NT[b]	NT
10^0	0/4	−	1/4	+	NT	NT

[a] Number of tubes showing EP cell growth/number of cultures.

[b] Not tested.

tive GMK lysates and incubated for 7 days. The cell-free lysates of the treated and untreated EP cells failed to yield detectable virus. These results indicate that, under these experimental conditions, subcellular components of GMK cells failed to induce virus synthesis in EP cells.

Parabiotic Cultures of EP and GMK Cells

The experimental evidence accumulated in these and our previous studies suggested that direct contact between tumor cell and indicator cell was necessary for the transfer of subviral infectivity. This point was investigated by means of parabiotic culture systems.

Fifty-millimeter petri dishes were divided by a strip of sintered glass filter, Corning medium grade, 3–4 mm thick, with a pore size of 10–15 μ (Fig. 3). The filter was annealed to the wall and bottom of the dish, and the lower 3 mm of the filter was heat-

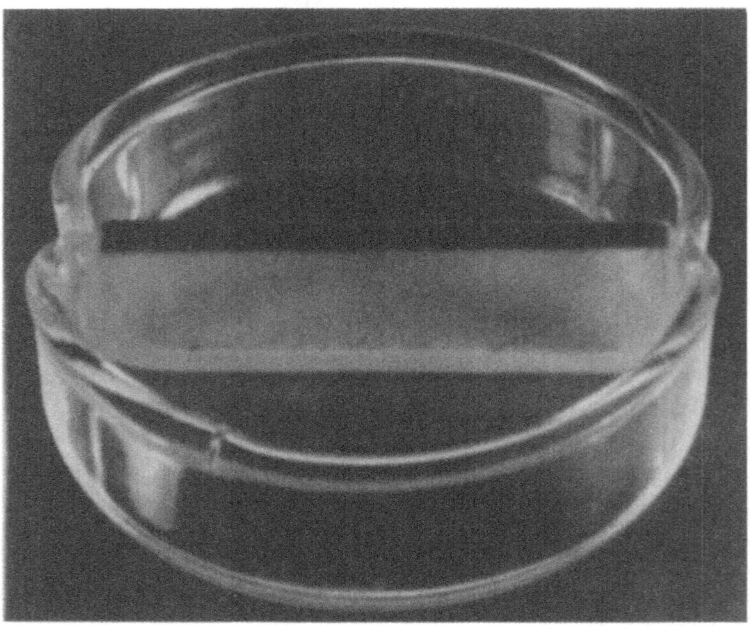

FIG. 3. Parabiotic culture dish

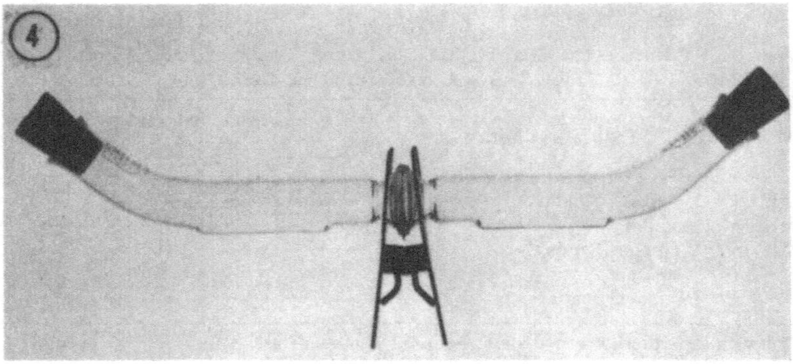

FIG. 4. Parabiotic culture tubes

sealed to avoid clogging by the cell sheet. Approximately 500,000 EP or GMK cells contained in 0.5 ml of medium were seeded in the respective halves of the culture dishes. One set was inoculated with a mixture of both cell types. After incubation at 37° for 4 hours, unattached cells were carefully removed and medium was added to each side to reach three-fourths of the height of the filter strip. Incubation was carried out in a humidified 5% CO_2 atmosphere at 37°. Samples of culture fluids were collected on the 10th day of incubation, and cell-free lysates were prepared on the 17th day. The results of two experiments summarized in Table 2 show the absence of virus in parabiotic cultures of EP and GMK cells as contrasted by the positive results obtained with mixed cultures. The data also demonstrate the permeability of these filters to SV40.

The parabiotic studies were repeated in chambers (Prier, 1963) separated by a Millipore membrane of 5 μ average porosity. The twin Leighton tube cultures separated by the membrane (Fig. 4) were initially incubated in the stationary position for 3 days, and afterward on a reciprocal rocking device to facilitate diffusion. Under these conditions, complete diffusion of 0.25% methylene blue was observed in 24 hours. The experimental results obtained with this system confirmed the observations of the preceding parabiotic culture experiments. These observations demonstrate the absence of any detectable effect on virus synthesis in EP cells by metabolic products derived from viable GMK cells in culture. Furthermore, these findings suggest that direct contact between tumor and indicator cell is essential for the transfer of subviral infectivity.

Relationship between Viability and Virogenicity of EP Cells

It was of interest to investigate the role of the physiologic state of the EP cells in their interaction with the indicator cells. EP cells were subjected to four kinds of physical treatments and were then seeded onto GMK cells and on glass to check their

TABLE 2

PARABIOTIC AND MIXED CULTURES OF EP AND GMK CELLS

Divided petri dish cultures	Virus assay of samples from both sides of dish	
	10-day fluids	17-day lysates
EP/Medium	–	–
GMK/Medium	–	–
EP/GMK	–	–
EP + GMK/Medium	+	+
SV40/GMK	+	+

TABLE 3

RELATIONSHIP BETWEEN VIABILITY AND VIROGENICITY OF TUMOR CELLS

Treatment	Tumor cells		Infectivity ($\log_{10}$ $TDID_{50}$/0.2 ml)	
	Viability	Virogenicity	SV40-DNA	Intact SV40
None	+	+	2.8	6.3
X-ray	+	+	3.0	6.2
UV, 30 sec	+	+	2.8	6.0
UV, 60 sec	–	–	3.0	6.2
56°C, 2 min	–	–	3.0	6.3
Sonic vibration	–	–	2.8	6.3

TABLE 4

EFFECT OF UV-IRRADIATED SENDAI VIRUS ON VIRUS RECOVERY FROM MIXED CULTURES OF TUMOR CELLS AND GMK CELLS

Cells	Preincubated with	SV40 isolation at indicated days of mixed cultures							
		1	2	3	4	5	6	7	8
EP + GMK	UV-inact. Sendai	−	+	+	+	+	+	+	+
EP + GMK	Medium	−	−	−	−	−	−	−	+
EP + GMK	UV-inact. Sendai, ether-treated[a]	−	−	−	−	−	−	−	+
EP	UV-inact. Sendai	−	−	−	−	−	−	−	−

[a] Treated with 20% diethyl ether for 18 hours at 4°.

viability. Ability of the treated cells to attach to glass and to exhibit metabolic activity were the criteria for viability. In addition, aliquots of SV40 virus or infectious viral DNA were treated in the same manner and infectivity assays were carried out in GMK cells. The experimental data are summarized in Table 3. Exposure of EP cells to X-irradiation up to 10,000 r, or to UV-irradiation up to 30 seconds, had no effect on the viability or virogenicity of the cells. On the other hand, longer exposure to UV, heating to 56° for 2 minutes, or ultrasonic treatment resulted in loss of cell viability and failure to demonstrate virogenicity. Since none of these treatments had any significant effect on the infectivity of SV40 DNA or intact virus, the results indicate that viability of EP cells is essential for the interaction with the indicator cells.

Influence of "Cell Fusion Factor" on Interaction between EP and GMK Cells

The time of appearance of detectable virus in mixed cultures of GMK and EP cells seems to depend on the number of tumor cells inoculated. Under optimal conditions the earliest detection of virus usually occurs after 7–10 days of mixed culture. Since most or possibly all EP cells carry the SV40 genome the late appearance of virus in mixed cultures suggests that the interaction between these two cell types occurs at a low frequency.

In an effort to enhance this process we studied the effect of noninfectious, UV-irradiated Sendai virus. This member of the parainfluenza group of viruses is reported to contain a factor present both in infectious and inactivated preparations which is capable of promoting fusion between cells of several continuous cell lines (Okada, 1958). Trypsin-dispersed GMK and EP cells were suspended in modified Hanks' solution at an approximate concentration of 10^7 cells/ml. To a mixture of 1.0 ml of each cell type, was added 2.0 ml of UV-irradiated Sendai virus containing 4000–8000 HA units/ml. This mixture was held at 4° for 15 minutes and then incubated at 37° for 20 minutes with frequent mixing. Appropriate controls were carried out in parallel. Aliquots of 0.5 ml were seeded in 2-ounce bottles. At daily intervals, one culture of each group was frozen for virus assay. Representative results of several experiments are summarized in Table 4. It can be seen that SV40 was not detectable until the 8th day in mixed cultures of EP and GMK cells in the absence of UV-inactivated Sendai virus or in the presence of ether-treated, UV-inactivated Sendai virus. On the other hand, SV40 could be demonstrated after 2 days in mixed cultures containing UV-inactivated Sendai virus.

Effect of Serial Passage of EP Cells in the Presence of Cytosine Arabinoside (CA)

In view of the known inhibitory effect of CA on the replication of some DNA-containing viruses, it was of interest to study the effect of this inhibitor on the virogenicity of EP cells. Cultures of EP cells were treated on alternate days for 24-hour periods with 5 μg/ml of CA during 5 successive cell passages and with 10 μg/ml of CA for an additional 8 passages. The surviving cells were then grown during 7 passages in the continuous presence of 10 μg/ml of CA.

Treated and untreated cells were examined for the presence of the "tumor antigen" ("T antigen") by the indirect

TABLE 5

EFFECT OF CYTOSINE ARABINOSIDE (CA) ON MULTIPLICATION OF SV40 IN $LLCMK_2$ CELLS

Treatment of SV40-infected cells	Extracellular virus	Intracellular virus
None	6.5[a]	5.5
CA, 7.5 μg/ml	<0 0	1.0

[a] Log_{10} $TCID_{50}$/0.5 ml.

fluorescent antibody staining technique (Pope and Rowe, 1964) and for virogenic activity The results obtained in repeated tests demonstrate the presence of the nuclear "T antigen" in all treated and untreated cells. Likewise, the CA-treated cells released infectious virus when plated on GMK cells. The effect of CA on the replication of SV40 in $LLCMK_2$ cells was tested in parallel experiments. These cells inoculated with 600 $TCID_{50}$ of SV40 were washed after a 1.5-hour adsorption period and then continously treated with 7.5 μg/ml of CA for 6 days with media changes on days 2 and 4. On the 7th day the supernatants of drug-treated and control tubes were pooled separately and the cell sheets were lysed by freezing and thawing in 0.5 ml of maintenance medium per tube. The extra- and intracellular virus content of the treated and control groups was determined. The results summarized in Table 5 indicate the complete inhibition by CA, of extracellular virus and only trace amounts of intracellular and/or cell-associated virus in the treated group. Initial treatments of EP cells with 10 μg/ml of CA caused considerable toxicity. Following intermittent treatments, the surviving cells tolerated continuous treatment without obvious cytotoxic changes. Treatment of the $LLCMK_2$ cells with 7.5 μg/ml of CA caused minimal cytotoxicity during the period of observation.

DISCUSSION

The phenomenon of cell-to-cell transfer of genetic information is not unique to the system reported here. Ochai *et al* (1959), studying multiple drug resistance in enteric bacteria observed that the resistance factors could be transferred *in vitro* from *Escherichia coli* to *Shigella* and vice versa. Cell-free filtrates of resistant strains failed to confer resistance upon sensitive strains and resistance could be transferred only by intact cells in direct contact with sensitive cells Subsequent studies by Watanabe and Fukasawa (1961) and Falkow *et al.* (1961) revealed that the transfer of resistance factors was episome-mediated and could also be demonstrated between members of widely separated families of bacteria.

Svoboda *et al.* (1963) investigated the virus-tumor relationship in rat tumor XC, produced by Rous virus. These workers observed that intact cells of tumor XC inoculated into chicks would give rise to virus-containing Rous sarcomas whereas, cell-free preparations of tumor XC consistently failed to be oncogenic in chicks. This problem was studied in tissue culture by Simkovic *et al.* (1962), who observed that mixing cultures of XC tumor cells and chick fibroblasts resulted in the release of Rous virus as evidenced by the oncogenic effect in chicks of cell-free culture fluids. This was in contrast to the lack of oncogenicity of cell-free fluids from either of these tissues alone.

Sabin and Koch (1963) studied the behavior of noninfectious SV40 viral genome in hamster tumor cells and confirmed our observations on the induction of viral synthesis in mixed culture of tumor cells and indicator cells. They applied the term "induction by association of viable tumor cells with sensitive indicator cells in tissue culture" to describe this phenomenon and suggested that "something may be transmitted from the sensitive cell to a tumor cell which then induces it to produce not more than a single, or at most, a few infectious particles of virus."

Black and Rowe (1963) analyzed the characteristics of hamster kidney cells transformed *in vitro* by SV40. After prolonged passage *in vitro* they failed to detect infectious virus in culture fluids or in disrupted cells. However, by means of mixed culture of transformed cells and indicator cells ("overlay method") they were able to induce viral synthesis and postulated that the transformed cells carried subviral infectious units. Black (personal communication) has cloned this transformed hamster cell line by isolating single cells in a microdrop in the presence of SV40 antiserum. Following three consecutive clonings of an isolated cell, he analyzed 12 clones and found that all contained the "T antigen" as determined by fluorescent antibody stain and that 11 of these clones gave rise to viral synthesis when cultured on sensitive indicator cells.

The experimental results reported here define some of the conditions required for the induction of viral synthesis in mixed cultures of virogenic hamster tumor cells and GMK cells. Viability of the tumor cells seems to be essential since prolonged UV-irradiation, heat, and sonic vibration abolished both cell viability and virogenicity while similar treatments of intact SV40 or infectious viral DNA had no significant effect on their infectivity. The requirement for direct contact of viable tumor cells with indicator cells was illustrated by the data obtained in parabiotic culture experiments. The interaction of tumor and indicator cells may involve either a transfer of subviral infectivity from the tumor cell to the indicator cell or, conversely, the transmission from the indicator cell of subcellular particles or metabolites essential for virus synthesis. The following evidence presented here appears to support the first alternative: (a) addition of concentrated lysates or subcellular fractions of GMK cells to viable cultures of tumor cells failed to induce release of virus; (b) tumor cells growing in parabiotic culture with GMK cells failed to release infectious virus.

The postulated transfer of genetic information could be mediated by any of several types of interaction between mammalian cells in culture. Barski (1960), and recently Harris and Watkins (1965) observed fusion of mammalian cells from the same or unrelated species. These workers showed karyotypic changes resulting from this interaction. It is also conceivable that adjacent cells in culture are capable of taking up molecular-sized particles by diffusion or active transport or are able to ingest subcellular structures by pinocytosis (Horkawa *et al.*, 1963).

The inhibitory effect of cytosine arabinoside on the replication of SV40 described here is in agreement with the findings of others (Rapp *et al.*, 1965). The absence of detectable effects on the virogenic properties of EP cells following prolonged serial passage in the presence of 10 μg/ml of CA is of interest. Additional studies are in progress to assess the significance of this finding.

The mechanism of the enhancing effect by UV-inactivated Sendai virus on the interaction of EP and GMK cells has not been elucidated. Further studies of this phenomenon are in progress.

Present evidence suggests that not every SV40-induced tumor or transformed cell line contains an infectious, inducible viral genome. One SV40-induced hamster tumor cell line and an *in vitro* transformed hamster embryo cell line have failed to be inducible when seeded onto sensitive indicator cells (Melnick *et al.*, 1964). Similar failures have been observed with SV40-transformed cells derived from human buccal mucosa (Habel, personal communication). Furthermore, all attempts to detect the presence of an infectious viral genome by mixed culture with sensitive cells or by chemical induction have failed in the case of polyoma virus or adenovirus type 12 transformed cells (Gerber, 1964). These findings suggest the possibility that transformation of hamster cells by SV40 virus may be associated with either an integration of the complete genome responsible for directing the synthesis of infectious virus or integration of only that portion of the genome involved in the synthesis of the new cellular antigen.

REFERENCES

Barski, G., Sorieul, S., and Cornefert, F. (1960). Production dans des cultures in vitro de deux souches cellulaires en association, de cellules de caractère "hybride." *Compt. Rend. Acad. Sci.* **251**, 1825–1827.

Black, P. H., and Rowe, W. P. (1963). An analysis of SV_{40}-induced transformation of hamster kidney tissue *in vitro*. I. General characteristics. *Proc. Nat. Acad. Sci. U.S.* **50**, 606–613.

Falkow, S., Marmur, J., Carey, W. F., Spilman, W. M., and Baron, L. S. (1961). Episomic transfer between *Salmonella typhosa* and *Serratia marcescens*. *Genetics* **46**, 703–706.

Gerber, P. (1963). Tumors induced in hamsters by simian virus 40: persistent subviral infection. *Science* **140**, 889–890.

Gerber, P. (1964). Virogenic hamster tumor cells: Induction of virus synthesis. *Science* **145**, 833.

Gerber, P., and Kirschstein, R. L. (1962). SV_{40}-induced ependymomas in newborn hamsters. I. Virus-tumor relationships. *Virology* **18**, 582–588.

HARRIS, H., and WATKINS, J. F. (1965). Hybrid cells derived from mouse and man: artificial heterokaryons of mammalian cells from different species. *Nature* **205,** 640–646.

HORKAWA, M., DOIDA, Y., and SUGAHARA, T. (1963). Preliminary studies on cytosis in mammalian cells cultured *in vitro*. *Exptl. Cell Res.* **32,** 404–407.

HULL, R. N., CHERRY, W. R., and TRITCH, O. J. (1962). Growth characteristics of monkey kidney cell strains LLC-MK_1, LLC-MK_2 and LLC-MK_3 (NCTC-3196) and their utility in virus research. *J. Exptl. Med.* **115,** 903–917.

MELNICK, J. L., KHERA, K. S., and RAPP, F. (1964). Papova virus SV40: failure to isolate infectious virus from tranformed hamster cells synthesizing SV40-induced antigens. *Virology* **23,** 430–432.

OCHAI, K., YAMANAKA, T., KIMURA, K., and SAWADA, O. (1959). Studies on the inheritance of drug resistance between *Shigella* strains and *Escherichia coli* strains. *Nihon Iji Shimpo* **1861,** 34–46.

OKADA, Y. (1958). The fusion of Ehrlich's tumor cells caused by HVJ virus *in vitro*. *Biken's J.* **1,** 103–110.

POPE, J. H., and ROWE, W. P. (1964). Detection of specific antigen in SV_{40}-transformed cells by immunofluorescence. *J. Exptl. Med.* **120,** 121 128.

PRIER, J. E. (1963). A simple parabiotic chamber for cell culture studies. *Am. J. Clin. Pathol.* **39,** 175–176.

RAPP, F., MELNICK, J. L., and KITAHARA, T. (1965). Tumor and virus antigens of simian virus 40: differential inhibition of synthesis by cytosine arabinoside. *Science* **147,** 625–627.

SABIN, A. B., and KOCH, M. A. (1963). Behavior of non-infectious SV_{40} viral genome in hamster tumor cells: Induction of synthesis of infectious virus. *Proc. Natl. Acad. Sci. U.S.* **50,** 407–417.

SIMKOVIC, D., VALENTOVA, N., and THURZO, V. (1962). An *in vitro* system for the detection of the Rous sarcoma virus in the cells of the rat tumor XC. *Neoplasm* **9,** 104–106.

SVOBODA, J., CHYLE, P., SIMKOVIC, D., and HILGERT, I. (1963). Demonstration of the absence of infectious Rous virus in rat tumor XC whose structurally intact cells produce Rous sarcoma when transferred to chick. *Folia Biol.* **9,** 77–81.

WATANABE, T., and FUKASAWA, T. (1961). Episome-mediated transfer of drug resistance in enterobacteriaceae. *J. Bacteriol.* **81,** 669–677.

ACTIVATION OF PRODUCTION OF INFECTIOUS TUMOR VIRUS SV40 IN HETEROKARYON CULTURES

H. Koprowski, F. C. Jensen, and Z. Steplewski

Infectious virus may be isolated from SV40-transformed human cells during the early stages of transformation.[1, 2] However, no virus can be detected in the majority of human cell lines after recovery from the crisis period.[3] In transformed green monkey kidney (GMK) cells, following abortive infection with SV40, infectious virus cannot be recovered at all.

Attempts to recover infectious virus from SV40-induced hamster tumors have been generally unsuccessful,[4–6] with the exception of hamster ependymomas from which SV40 was recovered after induction by chemical or physical means, or by fusion.[7] The virus has also been recovered from SV40-transformed hamster and mouse cells by placing the viable transformed cells in direct contact with susceptible indicator cells.[8, 9]

Since intimate cell-to-cell contact seems to facilitate recovery of the virus, an improved cell fusion technique, described recently by Harris *et al.*,[10] has made it possible to reinvestigate the presence of SV40 in these transformed cells after the production of artificial heterokaryons with normal cells.[11]

Materials and Methods.—Tissue cultures and their characteristics: Table 1 shows the SV40-transformed cell lines used in these experiments. GMK-EVa and GMK LL"E-46" are transformed cell lines derived from primary monkey kidney cultures. The GMK-EVa culture was transformed by exposure to a very low concentration of SV40 isolated from a human culture which continuously shed the virus for 24 months after infection. Eight clones were established from this line which, during continuous passage for 12 months, remained free of infectious virus. GMK LL "E-46" was transformed by the SV40 component of the adeno 7-SV40 hybrid, in which the infectivity of adeno 7 was destroyed by irradiation.[12] This line, which was kept in continuous cultivation for 8 months, did not demonstrate the presence of either infectious adeno 7 or SV40.

W18Va2 is an SV40-transformed culture derived from human buccal mucosa and kept in continuous passage since 1962.[13] This line has remained virus-free since January 1963. The W98Va line, derived from human skin, was transformed by SV40 in 1963.[3] Eight sublines of this culture were derived from a pool of 1×10^6 cells frozen after the culture became transformed but before it reached the crisis stage. After recovery from crisis, seven of these sublines stopped shedding infectious virus; in the remaining subline, W98VaE (see below), which continued to yield infectious SV40 for 19 months after crisis, no virus has been detected since May 1966.

F5-1 is a cell line derived from a hamster fibrosarcoma induced by inoculation of SV40 into a newborn hamster.[14]

As shown in Table 1, SV40-induced complement-fixation antigen (ICFA) detected by immunofluorescence (see below) is present in the nuclei of all cells of these cultures. However, none of the cells show presence of viral coat antigen (VP). The transformed cultures also show presence of SV40-induced transplantation antigen (ITA). They are resistant to superinfection with SV40, with the exception of the GMK LL"E-46" line.

Normal cells used for fusion with transformed cells were either of African green monkey kidney (AGMK) or hamster embryo (HE) origin; AGMK was used after one passage in tissue culture of primary monkey kidney suspension obtained from Flow Laboratories. Hamster embryo cultures were established from cell suspension of Syrian hamster embryos and used after one to five cell transfers.

Technique of cell fusion: The method used followed, essentially, that described by Harris *et al.*[10] A strain of Sendai virus obtained from Dr. G. Henle of the Children's Hospital, Philadelphia, Pa., was grown in the allantoic cavity of hens' eggs. The infectious allantoic fluids were harvested 48 hr after infection and, following overnight storage, were concentrated and inactivated by exposure to ultraviolet (UV) light for either 3 or 15 min.

In the actual fusion experiments, 5 million cells in 0.5 ml of each of the two cultures to be fused were mixed together in the presence of 1 ml of UV-inactivated Sendai virus containing 2000–4000 hemagglutinating units. The virus-cell mix was incubated for 10 min at +4°C and shaken for an additional 20 min in a water bath kept at 37°C. The fused cells were then centrifuged at 65 *g* for 3 min in an International centrifuge PR2 (Head 253), resuspended in growth medium, and seeded on Petri dishes containing coverslips or slides, and in milk dilution or Blake bottles. In other experiments, the fused cells were seeded on monolayers of primary GMK cultures. The ratio of heterokaryon cells was determined through labeling of cells of one parental line with H^3-thymidine (specific activity 14.5 c/mM) at a concentration of 0.2 μc/ml in 80 ml of medium. The cells were grown in presence of H^3-thymidine for 3 days before fusion. At the time of fusion, 100-fold excess of cold thymidine was added and the percentage of labeled nuclei determined by autoradiography of 24-hr cultures seeded on coverslips. This labeling technique resulted in labeling of 98.5–99.5% of parental cell nuclei. A heterokaryon cell was determined by the presence of at least one labeled nucleus in the same cell with one nonlabeled nucleus.

Isolation of infectious SV40: AGMK cultures were used throughout this work for isolation of SV40. Medium and cell extracts from heterokaryon cultures were passaged onto fresh AGMK monolayer.

SV40-induced complement fixation antigen (ICFA) and viral coat antigen (VP): Coverslips of heterokaryon cultures were washed with PBS, fixed in acetone at −24°C for 10 min, and then air-dried. Presence of ICFA or VP in the nuclei was then determined by staining the cells with fluorescein-isothiocyanate conjugated antibodies. Serum for staining of ICFA was obtained from tumor-bearing hamsters; for VP, from a baboon immunized against SV40.

Results.—Absence of infectious SV40 in transformed cell lines prior to fusion experiments: No infectious SV40 was demonstrated in extracts obtained by repeated freezing and thawing of 5 × 10^7 cells of the transformed cultures at different passage levels after transformation (Table 1). Furthermore, cocultivation of viable cells of these cultures with AGMK for varying time periods did not yield infectious SV40. Several of the cultures were subjected in various laboratories to experiments using methods which led to the demonstration of infectious SV40 in hamster transformed cells. However, no SV40 was isolated in these experiments. In addition, attempts were made to detect infectious SV40 in eight clonal sublines originating from single cells of GMK-EVa cultures, but without success.

Proportion of heterokaryons formed between various cell types: The number of cells participating in the formation of multinucleated cells in the presence and absence of Sendai virus is expressed in Table 2 as the percentage of nuclei in homo- and heterokaryons. This figure varied from 17.6 per cent for fused HE and GMK LL"E-46" cells to 61.9 per cent for fused HE and F5-1 cells, in the presence of Sendai virus. As expected, a certain proportion of multinucleated cells is represented by fused cells of the same parental lines (homokaryons).

The percentage of multinucleated cells produced 24 hours after cocultivation of the two cell types in the absence of Sendai virus is low, and the number of heterokaryons formed is minimal.

Figure 1 shows an autoradiograph of a trinucleated cell in which one nucleus (AGMK) is labeled and the other two are not (W18Va2). Figure 2 shows a multinucleated cell with several labeled nuclei (AGMK) and several unlabeled nuclei (W18Va2).

Activation of production of infectious SV40 in heterokaryon cultures: Results shown in Table 3 indicate that infectious SV40 was isolated in three separate ex-

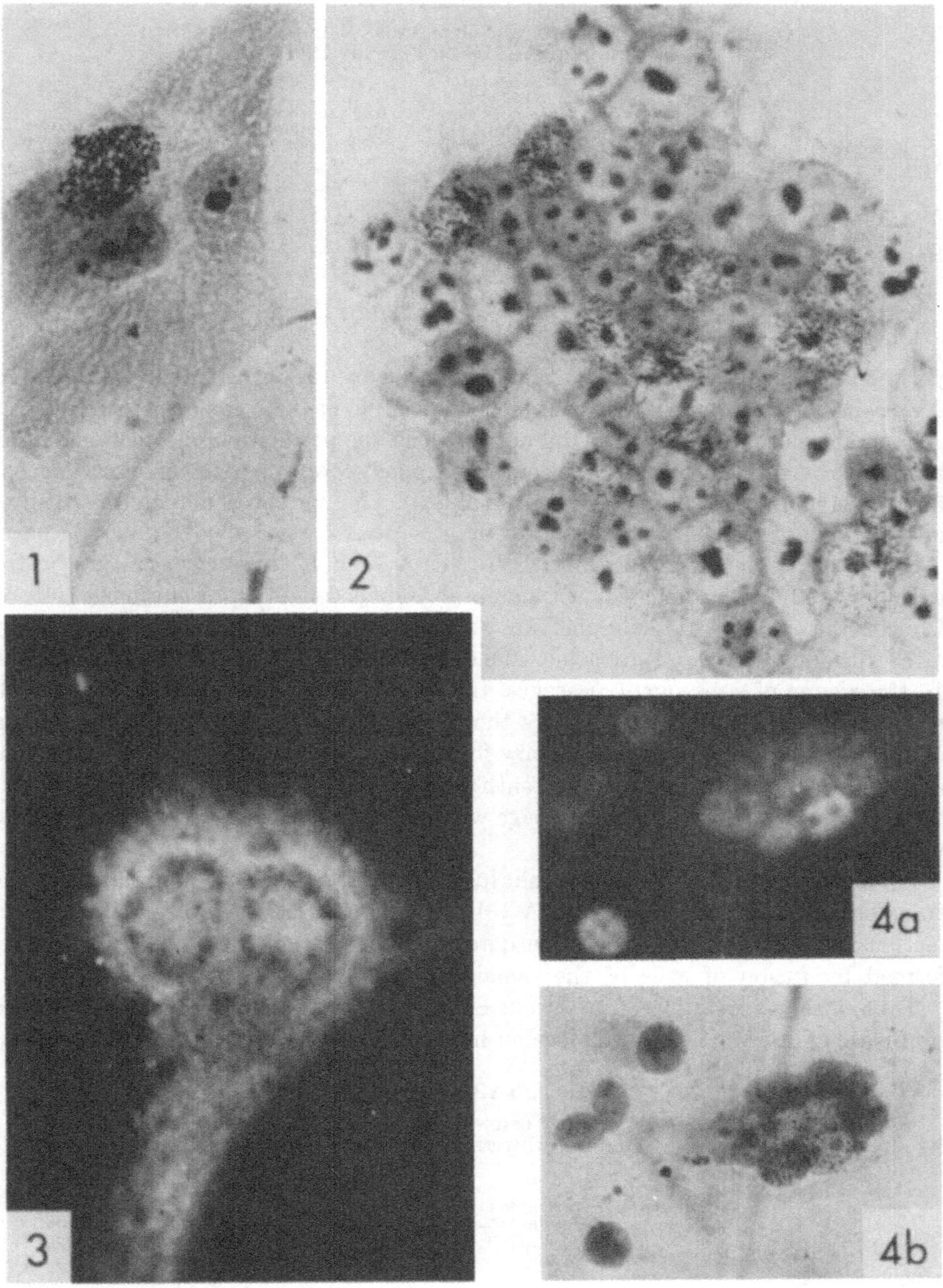

FIG. 1.—Heterokaryon of W18 Va2 and AGMK cells. One of the nuclei (labeled) is of AGMK origin.

FIG. 2.—Multinucleated cell showing the presence of several AGMK labeled nuclei. The non-labeled nuclei are W18 Va2.

FIG. 3.—Positive fluorescence in two of three nuclei stained for SV40 viral antigen with baboon anti-SV40 serum.

FIG. 4.—(*a*) Positive fluorescence in all nuclei of this heterokaryon stained for SV40-ICFA with hamster T-antisera. (*b*) Autoradiogram of the same heterokaryon as in (*a*) shows five labeled AGMK nuclei.

TABLE 1

ORIGIN OF SV40-TRANSFORMED CELL LINES USED IN FUSION WITH PRIMARY AGMK OR PRIMARY HE

Species	Cell line	Calculated number of cell generations since last detection of infectious SV40	Resistance to superinfection with SV40	Presence of SV40 VP	Presence of SV40 ICFA	Presence of SV40 ITA
Man	W18Va	1000	Yes	No	Yes	Yes
	W98Va*	150	Yes	No	Yes	Yes
African green	GMK-EVa	190	Yes	No	Yes	Yes
monkey	GMK LL "E-46"†	150	No	No	Yes	No?
Hamster	F 5-1‡	±1000	—	No	Yes	Yes

* Eight sublines.
† Transformed by exposure to irradiated adeno 7-SV40 hybrid (LL "E-46").
‡ Obtained from tumor induced by inoculation of SV40 into newborn hamster.

periments from heterokaryon cultures produced by fusion of transformed GMK-EVa cells with normal AGMK cells. Cocultivation of the transformed and normal cells in absence of Sendai virus did not result in the production of infectious SV40, but this is hardly surprising in view of the extremely low proportion of heterokaryons formed 24 hours after mixing the cells (Table 2). No virus was detected in cultures of fused cells of each of the parental strains.

SV40 was detected in the GMK-EVa plus AGMK heterokaryon culture by the cytopathic effect it produced on the parental AGMK cells still present in the culture. Concentration of SV40 in the medium obtained from this culture, on the 28th day after fusion when the cytopathic effect was observed, was $10^{8.5}$TCID$_{50}$/0.1 ml. In the course of subsequent passages, the level of infectious SV40 rapidly fell off to a level of $10^{0.5}$TCID$_{50}$/0.1 ml by the fourth transfer. Because of destruction of the parental AGMK cells necessary for its reproduction, no infectious virus was detected in the cultures from the tenth passage on. However, when cells of this culture were fused at the 15th passage with AGMK, activation of infectious SV40 was resumed.

Data shown in Table 4 indicate that fusion of cells derived from four clones of the transformed GMK-EVa line with AGMK cells resulted in the production of infectious SV40. No virus could be detected, however, in heterokaryon cultures formed by fusion of cells of the remaining four clones with AGMK.

Table 5 shows isolation of infectious SV40 from heterokaryon cultures, obtained by fusion of AGMK with transformed human cells (W18Va2), in which infectious

TABLE 2

HETEROKARYONS OBTAINED BY FUSION OF AGMK OR HE CELLS (PRELABELED WITH H^3-THYMIDINE) WITH SV40-TRANSFORMED CELLS

Parental Cell Types Fused (*A* and *B*): Normal (*A*)	Transformed (*B*)	Presence of Sendai virus	Percentage of Nuclei in Multinucleated Cells: Homo-karyons*	Hetero-karyons†	Remarks
AGMK	GMK-EVa	+	20.8	44.6	See Table 3
		−	4.3	0.8	
	W18 Va2	+	49.0	20.4	See Table 5
		−	6.5	0.6	
	W98 VaE	+	18.1	34.2	See Table 5
		−	7.5	0.5	
HE	F 5-1	+	25.2	61.9	
	GMK LL"E-46"	+	28.5	17.6	See Table 7
	W18 Va2	+	34.3	38.6	

* Homokaryon: a multinucleated cell containing nuclei of only one parental cell type.
† Heterokaryon: a multinucleated cell containing nuclei of both parental cell types.

TABLE 3

ACTIVATION OF INFECTIOUS SV40 PRODUCTION IN HETEROKARYON CULTURES PRODUCED BY FUSION OF GREEN MONKEY TRANSFORMED CELLS WITH PRIMARY AGMK CELLS

Expt. no.	Parental Cell Types Fused (*A* and *B*) *A*	*B*	Presence of Sendai virus	Isolation of infectious SV40
1	GMK LL "E-46"	AGMK	+	−
	GMK-EVa	AGMK	+	+
11	GMK-EVa	AGMK	+	+
		GMK-EVa	+	−
	AGMK	AGMK	+	−
25	GMK-EVa	AGMK	+	+
			−	−
		GMK-EVa	+	−
	AGMK	AGMK	+	−

SV40 had not been demonstrated for the past four years. Again, SV40 was not detected in cultures established by cocultivation in the absence of Sendai virus. This was consistent with the results shown in Table 2. Virus was also not detected in the fused cells of each of the parental lines.

Infectious SV40 was isolated from seven out of eight heterokaryon cultures produced through fusion, in the presence of Sendai virus, of cells of the eight sublines of transformed human skin W98Va with AGMK. Cultures established by cocultivation of four W98Va sublines with primary green monkey kidney in the absence of Sendai virus remained virus-free. Results of experiment 25 (Table 5) confirm activation of production of infectious SV40 in heterokaryon cultures formed by fusion of W98VaE cells with AGMK. No virus was found in the parental cultures.

SV40 plaque production by fused cells: Cells fused in the presence of Sendai virus in experiment 25 (Table 5) were resuspended after centrifugation and instead of being seeded in culture flasks, were plated directly on AGMK monolayers diluted to approximately 200 cells per plate (Table 6). Another portion of the fused cell suspension, in which one of the parental cell cultures (AGMK) was prelabeled with H^3-thymidine, was seeded on coverslips; the proportion of heterokaryons was determined by autoradiography performed 24 hours later.

The results of the plating experiment shown in Table 6 indicate that of the total number of 1.8×10^3 cells seeded on AGMK monolayers, two plaques were produced. Among these cells, only 252 were calculated to be true heterokaryons, approximately 1 per cent of which yielded infectious SV40.

Demonstration of VP antigen: No synthesis of VP was observed in the nuclei of the cell lines shown in Table 1, even after exposure to Sendai virus. However, when, as in experiment 25, transformed W98VaE cells were fused with AGMK cells, immunofluorescent staining of the heterokaryon cultures 96 hours after fusion revealed presence of VP in nuclei. Figure 3 shows presence of VP in two out of three nuclei of a multinucleated cell.

Internuclear transfer of ICFA: When heterokaryon cultures listed in Table 7 were examined 24 hours after fusion for presence of ICFA by immunofluorescent staining (see *Materials and Methods*), some multinucleated cells contained both

TABLE 4

ACTIVATION OF INFECTIOUS SV40 PRODUCTION IN HETEROKARYON CULTURES PRODUCED BY FUSION OF CELLS OF GMK-EVA CLONES WITH AGMK IN PRESENCE OF SENDAI VIRUS

Clone number:	2A-1	2B-1	5-1	6-1	7-1	8-1	9-1	10-1
Isolation of SV40:	−	−	+	+	+	−	−	+

TABLE 5

ACTIVATION OF INFECTIOUS SV40 PRODUCTION IN HETEROKARYON CULTURES PRODUCED BY FUSION OF HUMAN TRANSFORMED CELLS W18 Va2 AND W98Va WITH CELLS OF AGMK

Experiment no.	Parental Cell Types Fused (*A* and *B*) *A*	*B*	Presence of Sendai virus	Isolation of infectious SV40
11	W18 Va2	AGMK	+	+
			−	−
		W18 Va2	+	−
	AGMK	AGMK	+	−
	W98 Va-A		+	+
	B		+	+
	C		+	−
	D		+	+
	E		+	+
20		AGMK	−	−
	F		+	+
			−	−
	G		+	+
			−	−
	H		+	+
			−	−
	AGMK		+	−
25	W98 VaE	AGMK	+	+
			−	−
		W98 VaE	+	−
	AGMK	AGMK	+	−

fluorescing and nonfluorescing nuclei. Such cells were not found in cultures 72 hours after fusion; at this time, the multinucleated cells contained either all fluorescing nuclei or none, the latter being of AGMK origin.

Further confirmation of internuclear transfer of ICFA was obtained in experiments in which we observed the same heterokaryon cell for fluorescing nuclei (for ICFA) and, by autoradiography, for the presence of labeled nuclei.[15] Figure 4 shows such a multinucleated cell found in a culture produced after fusion of cells of the human transformed W18Va2 line with H^3-thymidine labeled normal AGMK. In this cell, five of the nine fluorescing nuclei are labeled. Since they are of AGMK origin, it is quite clear that transfer of ICFA from the unlabeled W18Va2 nuclei to the labeled AGMK took place.

Discussion.—These studies indicate that the genome of a DNA tumor virus, possibly integrated in a manner as yet undetermined with the genome of a transformed monkey or human cell, which could not be activated by known methods for detection of infectious SV40 in the transformed hamster or mouse cells,[2, 3] is nevertheless capable of becoming activated in heterokaryons to form a complete virus. It has been postulated that transformation of human and primate cells exposed to

TABLE 6

SV40 PLAQUE ASSAY OF HETEROKARYONS OF W98 EVA AND AGMK CELLS

Fused Cells Seeded on Primary AGMK Cells: Total number	Number per plate	Heterokaryon Cells: Ratio per total counted	Calculated number per 1.8 × 10^3 cells	Number of SV40 plaques
1.8 × 10^3	2 × 10^2	308/2214	252	2

TABLE 7

INTERNUCLEAR TRANSFER OF SV40-ICFA IN HETEROKARYON CELLS

Normal "recipient" cells	Transformed "Donor" Cells: W18 Va2	W98 VaE	GMK-EVa	GMK LL "E-46"	F 5-1
HE	+	+	+	+	+
AGMK	+	+	+	+	+

SV40 may occur after an abortive infection,[16, 17] i.e., without production of viral progeny. The results indicate that, at least in a fraction of the transformed cell population, SV40 may be present with a full complement of genetic material necessary for its synthesis, but may be unable (before fusion) to express its late function, such as synthesis of viral coat. The proportion of the cell population which contains the viral genome in such a form can only be determined by extension of the quantitative determinations of the ratio of heterokaryons releasing infectious SV40 (Table 6). This, in turn, is related to the estimate of the number of parental cells required to fuse for the activation of SV40 production. Experiments are in progress to determine the ratio of normal and transformed cells necessary to form virus-releasing heterokaryons.

It is possible that only a small fraction of the transformed cell population may yield infectious SV40 after heterokaryon formation. Moreover, there are transformed cultures, such GMK LL"E-46" or the four clones of GMK-EVa (Table 4), in which activation of SV40 synthesis through fusion with normal AGMK did not occur. If this assumption is correct, then it is logical to postulate that in these cells, the virus genome may be present in a form so defective that it cannot be rescued. Conversely, if complementation of the defective mutants may lead to the formation of infectious virus,[18] the fusion of cells infected with different defective mutants, such as "virus-negative" clones of GMK-EVa, may yield infectious virus. These problems are now under study.

Heterokaryons of transformed and normal cells may be useful for detection of hitherto undetectable tumor virus integrated in some manner into the genome of a malignant cell, and for the investigation of other aspects of carcinogenesis.

One of us (HK) is grateful to Professor Henry Harris of Sir William Dunn School of Pathology, University of Oxford, for his kindness in demonstrating technique of heterokaryon formation. The technical assistance of Mrs. Katherine M. Farrell is gratefully acknowledged.

This work was supported in part by USPHS grant 5-RO1-CA 04534-09 from the National Cancer Institute.

[1] Koprowski, H., J. Pontén, F. Jensen, R. G. Ravdin, P. S. Moorhead, and E. Saksela, *J. Cell Comp. Physiol.*, **59**, 281 (1962).
[2] Shein, H. M., and J. F. Enders, these PROCEEDINGS, **48**, 1164 (1962).
[3] Girardi, A. J., F. C. Jensen, and H. Koprowski, *J. Cell. Comp. Physiol.*, **65**, 69 (1965).
[4] Melnick, J. L., K. S. Khera, and F. Rapp, *Virology*, **23**, 430 (1964).
[5] Larson, V. M., P. A. Gosnell, and M. R. Hilleman, *Proc. Soc. Exptl. Biol. Med.*, **122**, 1182 (1966).
[6] Landau, B. J., V. M. Larson, G. A. Devers, and M. R. Hilleman, *Proc. Soc. Exptl. Biol. Med.*, **122**, 1174 (1966).
[7] Gerber, P., *Virology*, **28**, 501 (1966).
[8] Black, P. H., W. P. Rowe, and H. L. Cooper, these PROCEEDINGS, **50**, 847 (1963).
[9] Dubbs, D. R., and S. Kit, *Proc. of the Annual Meeting of the Am. Soc. Microbiol.* (1967), p. 143.
[10] Harris, H., J. F. Watkins, C. E. Ford, and G. I. Schoefl, *J. Cell Science*, **1**, 1 (1966).
[11] Koprowski, H., and F. C. Jensen, *Federation Proc.*, **26**, 313 (1967).
[12] Jensen, F. C., *Proc. Am. Assoc. Cancer Res.*, **8**, 34 (1967).
[13] Pontén, J., F. C. Jensen, and H. Koprowski, *J. Cell. Comp. Physiol.*, **61**, 145 (1963).
[14] Goldner, H., A. J. Girardi, and M. R. Hilleman, *Proc. Soc. Exptl. Biol. Med.*, **114**, 456 (1963).
[15] Sauer, G., H. Fischer, and K. Munk, *Virology*, **28**, 765 (1966).
[16] Koprowski, H., *Harvey Lectures*, Ser. 60 (1966).
[17] Sauer, G., and V. Defendi, these PROCEEDINGS, **56**, 452 (1966).
[18] Sauer, G., H. Koprowski, and V. Defendi, these PROCEEDINGS (in press).

PRODUCTION OF SV40 VIRUS IN HETEROKARYONS OF TRANSFORMED AND SUSCEPTIBLE CELLS

J. F. WATKINS AND RENATO DULBECCO

Cells transformed by DNA viruses do not produce progeny of the virus that caused the transformation. However, Gerber[1] and Sabin and Koch[2] found that small amounts of SV40 virus are produced if SV40-transformed cells are grown together with cells able to support SV40 multiplication. A clue to the possible mechanism of this activation was provided by Gerber[3] and by Koprowski and Jensen,[4] who observed that more virus is produced if UV-inactivated Sendai virus is added to the cell mixture. The inactivated Sendai virus promotes cell fusion;[5] it is possible, therefore, that SV40 virus is produced by heterokaryons resulting from fusion of transformed cells with virus-susceptible cells.

This article will give evidence in support of this hypothesis, and describe some of the characteristics of reactivation.

Materials and Methods.—Cell cultures: A line of SV40-transformed cells (SV3T3) derived from 3T3[6] cells was supplied by Dr. H. Green. It has been maintained for over a year in this laboratory by weekly transfer of about 10^4 cells in a reinforced Eagle's medium[7] with 10% calf serum (EC medium). Five other transformed lines, also from SV40-infected 3T3 cells, were produced by Dr. Marguerite Vogt in this laboratory. The SV40-susceptible line, BSC1, derived from African green monkey kidney cells (referred to as BSC), was maintained in reinforced Eagle's medium with 10% fetal bovine serum, and 10% tryptose phosphate (EFT medium).

Cells transformed by polyoma virus included a line of transformed 3T3 cells, a line of 3T3 cells doubly transformed with SV40 and polyoma virus (both supplied by Dr. H. Green), and a line of BHK[8] cells transformed by polyoma virus (Py 19).

Cell fusion: A modified version of the method described by Harris and Watkins[5] was used. Cells were diluted in Earle's saline[9] at 4°C. 0.5 ml of each cell suspension was placed in a 3 in. × $^1/_2$ in. test tube, and 1.0 ml of ultraviolet-inactivated Sendai virus (UVSeV), containing 400 hemagglutinating units, was added. The tube was stoppered and gently tilted horizontally until clumping of cells seemed maximal. It was then incubated in a 37° water bath, without further agitation, for 15 min. After gentle resuspensions of cells, 1.0 ml of the mixture was added to 5 ml of EFT medium in each of two 50-mm plastic Petri dishes, one of which usually contained several circular glass coverslips 11 mm in diameter. The dishes were incubated at 37° in a CO_2-flushed incubator. The medium was changed after 24 hr incubation.

Virus titration: Cells and medium were frozen and thawed three times. After centrifugation, 0.1 ml of supernatant was titrated by plaque formation on monolayers of BSC cells in 50-mm Petri dishes.[10] Virus yield was expressed in plaque-forming units (PFU).

Mixed immune hemadsorption[11] was carried out with a rabbit antiserum prepared against Ehrlich ascites tumor cells.[12] At a dilution of 10^{-4} this serum produced hemadsorption on all SV3T3 and 3T3 cells, and on no BSC cells.

Staining of coverslips: Coverslips were fixed in methyl alcohol, stained in 1:10 Giemsa at pH 6.7, dehydrated in acetone, cleared in xylene, and mounted in "Permount" (Fisher Scientific Co., New York).

Preparation of clonal cultures of SV3T3 cells: Sterile glass coverslips were broken into pieces a few millimeters across in 50-mm plastic Petri dishes. Five ml of EC medium containing about 500 cells was placed in the dish. After incubation for 18 hr at 37°, the coverslip segments were examined microscopically, and fragments bearing a single cell were transferred to separate Petri dishes (30-mm diameter), each containing 2 ml of EC medium. After 7 days' incubation at 37° in a CO_2-flushed incubator, fragments bearing colonies of SV3T3 cells were transferred to a drop (0.05 ml) of 0.15% trypsin in a separate Petri dish for each fragment, and placed at 37° for 15 min.

Two ml of EC medium was then added to each dish, and incubation was continued for 10 more days, when the cells were trypsinized and used in fusion experiments or for further passage.

Results.—Absence of SV40 infectivity in cell lines or Sendai virus preparation: No plaques were obtained on titration of the following materials: (*a*) supernatants after freezing and thawing up to 10^8 SV3T3 cells in 1.0 ml of EFT medium; (*b*) supernatants after freezing and thawing SV3T3 or BSC cells which had been treated one week previously with UVSeV; (*c*) UVSeV in Earle's saline at a concentration of 40,000 HAU/ml.

Production of virus after fusion of BSC and SV3T3 cells with UVSeV: After 10^6 BSC and 10^7 SV3T3 cells were fused, 0.2 ml (i.e., one tenth) of the suspension was placed in each of eight Petri dishes with 5 ml of medium; one of the Petri dishes contained coverslips. Each day one dish was removed for titration of SV40 and a coverslip for cytological examination. As a control, a similar experiment was performed without UVSeV. Table 1 shows that SV40 was detected from two days after fusion, always in much larger amounts than in the mixture without UVSeV. The time course of virus production corresponded closely with that found in SV40-infected BSC cells. The rise of virus titer at the seventh day probably originated from a second growth cycle for reasons given in a subsequent section. The recovered virus was neutralized by SV40-specific antiserum.

TABLE 1

COMPARISON OF SV40 YIELDS (PFU PER PETRI DISH) FROM MIXED SV3T3 AND BSC CULTURES WITH AND WITHOUT TREATMENT WITH UVSeV

Days after mixing cells	1	2	3	4	5	6	7
No Sendai virus added	<10	20	<10	120	<10	150	260
Sendai virus added	<10	1.8×10^3	2.2×10^4	1.6×10^5	1.7×10^5	1.7×10^5	3.0×10^5

Heterokaryons were readily detected on the coverslips one day after fusion, because BSC nuclei contain one or two roughly spherical nucleoli, and no heterochromatin, whereas SV3T3 nuclei have five or six irregularly shaped nucleoli, and small darkly staining areas of heterochromatin. Counts of the coverslip cultures showed that each dish contained approximately 4.4×10^4 multinucleated cells of which 2.0×10^4 (59%) were heterokaryons, in addition to many unfused BSC and SV3T3 cells.

Cytological evidence for SV40 multiplication in heterokaryons: Monolayers of BSC cells on coverslips, infected with SV40 and stained with Giemsa stain, showed three days later the alterations described[13] for SV40-infected green monkey kidney cells. The nuclei were enlarged and almost filled with purple homogeneous masses, often separated from the nuclear membrane by a slightly staining "halo." The nucleoli were enlarged and vacuolated. Similar changes, especially the "haloed" appearance of nuclei, were often seen in BSC/SV3T3 heterokaryons (Fig. 1). To see whether, in the heterokaryons, these appearances were diagnostic of SV40 infection, several cultures were prepared in which 10^5 3T3 or SV3T3 cells were fused with 10^6 BSC cells. In some cases, the 3T3 cells were infected in a monolayer with SV40 at a multiplicity of 1,000, incubated at 37° for four hours, washed three times in Earle's saline, trypsinized, washed again, and then fused; in other

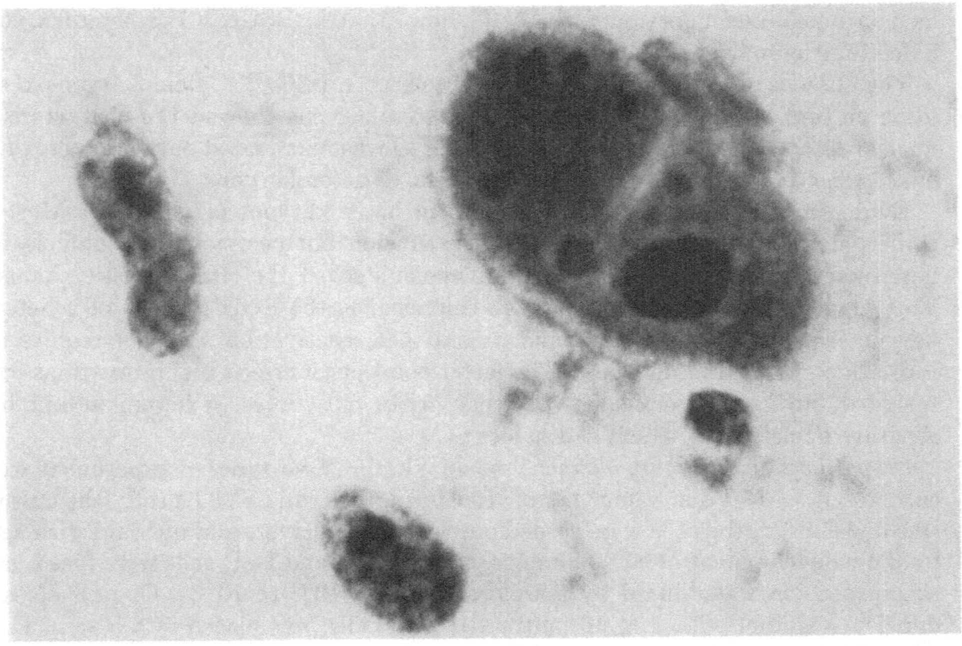

FIG. 1.—BSC/SV3T3 heterokaryon 4 days after fusion. Mouse surface antigen labeled with erythrocytes. The nucleus of the heterokaryon, formed by fusion of the original nuclei, shows a halo between the nuclear membrane and chromatin. An enlarged nucleolus is also present. Normal, unfused BSC cells are also in the field.

cases, cultures were infected after fusion. Other cultures were uninfected (Table 2). In all cases, the fused cells were placed in coverslip-containing Petri dishes with 5 ml of fresh medium. Four days after fusion, two coverslips were removed and, before staining, mixed immune hemadsorption for mouse antigens was carried out. Under these conditions, more than 99.5 per cent of all multinucleated cells containing mouse surface antigens were heterokaryons. On all preparations, the proportions of heterokaryons, homokaryons, and single BSC cells with "haloed" nuclei were determined. The results showed that haloed nuclei were seen in heterokaryons in all cases when the cultures were infected before or after fusion. They were not seen in BSC/3T3 heterokaryons or in single BSC cells, or in BSC homokaryons in any experiment unless the cultures had been infected after fusion with SV40. The perfect correlation between the haloed appearance of hetero-

TABLE 2

NUCLEAR ALTERATIONS IN HETEROKARYONS

Cells fused with BSC cells	Infection with SV40 24 hr after fusion	Heterokaryons per coverslip	Heterokaryons with haloed nuclei (%)	BSC homokaryons per coverslip	Homokaryons with haloed nuclei (%)	Unfused BSC cells with SV40 nuclear changes (%)
3T3	−	160	0	320	0	0
3T3	+	113	51	177	43	48
3T3 + SV40*	−	218	9	732	0	0
SV3T3	−	262	10	608	0	0
SV3T3	−	137	7	330	0	0
SV3T3	+	139	35	278	31	45

* 3T3 cells infected before fusion.

karyon nuclei and infection with SV40 shows that in BSC/SV3T3 heterokaryons haloed nuclei indicate SV40 development.

The absence of changes in isolated BSC cells or in BSC/BSC homokaryons in the same culture in which the BSC/SV3T3 heterokaryons showed the characteristic nuclear alteration is strong evidence that SV40 was reactivated only in heterokaryons. Reactivation occurred in 7–10 per cent of heterokaryons.

Evidence of SV40 replication was seen in heterokaryons of all sizes. Most of these heterokaryons had undergone nuclear fusion but occasionally a binucleated heterokaryon was seen, in which both nuclei showed the characteristic changes. This suggests that replication of SV40 can occur in the SV3T3 nuclei of a heterokaryon, as well as in the BSC nuclei, and that reactivation does not require the formation of a hybrid nucleus. The latter point is confirmed by observations with fusion of BSC and preinfected 3T3 cells, when most infected heterokaryons had separate nuclei, all of which had haloes.

Proportions of heterokaryons that produce SV40: Two types of experiment were carried out. (1) Twenty-four hours after fusing BSC and SV3T3 cells, the culture was trypsinized, the cells were seeded onto BSC monolayers and overlaid with agar to allow development of SV40 plaques. (2) SV3T3 and BSC cells were fused, and the suspension was diluted by tenfold steps from 10^{-1} to 10^{-3}. One ml of each dilution, together with 2×10^5 untreated BSC cells, was placed in a 50-mm Petri dish with 14 coverslips and 5 ml of EFT medium. After incubation for four hours at 37°, four coverslips were removed from each dish and stained to provide estimates of the average number of heterokaryons on each coverslip. Each of the remaining coverslips was placed in a separate 30-mm diameter Petri dish with 2 ml of EFT medium. The dishes were incubated, without being disturbed, for seven days at 37°, when all the coverslip cultures were stained with Giemsa stain. Microscopic examination showed that BSC cells with nuclear changes of SV40 infection were usually found in foci of between 10 and 50 contiguous cells. Occasionally a multinculeated cell with the nuclear changes of SV40 infection was seen in the center of such a focus. Many multinucleated cells were not associated with foci of infected BSC cells. It is most likely that the foci represented the sites of reactivation of SV40, since similar foci were seen in coverslip cultures of BSC cells infected with diluted SV40 and incubated for seven days. Some foci, however, may have originated from secondary infection early in incubation, although the time course of viral multiplication makes this unlikely. Therefore, counts of foci provide an upper limit for the proportions of heterokaryons yielding virus.

Both experiments were carried out with either 10^5 SV3T3 and 10^6 BSC cells, or 10^6 SV3T3 and 10^5 BSC cells. The populations of heterokaryons obtained in the two cases differed sharply in the predominant nuclear type. There were 15.8 BSC and 1.4 SV3T3 nuclei per heterokaryon, as average in the former case, and 2.0 BSC and 17.0 SV3T3 in the latter case.

The results of experiment 1 indicated that between 0.1 and 10 per cent of heterokaryons, depending on the proportions of the two kinds of nuclei per heterokaryon, were capable of giving rise to a plaque. Similar results were obtained in experiment 2 (Table 3). These data agree with the results of cytological observations (Table 2).

SV40 production in heterokaryons made with cloned SV3T3 populations: Ten

TABLE 3

FOCI OF SV40-INFECTED BSC CELLS IN RELATION TO NUMBERS OF HETEROKARYONS

Population	Heterokaryons per coverslip (mean of 4)	Foci of SV40-infected BSC cells per coverslip (mean of 10 coverslips)		Foci per coverslip as percentage of heterokaryons
BSC 10^6/SV3T3 10^5	230	Uncountable (>23)		>10
	35	7.2 (S.D. ± 3.6)		20
	7	1.0 (S.D. ± 0.57)		14
	1	0.1 (S.D. ± 0.36)		10
			Mean	15
BSC 10^5/SV3T3 10^6	480	3.9 (S.D. ± 2.0)		0.8
	28	0.8 (S.D. ± 0.42)		2.8
	5	0.2 (S.D. ± 0.42)		4.0
			Mean	2.5

clones of SV3T3 cells were grown for approximately 20 generations. Then 10^5 cells of each clone were fused with 10^6 BSC cells and incubated as usual. After four days' incubation at 37°, mixed immune hemadsorption was carried out on coverslips from each fused clonal culture. The coverslips were stained and examined. The cells in the second dish were assayed for SV40 after seven days' incubation. At the time of fusion, 10^4 cells of clone 3 were seeded into a dish. After seven days at 37° (about nine generations) ten samples of 10^5 cells from this dish were each fused with 10^6 BSC cells, and then examined in the same way as the cells from separate clones. Table 4 shows that virus was produced by nine of the

TABLE 4

PRODUCTION OF SV40 BY HETEROKARYONS MADE WITH CLONED SV3T3 POPULATIONS

Clone	Yield of virus seven days after fusion (PFU per Petri dish)
1	9.0×10^3
2	4.5×10^2
3	6.5×10^5
4	1.8×10^6
5	1.1×10^6
6	6.3×10^4
7*	<10
8	4.5×10^5
9	7.0×10^3
10	5.5×10^5

* This clone after 38 generations yielded 4.5×10^2 PFU and after 56 generations 2.5×10^4.

separate clones, and in highly variable amounts. In clones with higher yields, the virus arose not only from heterokaryons, but also from secondarily infected BSC cells, which could be detected cytologically. The yields from parallel samples of clone 3 were more uniform than from independent clones and ranged from 2.5 to 6.0×10^5 PFU. Once again, only a minority (never more than 5%) of heterokaryons had cytological signs of SV40 multiplication.

The negative clone 7 was examined further. Every seven days 10^4 cells were transferred to a 50-mm Petri dish with 5 ml of medium. Fusion of 10^5 cells with 10^6 BSC cells was carried out after 38 and 56 generations, when SV40 was recovered (Table 4).

In a second experiment, five SV3T3 clones were grown from single cells in medium containing SV40-specific antiserum at a dilution (1 in 50) able to neutralize about 99.5 per cent of the virus within 30 minutes at 37°. Seven days after fusion with

BSC cells, these clones all yielded SV40, between 2.5×10^4 and 2.5×10^6 PFU per culture.

These results show that most or all SV3T3 cells contain at least one complete SV40 genome, and that its presence is not dependent on external reinfection.

Generality of reactivation of viral genome by fusion of transformed cells with virus-susceptible cells: (a) *Reactivation in other SV3T3 lines:* With 10^5 cells of each of five lines of 3T3 cells, which had been independently transformed and carried for varying times, were fused 10^6 BSC cells, and the yield of SV40 determined after seven days' incubation. Infectious virus was obtained from all five lines in yields ranging from 5×10^4 to 5×10^6. No SV40 was recovered from 10^7 unfused cells of each line after freezing and thawing.

(b) *Failure to produce reactivation in lines transformed by polyoma virus:* No polyoma virus was detected by plaque titration on mouse embryo cells after seven days' incubation followed by freezing and thawing of polyoma-transformed 3T3 cells fused with either secondary mouse embryo cells or 3T3 cells. Fusion of a line of polyoma-transformed BHK cells (Py 19) with 3T3 cells also failed to induce the appearance of polyoma virus. In an experiment in which 3T3 cells doubly transformed with SV40 and polyoma virus were fused with BSC cells, SV40 was recovered but not polyoma virus.

Discussion.—3T3 cells transformed by SV40 do not produce infectious virus, but contain functioning viral genes; in fact, they synthesize virus-specific messenger RNA[14] and proteins (such as the T antigen). We have shown that when these cells are fused with BSC cells, which can support viral multiplication, infectious virus is produced in heterokaryons. This appears to be a general property of SV40-transformed cells of 3T3 origin. Furthermore, the study of clones of SV3T3 showed that most or all transformed cells contain the complete viral genome. These results raise two main questions: that of the mechanism preventing the expression of selected viral genes, and that of the state and location of the viral genome in the cells.

The lack of expression could be attributed either to the presence in the transformed cells of a specific repressor of viral functions (either viral, as in lysogeny, or cellular), or to lack of a factor required for the expression of these functions. These hypotheses are better defined and somewhat circumscribed by the results reported in this article. The following observations are especially relevant. (1) The ability to produce virus is dominant over nonproduction. (2) Virus production requires cytoplasmic but not nuclear fusion, and can occur even in dikaryons. (3) Both SV3T3 and BSC nuclei show cytological signs of viral multiplication in the heterokaryons. (4) The time course of virus production is similar to that observed in SV40-infected BSC cells. (5) 3T3 cells are unable to support SV40 multiplication, but the hybrid of BSC and infected 3T3 will do so. Then virus development in the 3T3 nuclei is revealed cytologically.

Since virus production is dominant, the BSC component of the heterokaryon contributes a positive factor, such as (*a*) repressor-free areas in which the SV40 genome can be totally expressed, (*b*) an antirepressor which neutralizes an SV3T3 repressor, or (*c*) a factor lacking in SV3T3 cells.

Considering the first point, it seems unlikely that a simple dilution of an SV3T3 repressor after heterokaryon formation could be responsible for the activation of the virus genome, because small heterokaryons (with two nuclei) show SV40 activation. It seems also unlikely that, as in zygotic induction of lysogenic bacteria, SV40 DNA becomes activated after passing from the SV3T3 part of the heterokaryon to a repressor-free space (for instance, a BSC nucleus), because virus-specific nuclear changes can occur in the SV3T3 nuclei of the heterokaryons. It must also be taken into account that SV40-infected untransformed 3T3 cells are unable to support SV40 multiplication. Thus, the repressor should be either very potent or very abundant, thus practically excluding hypothesis *a*.

There are no elements for discriminating between hypotheses *b* and *c*. If BSC cells contribute to the heterokaryons a factor lacking in either SV3T3 or 3T3 cells, this factor is cytoplasmic, since virus-specific changes occur in SV3T3 nuclei of heterokaryons without nuclear fusion. The factor could affect the translation of viral genes, which, as with other nuclear DNA viruses,[15] is likely to occur in the cytoplasm. The factor could be, for instance, a tRNA which recognizes a rate triplet; and the SV40 DNA could be viewed as carrying a mutation sensitive to a suppressor present in BSC but not in 3T3 or SV3T3 cells.

The similarity between the time course of viral multiplication after heterokaryon formation and in SV40-infected BSC cells shows that in SV3T3 cells viral development is arrested at an early step. This deduction is in agreement with the observation that in these cells the viral genome expresses only early functions, such as the synthesis of the T antigen, but not late functions, such as the synthesis of the capsid protein.

Concerning the state and location of the viral genome in SV3T3 cells, its presence in all or most cells suggests a rather stable association. Lack of reactivation in most heterokaryons could derive either from a heterogeneity of the SV3T3 component of heterokaryons or of the reactivation mechanism. That SV3T3 cells are heterogeneous is suggested by the differences of viral yields in different SV3T3 clones. Moreover, one clone became reactivable after many cell generations during which reactivation could not be demonstrated. The latter event strongly suggests that reactivability arises by a mutation-like event. A similar hypothesis could also explain the high variability of viral yields in the other clones. The event conferring reactivability would have to occur with a rather high probability, without, nevertheless, causing a progressive accumulation of reactivable cells. These two conditions would be met, for instance, if the event were a change of state of the viral genome, perhaps similar to that occurring in induction of lysogenic bacteria; and if the reactivable genome had a limited life. Other hypotheses are also possible, and further experiments are therefore required.

The observation that virus-specific changes occur in the BSC nuclei of heterokaryons shows that the viral DNA reaches them after fusion. If, in SV3T3 cells, the viral DNA were present exclusively in the nuclei, it would have to go through the cytoplasm in order to reach the BSC nuclei. Whether or not the viral DNA can leave the nucleus in which it resides is not known. An alternative possibility is that in the cells in which viral multiplication is activated by fusion, viral DNA exists free in the cytoplasm.

The failure to activate the viral genome in several lines of cells transformed by polyoma virus is unexplained. Owing to the small number of polyoma-virus-transformed cell lines studied so far, this result cannot yet be considered evidence that SV40 and polyoma virus establish different types of virus-cell relationships in transformation.[16]

Summary.—Evidence is provided that the reactivation of SV40 virus in mixtures of transformed cells (SV3T3) and susceptible cells (BSC) treated with UV-inactivated Sendai virus occurred only in heterokaryons. Most heterokaryons were non-productive. Fifteen SV3T3 clones and five independent SV3T3 lines all yielded SV40. The proportion of productive heterokaryons depended on the proportion of each kind of cell in a heterokaryon. It is suggested that the SV40 genome in SV3T3 cells may exist in either of two states, one reactivable, the other not. Attempts to reactivate polyoma virus from transformed cells by this technique were unsuccessful.

The authors thank Dr. Marguerite Vogt for providing the transformed 3T3 lines and for many discussions.

This work was supported by research grant CA-07952 from the National Cancer Institute.

The work reported in this paper was undertaken during the tenure of an Eleanor Roosevelt Cancer Fellowship of the American Cancer Society awarded by the International Union Against Cancer to J. F. Watkins.

[1] Gerber, P., and R. L. Kirschstein, *Virology*, **18**, 582 (1962).

[2] Sabin, A. B., and M. A. Koch, these PROCEEDINGS, **50**, 407 (1963).

[3] Gerber, P., *Virology*, **28**, 501 (1966).

[4] Koprowski, H., and F. Jensen, *Federation Proc.*, **26**, 313 (1967).

[5] Harris, H., and J. F. Watkins, *Nature*, **205**, 640 (1965).

[6] Todaro, G. J., and H. Green, *J. Cell Biol.*, **17**, 299 (1963).

[7] Dulbecco, R., and G. Freeman, *Virology*, **8**, 396 (1959).

[8] Macpherson, I., and M. Stoker, *Virology*, **16**, 147 (1962).

[9] Earle, W. R., *J. Natl. Cancer Inst.*, **4**, 165 (1943).

[10] Hatanaka, M., and R. Dulbecco, these PROCEEDINGS, **56**, 736 (1966).

[11] Espmark, J. A., and A. Fagraeus, *J. Immunol.*, **94**, 530 (1965).

[12] Watkins, J. F., and D. M. Grace, *J. Cell Sci.*, **2**, 193 (1967).

[13] Love, R., and M. V. Fernandes, *J. Cell Biol.*, **25**, 529 (1965).

[14] Benjamin, T., *J. Mol. Biol.*, **16**, 359 (1966).

[15] Fujinaga, K., and M. Green, these PROCEEDINGS, **55**, 1567 (1966); Fujiwara, S., and A. S. Kaplan, *Virology*, **32**, 60 (1967).

[16] When this manuscript had been completed, an article by H. Koprowski, F. C. Jensen, and Z. Steplewski (these PROCEEDINGS, **58**, 127 (1967)) contributed additional evidence on this subject.

VI
Transcription of the Viral Genome in Infected and Transformed Cells

Virus-Specific RNA in Cells Productively Infected or Transformed by Polyoma Virus

T. L. Benjamin

Virus-specific RNA is detected in mouse kidney cells infected by polyoma virus by means of hybridization between [^{3}H]thymidine-labeled virus DNA and ^{32}P pulse-labeled RNA. Small amounts of "early" RNA are observed before virus DNA synthesis begins; during the later phases of virus development, there is a sharp increase in the amount of virus-specific RNA. Synthesis of mouse RNA occurs, and is possibly stimulated, during the infection.

A small fraction of the pulse-labeled RNA from virus-free polyoma transformed cells is capable of hybridizing with polyoma DNA. Similarly, pulse-labeled RNA, from SV-40-transformed cells, is able to hybridize with SV-40 DNA. No hybridization is observed between RNA of cells transformed by one of these viruses and DNA of the other virus, or between viral DNA and RNA of normal or spontaneous malignant cells. These findings provide evidence for the persistence of the viral genome in transformed cells.

1. Introduction

The problem of demonstrating the presence of virus-related materials in mammalian cells transformed (i.e. rendered neoplastic, or "malignant") by viruses of the polyoma SV-40 group has been studied in recent years by means of several experimental approaches (Vogt & Dulbecco, 1962). Conclusive evidence on this point would be critical to the formulation of acceptable theories of viral carcinogenesis by these DNA tumor viruses.

Evidence for virus-specific complement-fixing antigens in tumor or transformed cells has been shown for polyoma (Habel, 1965), for SV-40 (Black, Rowe, Turner & Huebner, 1963) and for adenovirus types 12 and 18 (Huebner, Rowe, Turner & Lane, 1963). In addition, virus-specific transplantation antigens have been shown for polyoma and SV-40 (Sjogren, Hellstrom & Klein, 1961; Habel & Eddy, 1963; Koch & Sabin, 1963; Defendi, 1963). Other results have indicated that the genetic determinant for the polyoma transplantation antigen can be "rescued" by superinfection of transformed cells with appropriately marked virus strains (Ting, 1964). Although strongly suggestive of the continued presence of viral genes, these results alone are not directly conclusive in this regard.

Using techniques of nucleic acid hybridization, attempts have been made to reveal a complementarity of base sequence between DNA of polyoma-transformed cells and DNA of the virus (Axelrod, Bolton & Habel, 1964) or RNA made from the viral DNA by the DNA-dependent RNA polymerase from *Escherichia coli* (Winocour, 1965). Although a greater affinity of the viral DNA for DNA of transformed cells as compared to that of normal cells was observed, the results were somewhat obscured by a variable degree of homology with normal cellular DNA's. In the studies with synthetic viral RNA, no differences between normal and transformed cell DNA were observed; this would have been detected if roughly 20 viral DNA equivalents per cell were present.

In the present investigation, hybridization experiments are performed using purified viral DNA and RNA made *in vivo* by normal and transformed cells. As a preliminary step toward the detection of virus-specific RNA in transformed cells, conditions were first sought by which virus-specific RNA in cells infected by and releasing polyoma virus could be detected and characterized. The results show a progressive increase in the amount of virus-specific RNA during virus development. Using the same procedure, it was possible to detect a small fraction of virus-specific RNA in transformed cells, but not in normal or other malignant cells.

2. Materials and Methods

(a) *Preparation of [^{3}H]thymidine-labeled viral DNA*

Py was obtained from infected mouse kidney cultures as previously described (Winocour, 1963). On the 2nd or 3rd day after preparing the cultures, 10 ml. of fresh medium containing 2 μc/ml. of [^{3}H]thymidine (New England Nuclear Corp., specific activity $1 \cdot 4 \times 10^4$ μc/μmole) was added to each culture dish (100 mm $\times$ 20 mm). Cultures were harvested on the 5th or 6th day.

Concentration and purification of the virus from the crude lysates were carried out either by centrifugation in the S30 Spinco rotor for 2 hr at 28,000 rev./min, followed by treatment of the pellet with 0·5% sodium deoxycholate (Murakami, 1963), or by precipitation with absolute methanol (33% final vol.) in the cold. The methanol precipitate was taken up in a neutral tris buffer containing 0·025% trypsin, and treated with 0·5% sodium deoxycholate. The partially purified virus was then treated with RNase and DNase (10 μg/ml. each) for 1 hr at 37°C, spun down and resuspended in a CsCl solution of density 1·33 g/ml. The mixture was spun in the SW39 rotor for 18 to 24 hr at 30 to 35,000 rev./min. The virus band was collected, dialyzed against tris-buffered saline and stored for extraction at −70°C. The DNA of the virus was extracted with phenol (Weil, 1961).

Analyses of the DNA by band sedimentation through neutral or alkaline CsCl were performed in the Spinco model E ultracentrifuge to determine the distribution of the components (Vinograd, Lebowitz, Radloff, Watson & Laipis, 1965). Since even a minor contamination of the polyoma DNA by mouse DNA could obscure the significance of the hybridization results, the DNA preparations were fractionated on sucrose density-gradients at neutral pH in order to isolate the major viral components (I and II) from the small amount of heterogeneous component (III). In one preparation used in these studies, only components I and III were initially present (roughly 95% and 5%, respectively), and in the other preparation component II was present (roughly 20%) in addition to I and III. After fractionation, pure component I was obtained from the first preparation, and a mixture of components I and II from the second; no component III could be detected by analytical ultracentrifugation in either of the final preparations. The specific activity of the viral DNA, counted under the conditions used for hybrid detection, was 4·5 to $7 \cdot 3 \times 10^3$ cts/min/μg.

(b) *Preparation of ^{32}P-labeled RNA*

Baby mouse kidney cultures were incubated in phosphate-free Eagle's medium for 24 hr prior to infection. Cultures were infected at a multiplicity of 10 to 20 p.f.u.†/cell, washed, and incubated in a low-phosphate medium ($2 \cdot 5 \times 10^{-5}$ M, compared to 2×10^{-3} M for

† Abbreviation used: p.f.u., plaque-forming units.

normal medium). At different times after infection, 0·2 to 0·3 mc of carrier-free [^{32}P]orthophosphate was added to each culture containing 10 ml. of medium. Either 4 or 8 hr after addition of the isotope, the medium was removed, the cells washed with ice-cold tris-buffered saline, collected, rewashed, spun down and stored at −70°C.

Other normal or transformed cell lines were grown to near confluency, starved of phosphate, and pulse-labeled for 4 hr in the same manner as the infected cells.

RNA was extracted by a hot-phenol procedure. The cells were suspended in 27 vol. of a solution containing 4×10^{-3} M-$MgCl_2$, 0·5% sodium naphthalene disulfonate, and 1·25 mg/ml. bentonite. 3 vol. of 25% sodium dodecyl sulfate and 30 vol. of tris-buffered phenol (pH 7·2) containing 0·1% 8-hydroxyquinoline were added. The suspension was shaken vigorously in a 62°C water bath for 3 to 5 min, then quickly cooled in an ice–salt mixture. Following another cycle of heating, shaking and cooling, the mixture was centrifuged. The aqueous layer was removed and subjected to further deproteinization by shaking in the cold with an equal volume of the same phenol. The final aqueous phase was brought to 0·1 M-NaCl, and the RNA precipitated by adding 2 vol. of cold 100% ethanol. The RNA was dissolved in SSC/10 (0·015 M-NaCl, 0·0015 M-sodium citrate) containing 0·005 M-$MgCl_2$ and treated with 10 μg/ml. RNase-free DNase for 45 min at 37°C. The mixture was then shaken with phenol, and the aqueous phase passed through a 55 cm high column of Sephadex G100 equilibrated with SSC/10 to separate the RNA from products of DNase digestion. The leading radioactive peak (RNA) was collected, precipitated with ethanol, redissolved in $2 \times$ SSC and stored at −70°C. In some instances the material from the column was filtered through nitrocellulose filters (Schleicher & Schuell Co., type B-6) once or twice to help remove small amounts of denatured DNA. The ^{32}P in such preparations was rendered more than 99·9% trichloroacetic acid-soluble by alkali treatment (0·3 N-KOH, 20 hr, 30°C), and was insensitive to further attack by DNase. The concentration of RNA in these preparations varied from 50 to 200 μg/ml.

(c) *Hybridization*

The technique used here is similar to that described by Gillespie & Spiegelman (1965). To convert the closed circular form of polyoma DNA (component I) to a heat-denaturable form (component II), the DNA was incubated with pancreatic DNase (10^{-5} to 10^{-4} μg/ml.) at room temperature until roughly 90% of the preparation was present as component II (according to Vinograd *et al.*, 1965). The reaction was stopped by addition of EDTA, and the DNA dialyzed against SSC/10. The DNA was denatured at a concentration of 20 to 40 μg/ml. by heating in boiling water for 10 min, followed by rapid cooling in an ice–salt mixture. One-fourth volume of 2·5 M-KCl, 0·05 M-tris (pH 7·4) was then added. Appropriate amounts of the DNA were filtered under suction onto nitrocellulose filters presoaked for 5 to 10 min in 0·5 M-KCl, 0·01 M-tris (pH 7·4). (90 to 100% of the [^{3}H]polyoma DNA deposited on the filters was retained, and not more than 5% of the initially retained radioactive material was lost in the subsequent incubations and washings.) The filters were dried at room temperature for 1 hr, then in an oven at 80°C for 4 to 6 hr. In some cases, the filters were then placed in $2 \times$ SSC and incubated 10 to 14 hr at 60°C. This step was found to minimize the nonspecific attachment or background of ^{32}P in the subsequent incubation. For the formation of hybrids, the filters with DNA attached were placed singly in small chambers consisting of a plastic screw cap with fitting glass tube cut off just below the threads. The filter was placed inside the cap with the DNA-side exposed. The shortened glass tube was then screwed in place, serving as a well for the RNA solution. 0·4 ml. of the RNA was added, and the open end of the tube was sealed off with vacuum grease and a coverslip. The chambers were then incubated in a 60°C water bath for 18 to 24 hr. At the end of the incubation, the filters were removed, washed thoroughly in $2 \times$ SSC, and incubated for 1 hr at 30°C in $1 \times$ SSC containing 20 μg/ml. RNase. The RNase was previously heated at 80°C for 15 min to inactivate traces of DNase. The filters were finally washed by passing through 80 ml. of warm (56°C) 0·5 M-KCl, 0·01 M-tris. After drying thoroughly, the filters were placed in plastic vials for counting in a Packard Tri-Carb liquid-scintillation counter. The channels were set for simultaneous counting of ^{3}H and ^{32}P. Corrections for spillover between the channels were made whenever necessary. The approximate counting efficiencies were 15% for ^{3}H and 50% for ^{32}P for unquenched samples.

3. Results

(a) *Effects of phosphate-deprivation on the virus growth cycle*

The possibility that the low concentration of phosphate in the medium might prevent or alter the normal course of virus synthesis was examined by comparing the growth curves of virus under the normal and low-phosphate conditions. Replicate cultures were harvested at various times after infection, frozen and thawed three times, and assayed for total virus by hemagglutination (Eddy, Rowe, Hartley, Stewart & Huebner, 1958). The yield at the end of the cycle was also assayed by plaque formation (Dulbecco & Freeman, 1959) to assure that infectious particles were present. The results are presented in Fig. 1. It can be seen that both the time course of virus production and the final yield are the same for the two conditions.

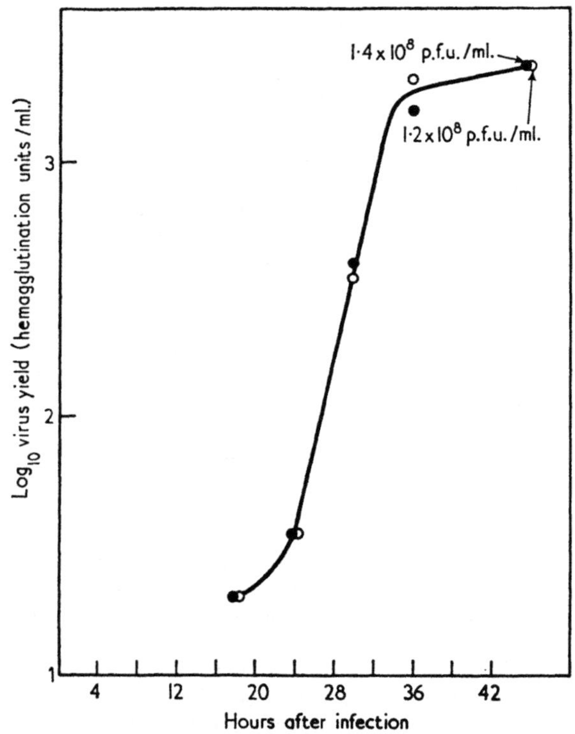

FIG. 1. Growth curves of virus. —●—●—, Normal phosphate conditions; —○—○—, low-phosphate conditions.

(b) *Detection of polyoma-specific RNA in infected mouse kidney cells*

To detect the presence of viral-specific RNA in the RNA extracted from infected cells, samples of the separate pulse-labeled RNA preparations from each eight-hour period from 0 to 32 hours after infection were pooled. A constant amount of the pooled RNA was added to a series of filters containing increasing amounts of polyoma DNA. Two controls were included: RNA from pulse-labeled uninfected mouse cells added to polyoma DNA, and infected-cell RNA added to *E. coli* DNA, which has a base composition close to that of polyoma DNA. The retention of *E. coli* DNA was determined by optical density measurements before and after filtration. The amounts of RNA used were 54 μg and 61 μg for infected and uninfected cell RNA, respectively. The fractions of radioactive material (counts) bound in the absence of DNA were $2{\cdot}7 \times 10^{-4}$ for infected-cell RNA and $1{\cdot}8 \times 10^{-4}$ for uninfected-cell RNA. Incubation was for 24 hours at 60°C. The results are shown in Fig. 2; each point represents the average of two determinations. It can be seen that only in the case of infected-cell

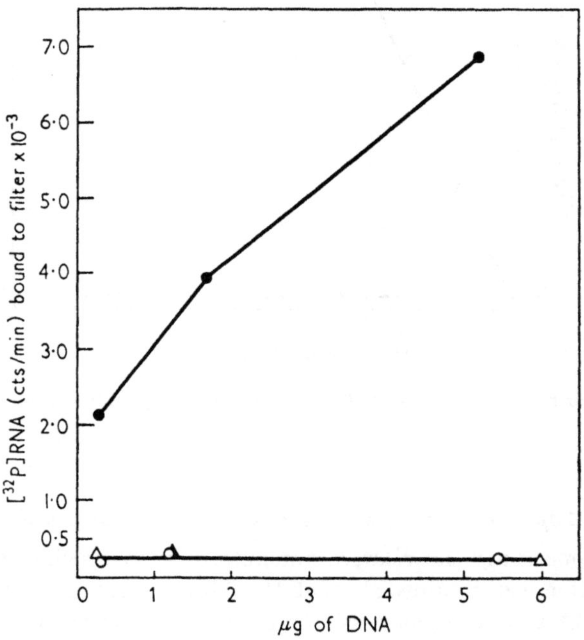

Fig. 2. Test for virus-specific RNA in infected mouse kidney cells. —●—●—, Infected-cell RNA, polyoma DNA; —○—○—, uninfected-cell RNA, polyoma DNA; —△—△—, infected-cell RNA, *E. coli* DNA.

RNA and polyoma DNA is there an increase in the amount of ^{32}P bound with increasing DNA on the filters. The controls show that the major factor in the rise of this curve cannot be due either to the presence of an RNA in uninfected mouse cells which is homologous to polyoma DNA, or to a nonspecific sticking to DNA by the infected-cell RNA.

The kinetics of formation of the hybrid at 60°C are shown in Fig. 3. Replicate hybrid chambers were set up with filters containing approximately 1·2 μg of polyoma DNA and a constant amount of infected-cell RNA. Duplicates were removed at the times indicated, and the amount of hybrid determined as described. Chambers containing filters without DNA and the same amount of RNA were included as background controls. It appears that no further hybrid is formed after about 16 hours, and the maximum amount of hybrid remains constant for up to 30 hours of incubation. In all subsequent experiments, the incubation time was between 18 and 24 hours.

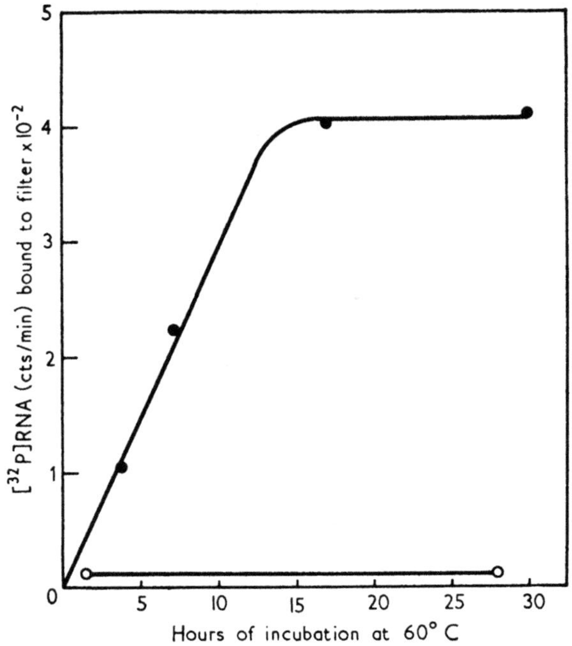

FIG. 3. Kinetics of hybrid formation at 60°C. —●—●—, Hybrid; —○—○—, background.

(c) *Appearance of virus-specific RNA during the growth cycle*

Information concerning the relative amounts of virus-specific RNA at different times after infection was obtained through the following experiments. Samples of the pulse-labeled RNA from each successive eight-hour period from 0 to 32 hours were diluted 40-fold from the initial concentration used above (section (b)), and incubated with filters containing increasing amounts of polyoma DNA in an attempt to "exhaust" the specific RNA in each of the RNA samples. The results are plotted as the fraction of the input of [^{32}P]RNA hybridized *versus* the amount of polyoma DNA (Fig. 4). The attainment of a plateau in such an experiment would indicate exhaustion, the level of the plateau giving directly the fraction of the labeled material capable of hybridizing to polyoma DNA. This would represent the actual virus-specific fraction, if all of the virus-specific RNA present was capable of hybridizing (i.e., if there was no appreciable secondary structure in the RNA to prevent hybridization).

In the cases of the early periods, 0 to 8 and 8 to 16 hours, as the amount of polyoma DNA is increased, a small but increasing fraction of the RNA is bound above the background value (binding in the absence of DNA), and no plateau is reached at the DNA levels employed. In the later periods, 16 to 24 and 24 to 32 hours, the relative amounts of virus-specific RNA are much greater, and plateaux are reached with less than 1 μg of DNA. The RNA of the latest period appears to be richest in specific RNA, although even in this case less than 1% of the labeled material hybridizes to virus DNA.

To verify that the RNA in the plateau region is actually depleted of virus-specific RNA and to rule out the possibility that the plateau is due to some limitation in the

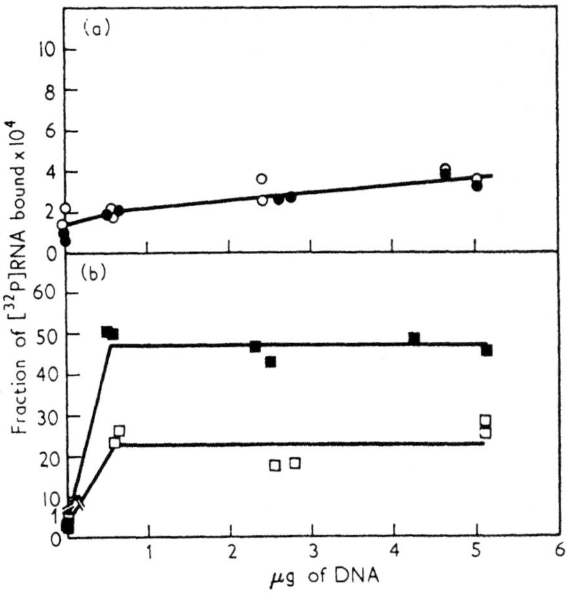

FIG. 4. Attempts to exhaust virus-specific RNA from RNA of different periods. (a) —○—○— 0 to 8 hr; —●—●—, 8 to 16 hr; (b) —□—□—, 16 to 24 hr; —■—■—, 24 to 32 hr.

availability of the DNA for hybridization, the "exhausted" RNA was incubated a second time with a new filter carrying polyoma DNA. The amount of ^{32}P bound in the second incubation was slightly above background, indicating that most, if not all, of the specific fraction was removed in the first incubation.

No direct comparison between the early and late periods, in terms of the relative amounts of specific RNA, can be made from these experiments, due to the inability to exhaust the early period RNA's. Nevertheless, it appears that a sharp increase in production of virus-specific RNA occurs sometime after 16 hours.

The converse type of experiment, in which increasing amounts of RNA are added to a constant amount of DNA, was performed using the same RNA preparations. Since the background is not strictly linear with the amount of RNA added, separate background controls were included for each input RNA concentration; these background values were then subtracted from the value for radioactive material bound with the same concentration of RNA in the presence of DNA. The results are plotted in Fig. 5 as the RNA hybridized (cts/min) per 0·1 μg DNA *versus* the RNA (cts/min) incubated. The actual amount of DNA was around 0·07 μg.

In principle, two kinds of information could be obtained from such an experiment. First, the initial slopes, in the region of DNA excess, are a measure of the relative amounts of virus-specific RNA, thus providing a common basis for comparing the RNA's from all periods. The consistency of this method with that of "exhaustion" can be seen by comparing results obtained by the two methods with the late-period RNA's: the ratio of the plateaux in Fig. 4(b) is in good agreement with the ratio of the initial slopes in Fig. 5(b) for the same RNA's, roughly 2 : 1 in each case. Second, the saturation of a low amount of DNA with RNA of any given period would allow the estimation of the fraction of the DNA involved in hybridization with the RNA of that period. It would appear from the results shown in Fig. 5 that the first type of information could be obtained reliably from measurements of the initial slopes; the second type of information could not be obtained for any but the latest-period RNA, which seems to give a saturation of the DNA.

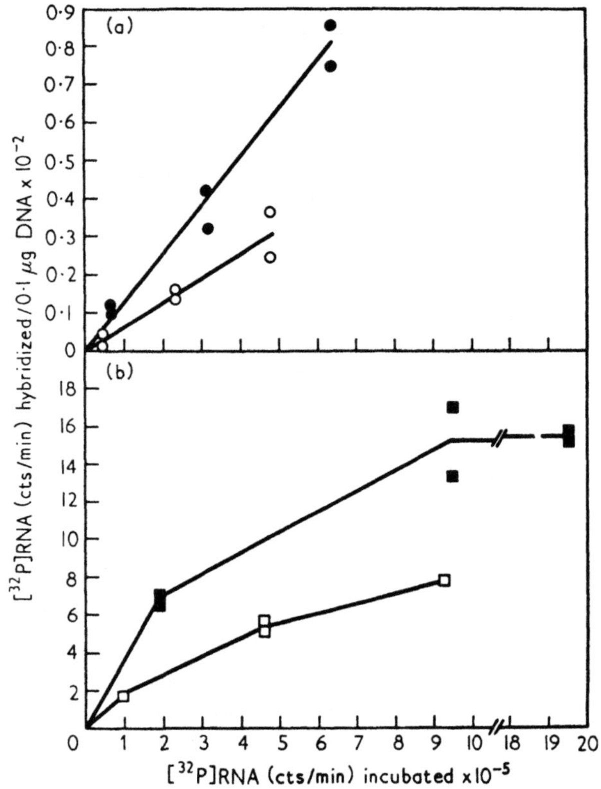

FIG. 5. Attempts to saturate viral DNA with RNA of different periods. (a) —○—○—, 0 to 8 hr; —●—●—, 8 to 16 hr; (b) —□—□—, 16 to 24 hr; —■—■—, 24 to 32 hr.

Experiments of this type were therefore repeated, using new preparations of RNA with successive 4-hour pulses from 4 to 32 hours after infection. RNA from each period was added in two concentrations to filters containing $0{\cdot}5 \pm 0{\cdot}04$ μg DNA. The relative amount of virus-specific RNA was calculated for each period by dividing the increase in counts of ^{32}P hybridized by the increase in counts of ^{32}P incubated for the two concentrations. Since the amount of DNA used was in excess for all of the RNA's, the initial slopes were being measured. These values, plotted against the time after infection, show the relative rate of synthesis of virus-specific RNA during the virus growth cycle (Fig. 6). The uptake of [^{14}C]thymidine into trichloroacetic acid-insoluble material by infected mouse kidney cultures is plotted for comparison. The increase in the fraction of hybridizable RNA after 16 hours is undoubtedly due in part to an increase in absolute amount of virus-specific RNA at these times; however, unknown factors, such as changes in the rate of incorporation of ^{32}P into other RNA (see below;

since the over-all rate of incorporation of ^{32}P into RNA also increases during the late period, this factor seems somewhat unlikely), and changes in the specific activity of the phosphate pool with time, do not allow the exact determination of the absolute rate of synthesis of virus-specific RNA.

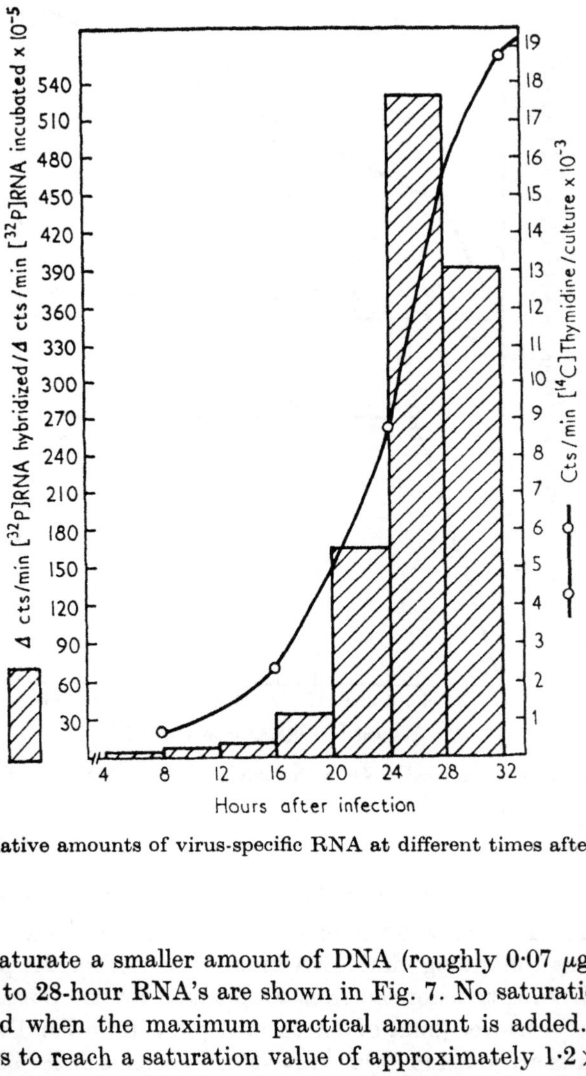

Fig. 6. Relative amounts of virus-specific RNA at different times after infection.

Attempts to saturate a smaller amount of DNA (roughly 0·07 μg) with the 12- to 16-hour and 24- to 28-hour RNA's are shown in Fig. 7. No saturation with the early RNA is observed when the maximum practical amount is added. The late RNA, however, appears to reach a saturation value of approximately $1{\cdot}2 \times 10^4$ cts/min per 0·1 μg DNA. Assuming the RNA has the same specific activity as the medium (corrected for decay), this value corresponds to the binding to roughly 25 to 35% of a single viral DNA strand.

(d) *Hybridization of infected and uninfected cell RNA with mouse DNA*

Pulse-labeled RNA's from infected (16 to 24 and 24 to 32 hours, pooled) and uninfected cells were incubated with filters containing no DNA, 5 μg of [^{3}H]polyoma DNA, or 5 μg of [^{3}H]mouse DNA. The results are shown in Fig. 8. As expected, only the infected-cell RNA forms a significant amount of hybrid with the viral DNA. Both

RNA's, however, hybridize with mouse DNA, the infected cell RNA to a greater extent than the uninfected.

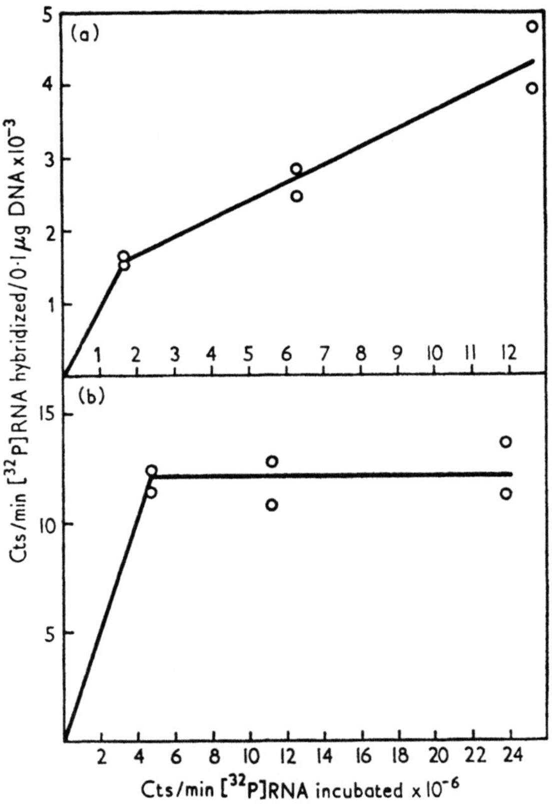

FIG. 7. Attempts to saturate viral DNA with early and late RNA. (a) 12- to 16-hr RNA, (b) 24- to 28-hr RNA.

(e) *Detection of virus-specific RNA in transformed cell lines*

Pulse-labeled RNA's from a series of normal and transformed cell lines were incubated with filters containing viral DNA in order to test for the presence of virus-specific RNA. All virus-transformed lines were tested for the presence of virus and were found to be virus-free (limit of detection: 1 p.f.u./10^6 cells). Descriptions of the cell lines and results are given in Table 1. Results with some of the lines are presented graphically in Fig. 9. Small but significant fractions of the labeled RNA from polyoma-transformed lines hybridize with polyoma DNA. Virus-specific RNA can be detected in cells of three different species transformed *in vitro* by polyoma; no such RNA is found in either normal, SV-40-transformed or spontaneously malignant cells. However, a fraction of the RNA of SV-40-transformed cells can hybridize with DNA of SV-40 virus. (The SV-40 DNA, made available by Dr J. Vinograd and Phil Laipis, was largely component II. Since this material was not radioactively labeled, no direct measurement of retention was possible; the values listed in Table 1 represent the weight in μg of DNA applied to the filters.)

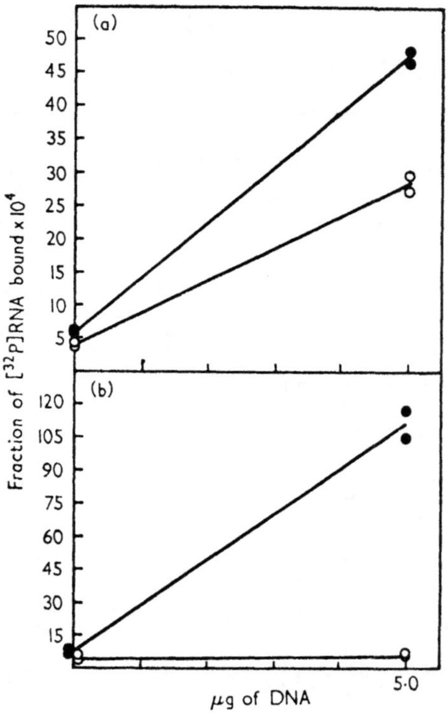

FIG. 8. Hybridization of infected- (—●—●—) and uninfected- (—○—○—) cell RNA with mouse DNA (a) or polyoma DNA (b).

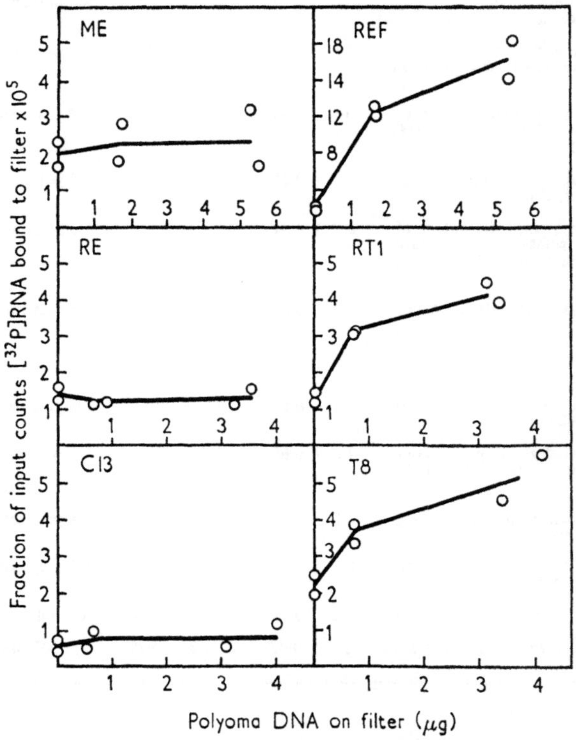

FIG. 9. Test for virus-specific RNA in normal and transformed cells (see Table 1).

TABLE 1

Tests for virus specific [^{32}P]RNA in various normal and transformed cell lines

Cell line	Origin and description	Fraction† of ^{32}P bound to filters $\times 10^5$ −DNA	Fraction† of ^{32}P bound to filters $\times 10^5$ +DNA	μg of virus DNA on filters	% increase of fraction bound +DNA over −DNA
(a) Mouse					
ME	Normal embryo secondary cultures	2·0 ± 0·26	2·3 ± 1·2	5·3 polyoma	+ 15
3T3	Normal established line (Todaro & Green, 1963)	4·0 ± 0·68 3·7 ± 0·67	2·8 ± 0·67 5·1 ± 0·65	1·5 polyoma 1·3 SV-40	− 30 + 37·8
Py-3T3-1	Py transformed derivative of 3T3 (Todaro & Green, 1965)	3·3 ± 0·36	9·5 ± 1·4	1·8 polyoma	+ 188
Py-3T3-11	Py transformed, similar to Py-3T3-1	3·1 ± 0·19	17·6 ± 3·2	1·6 polyoma	+ 467
3T3-Py-1	Py transformed derivative of 3T3, isolated in agar (Benjamin, unpublished observation)	2·4 ± 0·34	7·5 ± 0·83	1·6 polyoma	+ 213
3T3-Py-6	Py transformed, similar to 3T3-Py-1	1·1 ± 0·10 1·4 ± 0·30	6·8 ± 0·89 1·5 ± 0·30	1·7 polyoma 1·3 SV-40	+ 568 + 7·2
SV-3T3-47	SV-40 transformed derivative of 3T3 (Todaro & Green, 1965)	1·4 ± 0·27 1·2 ± 0·10	0·9 ± 0·28 5·2 ± 0·80	1·7 polyoma 1·3 SV-40	− 35·5 + 333
SVPY-3T3-11	SV-40, Py doubly transformed (Todaro, Habel & Green, 1965)	2·1 ± 0·78 2·8 ± 0·31	19·2 ± 0·38 9·2 ± 0·40	1·6 polyoma 1·3 SV-40	+ 813 + 221
(b) Hamster					
C13	Normal baby hamster kidney (Stoker & Abel, 1962)	0·6 ± 0·12	0·8 ± 0·30	3·6 polyoma	+ 33
S1	Non-transformed spontaneous mutant of C13 that grows in agar	3·8 ± 0·80	2·5 ± 0·43	1·6 polyoma	− 34·2
Ha 442	Spontaneous malignant derivative of C13 (see Bases, 1963)	3·5 ± 0·11	2·7 ± 0·19	1·6 polyoma	− 22·9
Ha 651	Spontaneous malignant derivative, similar to Ha 442	4·2 ± 0·25	3·8 ± 0·61	4·6 polyoma	− 9·6
T1	Py transformed derivative of C13 isolated in agar (see Macpherson & Montagnier, 1964)	2·8 ± 0·81	11·1 ± 0·11	1·6 polyoma	+ 296
T8	Py transformed, similar to T1	2·1 ± 0·21	5·1 ± 0·60	1·6 polyoma	+ 143
(c) Rat					
RE	Normal line of rat embryo fibroblasts	1·4 ± 0·20	1·3 ± 0·20	3·3 polyoma	− 7·1
RS2	Non-transformed spontaneous mutant of RE that grows in agar	4·8 ± 0·86	8·3 ± 2·4	1·3 polyoma	+ 52
REF	Py transformed "focal" line, derivative of RE	2·0 ± 0·10	17·2 ± 1·0	5·4 polyoma	+ 720
RT1	Py transformed derivative of RE isolated in agar	1·3 ± 0·14	4·2 ± 0·23	3·2 polyoma	+ 223
RT2	Py transformed, similar to RT1	2·2 ± 0·22	10·8 ± 0·94	1·3 polyoma	+ 391
RT4	Py transformed, similar to RT1	1·0 ± 0·07	3·2 ± 0·43	1·3 polyoma	+ 220

† Average of two determinations.

4. Discussion

(a) *Virus-specific RNA in infected mouse kidney cells*

A fraction of the ^{32}P pulse-labeled RNA extracted from mouse kidney cells infected by polyoma virus is capable of hybridizing specifically with polyoma DNA. The size of this fraction depends on the time during the virus growth cycle at which the ^{32}P is given. The curve showing the size of the virus-specific fraction as a function of time follows fairly closely the uptake of [^{14}C]thymidine into DNA in the infected cultures (Fig. 6). Since part of the DNA synthesized after infection is known to be mouse DNA, a better comparison would be the time of appearance of infectious viral DNA (Dulbecco, Hartwell & Vogt, 1965). DNA infectivity begins to appear around 22 to 24 hours and increases rapidly in parallel with the increase of virus-specific RNA. These results suggest that progeny viral DNA acts as template for RNA synthesis.

The inability to exhaust virus-specific RNA from the early RNA's with amounts of viral DNA 3 to 4 times greater than that required to exhaust the late RNA's (Fig. 4(a) and (b)) may be related either to the smaller amount of the early RNA, or to its homology with a relatively smaller fraction of the viral genome. From the saturation experiment with 24 to 28-hour pulse-labeled RNA (Fig. 7(b); Results, part (c)) it would appear that not all of the virus DNA is involved in hybridization with the late RNA; however, the calculation involves the untested assumption that all of the DNA on the filter is available for hybridization. The present results, therefore, give no clear indication as to the specificity of transcription of DNA sites during the course of virus development. It is hoped that such information can be obtained from further experiments, now in progress, involving competition between the early and late RNA's.

(b) *Effect of infection on the capacity of pulse-labeled RNA to hybridize with mouse DNA*

During the period of maximum synthesis of virus-specific RNA (24 to 28 hours), not more than 1% of the labeled RNA can hybridize to polyoma DNA. Even allowing for the fact that only half (roughly) of the cells in the culture are infected, it would appear that infection by polyoma does not block the uptake of ^{32}P into cellular RNA. In fact, comparison of the capacities of infected and uninfected cell RNA to hybridize with cellular DNA shows a stimulation during infection (Fig. 8). These results could represent a "cross-hybridization" between virus-specific RNA and mouse DNA, which would be predicted from the partial homology observed between synthetic polyoma RNA and mouse DNA (Winocour, 1965). On the other hand, they could indicate the actual stimulation of host-cell RNA synthesis by the virus; this would be in keeping with the observations made on the same system (Dulbecco *et al.*, 1965) of a stimulation (or de-repression) of mouse DNA synthesis. Evidence of a stimulation of cellular RNA synthesis has also been obtained in polyoma-infected mouse embryo fibroblasts (Benjamin, unpublished results).

(c) *Presence of virus-specific RNA in transformed cells*

The results shown in Table 1 demonstrate the existence of a fraction of the RNA from virus-free transformed cells which is capable of hybridizing with the virus DNA. This hybridization is characteristic of virus transformed cells, since no such hybridization is observed using RNA from normal cells, or spontaneous malignant cells, or mutant cell types selected on the same basis as virus-transformed cells by their enhanced ability to grow in agar. Furthermore, among the cell lines transformed by polyoma or SV-40, no "cross-hybridization" is observed between RNA of cells transformed by one of these viruses and DNA of the other virus. RNA from a doubly transformed line, however, is capable of hybridizing with both viral DNA's.

From the amount of [^{32}P]RNA extracted from a given number of polyoma-transformed cells which was observed to hybridize with polyoma DNA, a rough estimate can be made of the amount of virus-specific RNA per cell. Assuming the RNA synthesized during the pulse is of the same specific activity as the external phosphate pool, the calculated values run from 1 to 5 half-polyoma DNA equivalents of RNA per cell. These estimates are undoubtedly low, since the amount of RNA hybridized is probably less than the amount of virus-specific RNA present (i.e., the specific fraction is not exhausted by the amounts of viral DNA used here), and since both the recovery of RNA during the extraction and the efficiency of hybridization are less than 100%. Thus, the average number of virus equivalents of RNA per cell is probably considerably greater than 1, making it likely that all of the cells in the population are producing the virus-specific RNA.

These results offer strong evidence for the persistence of at least a part of the viral genome in transformed cells. It would seem reasonable, furthermore, to assume that the virus-specific RNA functions *in vivo* as a messenger RNA; whether or not the corresponding protein is required for maintenance of the transformed state is still a matter of speculation.

I wish to express my gratitude to Dr Giuseppe Attardi for his sound advice and encouragement. I also thank Dr Jerome Vinograd and his associates for their help with the preparation and analysis of polyoma DNA and for the gift of the SV-40 DNA, and Dr George Todaro for the gift of several of the cell lines. The comments and criticisms of Dr Robert Sinsheimer were helpful in the preparation of the manuscript. Finally, I wish to acknowledge the assistance of Mrs Laverne Wenzel and Mrs Arger Drew in the tissue culture aspects of this work. This investigation has been supported by U.S. Public Health Service Training Grant 5T1-GM-86.

REFERENCES

Axelrod, D., Bolton, E. T. & Habel, K. (1964). *Science*, **146**, 1466.

Bases, R. (1963). *Cancer Res.* **24**(2), 1216.

Black, P. H., Rowe, W. B., Turner, H. C. & Huebner, R. J. (1963). *Proc. Nat. Acad. Sci., Wash.* **50**, 1148.

Defendi, V. (1963). *Proc. Soc. Exp. Biol. Med.* **113**, 12.

Dulbecco, R. & Freeman, G. (1959). *Virology*, **8**, 396.

Dulbecco, R., Hartwell, L. H. & Vogt, M. (1965). *Proc. Nat. Acad. Sci., Wash.* **53**, 403.

Eddy, B. E., Rowe, W. B., Hartley, J. W., Stewart, S. E. & Huebner, R. J. (1958). *Virology*, **6**, 290.

Gillespie, D. & Spiegelman, S. (1965). *J. Mol. Biol.* **12**, 829.

Habel, K. & Eddy, B. E. (1963). *Proc. Soc. Exp. Biol. Med.* **113**, 1.

Habel, K. (1965). *Virology*, **25**, 55.

Huebner, R. J., Rowe, W. P., Turner, H. C. & Lane, W. T. (1963). *Proc. Nat. Acad. Sci., Wash.* **50**, 379.

Koch, M. A. & Sabin, A. B. (1963). *Proc. Soc. Exp. Biol. Med.* **113**, 4.

Macpherson, I. & Montagnier, L. (1964). *Virology*, **23**, 291.

Murakami, W. (1963). *Science*, **142**, 56.

Sjogren, H. O., Hellstrom, I. & Klein, G. (1961). *Cancer Res.* **21**, 239.

Stoker, M. & Abel, P. (1962). *Cold Spr. Harb. Symp. Quant. Biol.* **27**, 375.
Ting, R. (1964). *Virology*, **24**, 227.
Todaro, G. J. & Green, H. (1963). *J. Cell. Biol.* **17**, 299.
Todaro, G. J. & Green, H. (1965). *Science*, **147**, 513.
Todaro, G. J., Habel, K. & Green, H. (1965). *Virology*, **27**, 179.
Vinograd, J., Lebowitz, J., Radloff, R., Watson, R. & Laipis, P. (1965). *Proc. Nat. Acad. Sci., Wash.* **53**, 1104.
Vogt, M. & Dulbecco, R. (1962). *Virology*, **16**, 41.
Weil, R. (1961). *Virology*, **14**, 46.
Winocour, E. (1963). *Virology*, **19**, 158.
Winocour, E. (1965). *Virology*, **25**, 276.

Characterization of the Simian Virus 40-specific RNA in Virus-yielding and Transformed Cells

YOSEF ALONI, ERNEST WINOCOUR AND LEO SACHS

The SV40†-specific RNA in virus-yielding and in transformed cells has been characterized by DNA–RNA hybridization and by base ratio determinations on the hybridized RNA.

In virus-yielding monkey cells (BS-C-1 line) the viral messenger RNA that is synthesized before replication of virus DNA ("early" RNA) was shown, both by competition experiments and by base ratio analysis, to be different from the viral messenger RNA that is present after the onset of virus DNA synthesis ("late" RNA). The amount of late RNA synthesized was found to be approximately 40 times greater than the amount of early RNA.

The detection of SV40-specific RNA in SV40-transformed (but virus-free) mouse cells, of the 3T3 line, was greatly enhanced by means of two successive hybridization steps. The amount of virus-specific RNA found in the transformed cells was comparable to the amount of early RNA detected in virus-yielding cells. Competition experiments between the virus-specific RNA from transformed cells and the late RNA from virus-yielding cells indicated that only part (about one-third) of the SV40 genome was transcribed in the transformed cells. The base composition of the virus-specific RNA from transformed cells was not identical to the base composition of either the early or late virus-specific RNA present in virus-yielding BS-C-1 cells.

1. Introduction

The small DNA viruses polyoma and SV40 can induce the transformation of normal cells into tumour cells, and these viruses have been extensively used for *in vitro* studies on the mechanism of neoplastic cell transformation (see reviews in Sachs, 1965,1967; Dulbecco, 1966). One of the basic questions relating to this mechanism is whether the process of virus-induced neoplasia is always accompanied by an incorporation of viral genes into the cellular genome, and if so, what part of the viral genome is transcribed in the transformed cells. Regarding the incorporation of viral genes, it has been reported that virus-specific RNA can be detected by DNA–RNA hybridization techniques in virus-free cells transformed by polyoma (Benjamin, 1966), SV40 (Benjamin, 1966; Reich, Black & Weissmann, 1966), and the carcinogenic adenoviruses (Fujinaga & Green, 1966,1967). The present studies with SV40 were undertaken in order to improve the sensitivity of the hybridization method for detecting virus-specific RNA in transformed cells, and to characterize this RNA by comparison with the viral messenger RNA's in virus-yielding cells using both nucleic acid hybridization techniques and base ratio determinations.

† Abbreviations used: SV40, simian virus 40; MAK, methylated albumin-kieselguhr; PCA, perchloric acid; p.i. post-infection.

2. Materials and Methods

(a) *Preparation of SV40 DNA*

SV40 (strain 777) was grown on monkey cells of the BS-C-1 line, as previously described (Gershon, Sachs & Winocour, 1966). Concentration and purification of the virus from the tissue culture lysates were carried out by equilibrium centrifugation in caesium chloride density-gradients by procedures reported elsewhere (Black, Crawford & Crawford, 1964). The DNA was extracted from the purified virus with phenol (Weil, 1961) and dialysed against 0·1 × SSC. (SSC is 0·15 M-NaCl–0·015 M-sodium citrate.) To purify the virus DNA further, the dialysed extracts were subjected to a brief heat-treatment (boiled for 5 min followed by rapid cooling in ice-water) and fractionated on a 3-layer MAK column (Mandell & Hershey, 1960) using a linear elution gradient from 0·4 M-NaCl. to 0·9 M-NaCl. This step effectively separates the rapidly renaturing SV40 DNA molecule from linear DNA molecules of cellular origin (Gershon *et al.*, 1966), and was included in the purification procedure since experiments with polyoma virus have shown that host cell DNA is incorporated into some virus particles (Winocour, 1967*a*,*b*; Kaye & Winocour, 1967).

To prepare [^{3}H]thymidine-labelled viral DNA, 2 μc of [^{3}H]thymidine (The Radiochemical Centre, Amersham; specific activity 17,100 mc/m-mole) were added/ml. of culture supernatant after virus infection. Seven to 9 days post-infection, the virus was harvested, purified, and the labelled-virus DNA extracted and purified as above.

(b) *Infection of cultures and preparation of ^{32}P-labelled cell RNA*

BS-C-1 cells were incubated in phosphate-free Eagle's medium (containing a fourfold concentration of amino acids and vitamins) with 10% dialysed calf serum for 24 hr prior to infection. The cultures (5×10^6 cells) were infected with 0·8 ml. of a single stock of SV40 (10^8 plaque-forming units/ml.). After a 2-hr period for virus adsorption, the cultures were incubated with Eagle's medium containing 0·01 the normal concentration of phosphate and 10% dialysed calf serum. For various time periods after virus infection (see Results), 0·5 to 1·0 mc of [^{32}P]orthophosphate (50 to 100 c/mg P; The Radiochemical Centre, Amersham) was added to each culture. At the end of the labelling period, the radioactive medium was removed, the cell monolayers were washed twice with cold Tris-buffered saline, and the cells harvested by scraping. The cells were re-washed by low-speed centrifugation and stored as a dry pellet at −20°C. To test the proportion of cells infected under the above-mentioned conditions, an immuno-fluorescent test for the SV40-induced nuclear tumour antigen was carried out as described elsewhere (Gershon *et al.*, 1966). The results showed that at 48 hr p.i., 70% of the cells displayed fluorescent nuclei.

SV40-transformed cells of the mouse line 3T3 (Todaro, Green & Goldberg, 1964) were grown to near confluency, starved of phosphates for 24 hr, and labelled with [^{32}P]orthophosphate for time periods as indicated in Results. The cells were collected and washed in the same manner as the infected BS-C-1 cells. The line of SV40-transformed 3T3 cells used in the present experiments was clone 1 from SV-3T3-10 kindly supplied by Drs H. Green and G. J. Todaro.

The RNA was extracted from the cell pellets by a slight modification of the procedure reported by Benjamin (1966). The cell pellets were homogenized in 27 vol. of a solution containing 4×10^{-3} M-$MgCl_2$, 0·5% sodium naphthalene disulphonate and 1·25 mg/ml. bentonite. 3 vol. of 5% sodium dodecyl sulphate and 30 vol. of Tris-buffered phenol (pH 7·2) were then added. The mixture was vigorously shaken in a 62°C water bath for 5 min and then cooled rapidly in ice-water. After a second cycle of shaking at 62°C followed by cooling, the mixture was centrifuged. The aqueous layer was removed and subjected to further deproteinization by shaking in the cold with an equal vol. of Tris-buffered phenol. The final aqueous phase was brought to 0·1 M-NaCl and the RNA precipitated by adding 2 vol. of cold 100% ethanol. The precipitated RNA was collected by centrifugation, re-dissolved in 0·1 × SSC containing 0·005 M-$MgCl_2$, and incubated for 60 min at 37°C with 20 μg DNase (electrophoretically purified and free of RNase)/ml. The mixture was re-extracted with phenol, centrifuged, and the aqueous phase brought to 0·2 M-NaCl, in 0·05 M-sodium phosphate buffer pH 6·8). This material was then applied to a one-layer MAK column (Spiegelman, personal communication) previously washed with 100 bed-vol.

of 0·2 M-NaCl–0·05 M-sodium phosphate. The loaded column was washed extensively with the same buffer until the washings contained no detectable radioactivity. The [^{32}P]RNA was eluted from the column with 1·8 M-NaCl in 0·05 M-sodium phosphate buffer and dialysed against 2×SSC. These dialysed preparations were further purified by passage through nitrocellulose membrane filters. The specific activity of the [^{32}P]RNA preparations was determined from their optical density at 260 mμ and the amount of radioactive material in the acid-insoluble fraction. Essentially 100% of the radioactive material was sensitive to alkali degradation.

RNA from unlabelled cells was extracted as described above except that the MAK purification step was omitted.

(c) *DNA–RNA hybridization*

The technique used was similar to that described by Gillespie & Spiegelman (1965). The DNA in 0·01×SSC was denatured by boiling for 15 min followed by rapid cooling in ice-water. The denatured DNA solutions were diluted to 5 ml. with 2×SSC and passed slowly under suction through nitrocellulose membrane filters which had been soaked in 2×SSC for 10 min and washed with 25 ml. of the same buffer. The DNA filters were then washed with 100 ml. of 2×SSC, dried at room temperature over night, and baked at 80°C for 4 hr. 90 to 100% of the ^{3}H-labelled SV40 DNA deposited on the filters was retained, and not more than 5% of the DNA initially retained was lost in the subsequent incubations and washings. Hybrids were formed by immersing the DNA filters in scintillation vials containing 5 ml. of [^{32}P]RNA in 2×SSC and incubating the vials at 66°C for 36 hr (see under Results). After incubation, the filters were removed from the vials, washed by suction filtration on each side with 100 ml. of 2×SSC and incubated with 5 ml. of 2×SSC containing 20 μg of boiled RNase/ml. for 1 hr at room temperature. The filters were then re-washed on each side with 100 ml. of 2×SSC, dried, and assayed for radioactivity in a scintillation spectrometer.

For isolation of the [^{32}P]RNA hybridized to DNA, the filters were removed from the scintillation fluid (after counting in the scintillation spectrometer) washed extensively in toluene, and the toluene removed by evaporation at room temperature. The filters were then boiled for 3 min in 3 ml. of 0·01×SSC and rapidly cooled. 90% of the [^{32}P]RNA was eluted by this step. The base ratio determinations were carried out on the [^{32}P]RNA eluted into the solution by this method. For rehybridization experiments, the eluted [^{32}P]RNA was passed 5 times through nitrocellulose membrane filters in order to remove traces of unlabelled DNA which was eluted from the filters together with the [^{32}P]RNA.

(d) *DNA–DNA hybridization*

A modification of the methods reported by Denhardt (1966), Warnaar & Cohen (1966) and Richards (1967) was used. Nitrocellulose membrane filters containing immobilized denatured DNA (as for the RNA–DNA hybridization experiments) were incubated at 60°C for 6 hr in vials with 1·5 ml. of a solution containing 3×SSC and 0·04% bovine serum albumin. Small portions of fragmented (10 min sonic vibration at full power in a 10 kc/s Raytheon sonic vibrator), denatured, labelled DNA were then introduced into the vials and incubation continued for another 23 hr at 60°C. The filters were then removed from the vials, rinsed briefly in 0·003 M-Tris buffer (pH 9·4), and washed on both sides under suction with 100 ml. of the same buffer. The background value of labelled DNA non-specifically absorbed to blank filters, or to filters containing heterologous DNA, was found to be only 0·01 to 0·05% of the total radioactive DNA incubated.

(e) *Base composition of hybridized [^{32}P]RNA*

Base ratios were determined according to the method of Sebring & Salzman (1964). The [^{32}P]RNA eluted from the filters after hybridization was mixed with 0·5 mg of carrier RNA and precipitated by the addition of an equal vol. of 5% PCA. The precipitated RNA was collected by centrifugation, washed with 2 ml. of 2·5% PCA, and then hydrolysed by incubating in 0·25 ml. of 0·3 M-KOH for 16 to 18 hr at 37° C. The digest was adjusted to pH 3·5 with PCA, chilled, and the precipitated $KClO_4$ removed by centrifugation. The supernatant fluid was then applied to a sheet of Whatman no. 1 paper and the [^{32}P]mononucleotides separated by electrophoresis at 3000 v for 70 min in pyridine acetate buffer (pH 3·5). To remove the ultraviolet-absorbing solvent, the chromatogram was dried overnight in a ventilated hood. The chromatogram was cut according to the positions of the 4 nucleotide bands located under ultraviolet light, and the amount of radioactivity in each band measured by scintillation spectrometry. In all cases, the resolution of the 4 nucleotide bands identified under ultraviolet light was identical to the peaks of ^{32}P-radioactivity obtained by scanning the entire chromatogram in a Vanguard Autoscanner 880. 100% of the radioactivity applied to the Whatman paper was recovered in the 4 nucleotide bands.

Samples of each [^{32}P]RNA preparation (eluted from the filters after hybridization), were also hydrolysed directly in 0·3 M-KOH without prior precipitation of the RNA in 2·5% PCA. In this case, the alkali digest was adjusted to pH 3·5 with PCA, chilled, the precipitated $KClO_4$ was removed by centrifugation, and the volume of the supernatant was reduced to 0·25 ml. by evaporation under a stream of nitrogen. The subsequent separation of the [^{32}P]mononucleotides was carried out as described above. Identical results were obtained by both methods.

3. Results

(a) *SV 40-specific RNA in infected BS-C-1 cells*

(i) *Detection of virus-specific RNA in infected BS-C-1 cells*

To detect the presence of virus-specific RNA in infected BS-C-1 cells undergoing the lytic interaction, [^{32}P]RNA from infected cultures, labelled from the end of virus adsorption to 72 hours p.i., was incubated with filters containing SV40 DNA, or *E. coli* DNA as control filters. The ability of [^{32}P]RNA from uninfected BS-C-1 cells to hybridize with SV40 DNA was also examined. The results are shown in Table 1.

TABLE 1

Virus-specific RNA in BS-C-1 cells infected with SV40

Expt	[^{32}P]RNA from	DNA on filter	μg DNA/ filter	Cts/min bound to filter	Cts/min bound as % of input
1	infected BS-C-1 cells	{ SV40	5	16,006	0·160
		{ *E. coli*	5	294	0·003
		{ SV40	25	25,232	0·252
		{ *E. coli*	25	406	0·004
2	uninfected BS-C-1 cells	{ BS-C-1 cell	10	10,003	0·200
		{ *E. coli*	10	892	0·018
		{ BS-C-1 cell	10	10,520	0·210
		{ SV40	10	1003	0·020
		{ SV40	5	937	0·019
		{ None	0	920	0·018

In experiment 1, BS-C-1 cells were infected as described under Materials and Methods. After virus adsorption, 0·5 to 0·8 mc of ^{32}P (50 to 100 c/mg P) were added to each culture. 72 hr p.i., the [^{32}P]RNA was extracted from the cells, purified and tested for its ability to hybridize with SV40 DNA as described under Materials and Methods. Each hybrid-forming reaction contained 1×10^7 cts/min [^{32}P]RNA (100 μg) in 5 ml. of 2 × SSC, one filter with SV40 DNA, and one filter with *E. coli* DNA at the indicated amounts (the filters included in the same reaction vial are bracketed).

In experiment 2, uninfected BS-C-1 cells were labelled for 72 hr as above, and the [^{32}P]RNA extracted as described under Materials and Methods. Each hybrid-forming reaction contained 5×10^6 cts/min [^{32}P]RNA (52 μg) in 5 ml. of 2 × SSC and two filters as indicated above (bracketed).

It will be seen that specific hybridization was obtained between [^{32}P]RNA from infected cells and SV40 DNA; no hybridization between SV40 DNA and [^{32}P]RNA from uninfected BS-C-1 cells was detected in these experiments. The kinetics of hybrid formation at 66°C between SV40 DNA and [^{32}P]RNA from infected BS-C-1 cells are shown in Fig. 1.

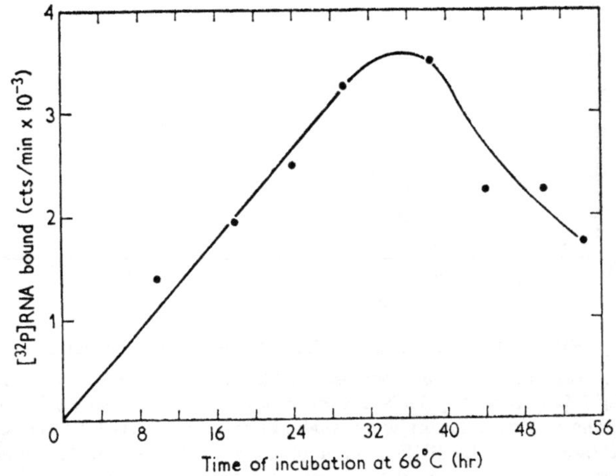

FIG. 1. Kinetics of hybrid formation at 66°C.

Each hybrid-forming reaction contained $6{\cdot}2 \times 10^6$ cts/min (30 μg) of [^{32}P]RNA (from infected BS-C-1 cells labelled for 72 hr) in 5 ml. of 2 × SSC, one filter with 0·05 μg of SV40 DNA and one blank filter. The amount of hybridizable [^{32}P]RNA was in excess of the amount of SV40 DNA on the filter. The background value (0·003% of the input) of cts/min [^{32}P]RNA bound to the blank filter is subtracted in each case from the cts/min [^{32}P]RNA bound to the SV40 DNA filter. At the indicated times of incubation, hybrid formation was stopped by immersing the reaction vials in ice-water.

(ii) *Amount of virus-specific RNA in BS-C-1 cells at different times after infection*

In order to correlate the amounts of virus-specific RNA detected in the infected cells with the replication of SV40 DNA, experiments were first performed to determine the time of viral DNA replication in cells infected under the same conditions as those used for extraction of the virus-specific RNA. This experiment was performed by hybridization tests between [^{3}H]thymidine-labelled DNA extracted from the cells at different times after infection, and SV40 DNA. The results in Fig. 2 show that viral DNA synthesis was first detected in cultures labelled from 22 to 25 hours p.i. After this time, there was a sharp increase in viral DNA replication, which reached a peak at 48 hours p.i., and then declined.

In view of the above results, virus-specific RNA which was extracted from the infected cells before the time of replication of SV40 DNA (up to 22 hours p.i.) is designated as "early" RNA; SV40-specific RNA which was extracted from cells at any time after the onset of viral DNA replication (after 22 hours p.i.) is designated as "late" RNA (these preparations may also contain early RNA).

To determine the relative amounts of early and late RNA, [^{32}P]RNA was extracted from BS-C-1 cells labelled from 2 to 22 hours p.i., and from 50 to 70 hours p.i., and incubated with filters containing increasing amounts of SV40 DNA. The results in Fig. 3 show the "exhaustion" levels of virus-specific RNA in these preparations

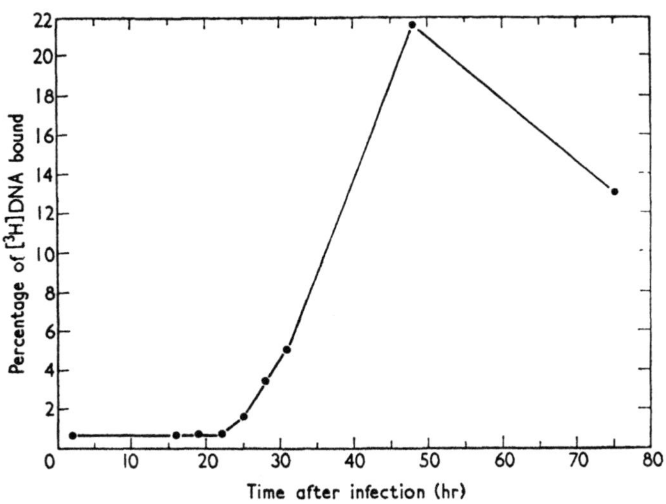

FIG. 2. Time of virus DNA synthesis in BS-C-1 cells.

Confluent cultures were infected as described in Materials and Methods. 3 hr before the time indicated, 50 μc of [^{3}H]thymidine (17,000 mc/m-mole) were added to each infected culture. At the indicated times, the cell monolayers were washed twice with Tris-buffered saline and the total cell nucleic acid extracted by the SDS–phenol method of Dulbecco, Hartwell & Vogt (1965). RNA was then removed by treatment with RNase, and the specific activity of the [^{3}H]thymidine-labelled DNA was determined. The amount of virus DNA in the labelled DNA extracted from the infected cells at the indicated times was determined by DNA–DNA hybridization as described in Materials and Methods. Each reaction contained 10 μg of purified, denatured SV40 DNA immobilized on a filter, or a blank filter, and 6000 to 31,000 cts/min of fragmented, denatured [^{3}H]DNA extracted from the infected cells at the indicated times. Less than 0·02% of the total radioactive DNA incubated was bound to blank filters. Each hybridization reaction was performed in duplicate.

after hybridization with increasing amounts of SV40 DNA. To verify that the non-hybridized RNA remaining after the attainment of the plateaux in Fig. 3 is truly depleted of virus-specific RNA, the exhausted RNA preparations were incubated a second time with filters containing SV40 DNA; the amount of [^{32}P]RNA bound in this second incubation was found to be no higher than the background value. Furthermore, the exhaustion of the virus-specific RNA cannot be due to degradation, since the residual RNA fraction was fully capable of hybridizing with cellular DNA, and sucrose gradient centrifugation of the preparation after six days at 66°C showed that 80% of the radioactive material sedimented with a value greater than 16 s. From Fig. 3, it will be seen that the amount of virus-specific RNA present in the cells labelled from the fiftieth to seventieth hour p.i. (curve A) is 40 times greater than that present in cells labelled from 2 to 22 hours p.i. (curve B); that is, before the onset of viral DNA replication. Two additional exhaustion experiments with late RNA, in which the fraction of virus-specific RNA present in cells labelled from 66 to 72 hours and 2 to 72 hours p.i. was determined, produced similar results; indicating that the amount of virus-specific RNA present after SV40 DNA replication was about 40 times greater than the amount present before the onset of viral DNA replication.

(iii) *Competition experiments between early and late virus-specific RNA*

The experiments described in the previous section indicated differences in the relative amounts of virus-specific RNA present during the early and late periods of the

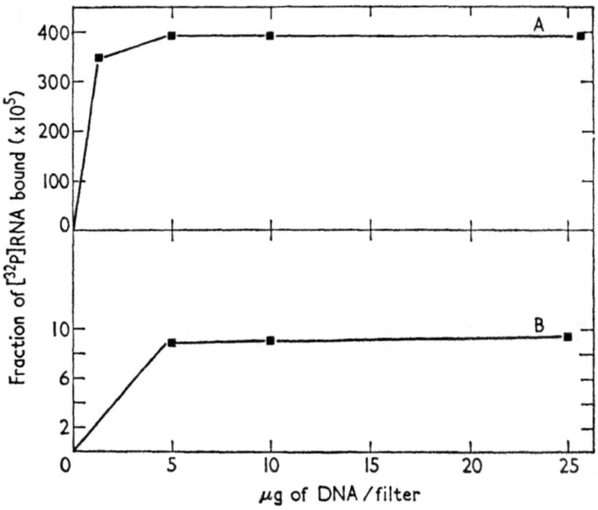

FIG. 3. Hybridization between increasing amounts of SV40 DNA and late and early virus-specific RNA from infected BS-C-1 cells.

Curve A, the [^{32}P]RNA was extracted from infected BS-C-1 cells labelled from 50 to 70 hr p.i., and is thus designated late RNA. Each hybrid-forming reaction contained 5×10^6 cts/min [^{32}P]RNA (25 μg) in 5 ml. $2 \times$SSC, one filter containing the amount of SV40 DNA indicated, and one blank filter. The experimental background (cts/min [^{32}P]RNA bound to blank filter) was 0·003% of the input and has been subtracted from the values plotted.

Curve B, the [^{32}P]RNA was extracted from BS-C-1 cells labelled from the end of the virus absorption period to 22 hr p.i. (designated early RNA). Each reaction contained 26×10^6 cts/min (= 70 μg) [^{32}P]RNA in 5 ml. $2 \times$SSC, one filter with the indicated amount of SV40 DNA, and one blank filter. The experimental background was 0·0029% of the input and this value has been subtracted from the results plotted.

infectious cycle. To determine if there are qualitative differences in the species of RNA molecules present during these stages of the infectious cycle, competition experiments were performed between unlabelled early RNA (extracted at 22 hours p.i.), unlabelled late RNA (extracted at 72 hours p.i.), and late [^{32}P]RNA (labelled from 2 to 72 hours p.i.).

As a preliminary to the competition experiments, it was first necessary to determine the conditions under which the majority of hybridizing sites on SV40 DNA can be saturated with late [^{32}P]RNA. Accordingly, small amounts of ^{3}H-labelled SV40 DNA immobilized on filters were incubated with increasing amounts of late [^{32}P]RNA. Figure 4 shows the results of saturation experiments using 0·044 μg of SV40 DNA immobilized on the filter. It is evident that all available sites on the virus DNA molecule have been saturated. The fraction of the virus DNA molecule that is saturated was estimated as follows. During the 72-hour period of labelling, it is assumed that all species of RNA synthesized have the same specific activity, which was found to be $1{\cdot}8 \times 10^5$ cts/min/μg. The 3000 cts/min of [^{32}P]RNA hybridized at the saturation plateau in Fig. 4 is thus equivalent to 0·0167 μg RNA. This amount of RNA was hybridized by 500 cts/min of ^{3}H-labelled SV40 DNA (=0·044 μg, from the known specific activity of the virus DNA). Thus it is estimated that 76% of the viral genome was saturated by the late [^{32}P]RNA.

Figure 5 shows the results of the competition experiments between early and late RNA for sites on the SV40 DNA molecule. It is clear that the early and late RNA are

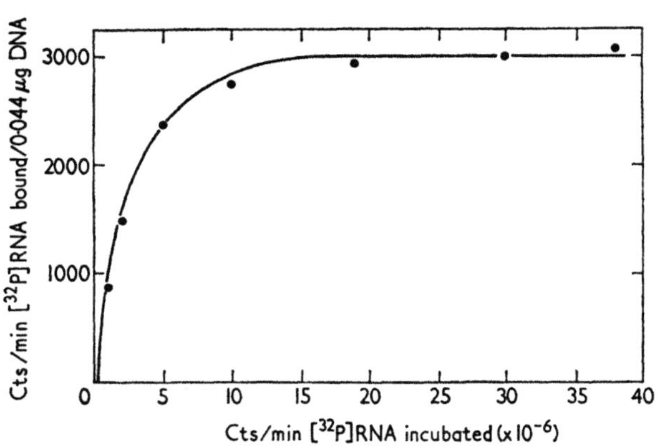

Fig. 4. Saturation of SV40 DNA with increasing amounts of late [^{32}P]RNA from infected BS-C-1 cells.

The late [^{32}P]RNA was extracted from BS-C-1 cells labelled from 2 to 72 hr p.i. The specific activity of this RNA preparation was $1 \cdot 8 \times 10^5$ cts/min/μg. Each reaction vial contained one filter with 0·044 μg of ^{3}H-labelled SV40 DNA (500 cts/min), one blank filter, and increasing amounts of [^{32}P]RNA as indicated. The experimental background (cts/min [^{32}P]RNA bound to the blank filter) has been subtracted from the cts/min [^{32}P]RNA bound to the SV40 DNA filter in each reaction.

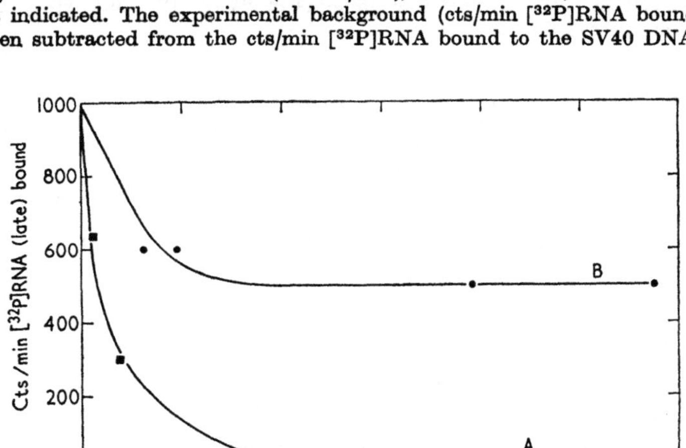

Fig. 5. Competition between late [^{32}P]RNA and unlabelled late and early RNA for hybridizing sites on SV40 DNA.

Each reaction mixture contained (in 5 ml. of $2 \times$SSC): one filter with 0·044 μg SV40 [^{3}H]DNA (500 cts/min), one blank filter, 4×10^6 cts/min (= 30 μg) of late [^{32}P]RNA (from infected BS-C-1 cells labelled from 2 to 72 hr) and increasing amounts of unlabelled RNA extracted from BS-C-1 cells 22 hr p.i. (curve B = early RNA) or 72 hr p.i. (curve A = late RNA). The experimental backgrounds (0·003% of input) have been subtracted in each case. In reactions containing no unlabelled RNA, 1000 cts/min of the late [^{32}P]RNA were hybridized by 0·044 μg of SV40 DNA in the filter; this is a saturation level.

qualitatively different, in that a fraction of the late RNA (curve A) binds to sites not shared by the early RNA (curve B). The specificity of the competition experiment was tested by adding increasing amounts of the same unlabelled early and late virus-specific RNA preparations to the hybridization reaction between wheat germ DNA and wheat germ [^{32}P]RNA, and by adding increasing amounts of unlabelled *E. coli* RNA, or normal BS-C-1 cell RNA, to the reaction mixture described under Fig. 5

(as well as to the heterologous hybridizing system). In each case non-specific competition to the extent of 20% was observed, and this factor has been taken into account in the ensuing calculations. All the unlabelled RNA preparations used in the competition experiments possessed no detectable RNase activity.

Assuming that all the [^{32}P]RNA's synthesized in infected BS-C-1 cells have equal specific activity, it may be estimated from Fig. 5 that about 37% of the SV40 DNA sites which bind late RNA bind early RNA. Confirmation of the qualitative differences between the early and late virus-specific RNA's has been obtained by base-ratio determinations on these RNA's (see section (c) below).

(b) *SV40-specific RNA in transformed 3T3 cells*

(i) *Detection of virus-specific RNA in 3T3 cells transformed by SV40*

To detect the presence of SV40-specific RNA in transformed cells of the mouse line 3T3, cultures of these cells were labelled with ^{32}P for different time periods, and the [^{32}P]RNA was extracted and tested for its capacity to hybridize specifically with SV40 DNA. The results in Table 2 show that the [^{32}P]RNA in four preparations, extracted after labelling periods of 6, 22 and 96 hours, hybridized specifically with

TABLE 2

Detection of virus-specific RNA in 3T3 cells transformed by SV40

Expt	DNA on filter	μg DNA/ filter	Cts/min bound to filter(average)	Cts/min bound as % of input	No. of replicates
	SV40	1	6820±2%	0·0087	2
1	SV40	5	11,096±3%	0·0142	4
	None	0	2227±11%	0·0028	4
	SV40	5	13,760±9%	0·0088	2
2	SV40	10	14,475±4%	0·0093	2
	SV40	25	15,922	0·0104	1
	None	0	2688±2%	0·0017	2
	SV40	3	6090±6%	0·0043	5
3	None	0	2475±9%	0·0017	2
	Polyoma	3	1659±38%	0·0012	2
4	SV40	20	80,982±5%	0·0252	2
	None	0	20,480±10%	0·0064	2
5	SV40	20	40,462±8%	0·0126	2
	None	0	5142±10%	0·0016	2

3T3 cells transformed by SV40 were labelled with ^{32}P (0·5 to 0·8 mc/plate; specific activity 75c/mgP) for 6 hr (expt 1), 22 hr (expts 2, 4 and 5) and 96 hr (expt 3). The [^{32}P]RNA was extracted, purified and incubated with filters containing the indicated amount of SV40 DNA, polyoma DNA, or blank filters as described under Materials and Methods. Each reaction contained 5 ml. of [^{32}P]RNA in 2 × SSC containing: 78×10^6 cts/min, $3{\cdot}6 \times 10^5$ cts/min/μg (expt 1); 156×10^6 cts/min, $4{\cdot}3 \times 10^5$ cts/min/μg (expt 2); 140×10^6 cts/min, $8{\cdot}7 \times 10^5$ cts/min/μg (expt 3); 318×10^6 cts/min, $8{\cdot}2 \times 10^5$ cts/min/μg (expts 4 and 5). In experiments 1, 2, 4 and 5, each vial contained one SV40 DNA filter and one blank filter; in experiment 3 each vial contained one SV40 DNA filter, and either one polyoma DNA filter or one blank filter. In experiments 4 and 5, the same [^{32}P]RNA preparation was used, but in experiment 4 the RNase step was omitted.

SV40 DNA (the maximum amount of [^{32}P]RNA bound to SV40 DNA filters was seven- to eightfold greater than the amount bound to control filters). The binding of [^{32}P]RNA from non-transformed 3T3 cells (labelled for six hours; specific activity = 131,000 cts/min/μg total RNA) to filters containing 3 μg SV40 DNA was no higher than the binding to filters containing 3 μg of polyoma DNA.

When the transformed cell [^{32}P]RNA bound to the SV40 DNA filters was eluted and tested for its capacity to rehybridize with SV40 DNA, the significance of the above results, indicating the presence of virus-specific nucleotide sequences, was considerably enhanced; Table 3 shows that the amount of eluted and rehybridized [^{32}P]RNA bound to SV40 DNA filters was 30 to 100 times greater than the amount bound to polyoma DNA, *E. coli* DNA or blank filters. To examine the specificity of the

TABLE 3

Rehybridization of virus-specific [^{32}P]RNA from 3T3 cells transformed by SV40

Expt	DNA on filter	μg DNA/ filter	Cts/min bound to filter	Cts/min bound as % of input
1	SV40	5	2820	49·0
	SV40	5	2670	47·0
	None	0	28	0·5
	None	0	50	0·9
2	SV40	10	3750	58·0
	SV40	10	4270	66·0
	Polyoma	10	60	0·9
	Polyoma	10	55	0·8
	3T3 cell	3	160	2·5
	None	0	71	1·1
3	SV40	3	244	20·0
	SV40	3	206	17·0
	E. coli	3	0	0
	E. coli	50	0	0
	3T3 cell	3	8	0·7
	3T3 cell	50	17	1·4
4	SV40	20	9811	14·1
	SV40	20	10,630	15·2
	E. coli	20	363	0·5
	Polyoma	20	449	0·6
	None	0	300	0·4
	None	0	369	0·5

The virus-specific [^{32}P]RNA bound to filters containing SV40 DNA in experiments 1, 2, 3 and 4 of Table 2 was eluted from the filters, purified and rehybridized as described under Materials and Methods.

In experiment 1, each reaction contained 5750 cts/min [^{32}P]RNA (from SV40-transformed 3T3 cells labelled for 6 hr) in 5 ml. of 2 × SSC, one SV40 DNA filter, and one blank filter. In experiment 2, each reaction contained 6450 cts/min [^{32}P]RNA (labelling time was 22 hr), one SV40 DNA filter, one polyoma DNA filter, and one blank or uninfected 3T3 cell DNA filter. In experiment 3, each vial contained 1220 cts/min [^{32}P]RNA (labelling time was 96 hr), one SV40 DNA filter, one *E. coli* DNA filter, and one 3T3 cell DNA filter. In experiment 4, each reaction contained 70,000 cts/min [^{32}P]RNA (labelling time was 22 hr), one SV40 DNA filter, one polyoma DNA filter or *E. coli* DNA filter, and one blank filter. All reactions were carried out in duplicate.

rehybridization method, a parallel control experiment was performed (under identical conditions) in which [^{32}P]RNA's from *normal* 3T3 cells, bound to filters containing SV40 and polyoma DNA, were eluted and tested for their ability to rehybridize with filters containing SV40, polyoma and no DNA. The amount of [^{32}P]RNA bound to these filters was found to be essentially the same in each case.

In experiments to determine the amount of virus-specific [^{32}P]RNA in transformed cells, [^{32}P]RNA (from cells labelled for 22 hours) was hybridized with increasing amounts of SV40 DNA. The results (Fig. 6) show that 5 μg of SV40 DNA exhausted all the virus-specific RNA (0·0078% of the total cts/min incubated) in this preparation. The fractions of [^{32}P]RNA from transformed cells, labelled for 6 and 96 hours, which bound to 5 μg SV40 DNA were found to be 0·018 and 0·0036%, respectively, of the total [^{32}P]RNA incubated. The amount of virus-specific RNA in the transformed cells is thus of the same order of magnitude as that found before the onset of viral DNA replication in infected BS-C-1 cells.

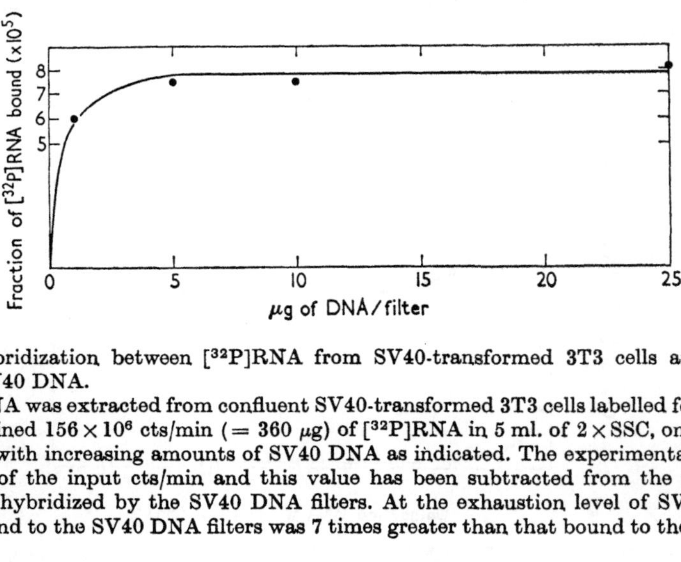

FIG. 6. Hybridization between [^{32}P]RNA from SV40-transformed 3T3 cells and increasing amounts of SV40 DNA.

The [^{32}P]RNA was extracted from confluent SV40-transformed 3T3 cells labelled for 22 hr. Each reaction contained 156×10^6 cts/min (= 360 μg) of [^{32}P]RNA in 5 ml. of 2 × SSC, one blank filter, and one filter with increasing amounts of SV40 DNA as indicated. The experimental background was 0·0016% of the input cts/min and this value has been subtracted from the percentage of input cts/min hybridized by the SV40 DNA filters. At the exhaustion level of SV40 DNA, the [^{32}P]RNA bound to the SV40 DNA filters was 7 times greater than that bound to the blank filters.

The above results, indicating the presence of virus-specific nucleotide sequences in 3T3 cells transformed by SV40, are thus in agreement with previous reports by Benjamin (1966) and Reich *et al.* (1966). To verify that the line of SV40-transformed 3T3 cells used in the present studies was free of infectious SV40, 45×10^6 transformed 3T3 cells were sonicated in 5 ml. of medium and 0·5-ml. samples of the sonicated cells were plated on nine BS-C-1 monolayers in a plaque assay procedure; no plaques appeared during 30 days of incubation.

(ii) *Competition experiments between virus-specific RNA from SV40-transformed 3T3 cells and virus-specific RNA from infected BS-C-1 cells*

A question of interest is whether the virus-specific RNA detected in the SV40-transformed 3T3 cells is complementary to the entire SV40 genome. To study this point, competition experiments were carried out between unlabelled RNA from SV40-transformed 3T3 cells, unlabelled late RNA from infected BS-C-1 cells, and late [^{32}P]RNA from infected BS-C-1 cells. The competition experiments were carried out under the same conditions as those described in section (a, iii), above; that is, the amount of late [^{32}P]RNA incubated was in excess of that required to saturate the amount of SV40 DNA immobilized on the filter, the hybrid-forming reactions were incubated for a time sufficient to ensure completion of the reaction, and all the unlabelled RNA preparations used were tested for non-specific competition in a heterologous hybridizing system (wheat germ DNA and wheat germ [^{32}P]RNA). The results are presented in Fig. 7, where it will be seen that a fraction of the late RNA from lytically infected BS-C-1 cells binds to SV40 DNA sites which are not represented in the virus-specific RNA extracted from transformed 3T3 cells. The SV40-specific RNA from transformed cells thus resembles the virus-specific early RNA from lytically infected BS-C-1 cells in that it is devoid of sequences present in the late RNA from BS-C-1 cells. If we assume, as before, that all the [^{32}P]RNA's synthesized in infected BS-C-1 cells have equal specific activity, it may be estimated from Fig. 7 that 31 to 37% of the SV40 DNA sites which bind late RNA from BS-C-1 cells, bind virus-specific RNA from transformed 3T3 cells (corrected for non-specific competition to the extent of 20%).

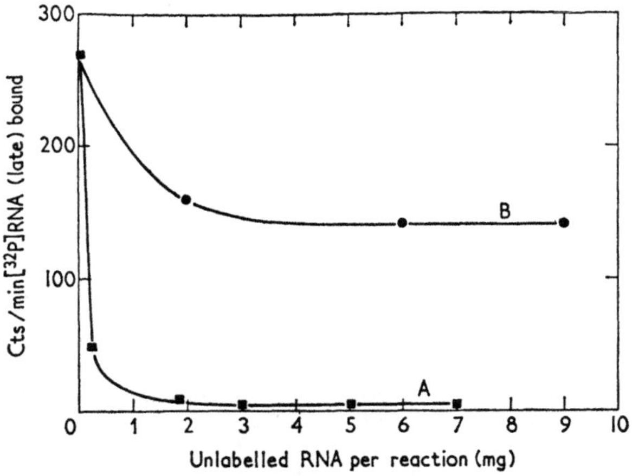

FIG. 7. Competition between [^{32}P]RNA from infected BS-C-1 cells labelled from 2 to 72 hr p.i., unlabelled RNA from infected BS-C-1 cells (extracted at 72 hr p.i.), and unlabelled RNA from SV40-transformed 3T3 cells.

Each reaction mixture contained, in 5 ml. of 2 × SSC: one filter with 0·044 μg [^{3}H]SV40 DNA, one blank filter, 5×10^{6} cts/min (30 μg) late [^{32}P]RNA (labelled from 2 to 72 hr p.i.), and increasing amounts of unlabelled RNA from BS-C-1 cells (72 hr p.i.) (curve A) or unlabelled RNA from SV40-transformed 3T3 cells (curve B). The amount of [^{32}P]RNA bound to the blank filters has been subtracted from the values plotted for the amounts bound to the SV40 DNA filter. The amount of [^{32}P]RNA bound to SV40 DNA in the absence of unlabelled RNA was 270 cts/min (this is a saturation level).

(c) *Base composition of virus-specific [^{32}P]RNA's*

To confirm the results obtained in the competition experiments, which indicated qualitative differences between the early and late RNA from lytically infected BS-C-1 cells, and between the late RNA and the virus-specific RNA from transformed 3T3 cells, the base ratios of the early, late and transformed cell virus-specific RNA's were determined. The virus-specific RNA's were obtained from hybridization experiments with SV40 DNA at an exhaustion level. The amount of [^{32}P]RNA bound to blank filters was no higher than 15% of the amount bound to SV40 DNA filters. The [^{32}P]RNA was eluted from the filters, hydrolysed to the mononucleotides with alkali, and the four [^{32}P]mononucleotides isolated by electrophoresis as described in Materials and Methods.

The results in Table 4 show that the three types of virus-specific RNA differ in their base composition. The virus-specific early RNA from infected BS-C-1 cells and that from transformed 3T3 cells are richer in purines and poorer in uridine in comparison to the virus-specific late RNA from infected BS-C-1 cells. Furthermore, the virus-specific RNA from transformed cells appears to be richer in adenine and poorer in guanine in comparison to the early RNA from infected BS-C-1 cells.

4. Discussion

The analysis of the virus-specific RNA which appears during the lytic interaction between BS-C-1 cells and SV40 has indicated the presence of at least two distinct classes of viral messenger RNA: an "early" RNA that is synthesized before the replication of viral DNA, and a "late" RNA that is present after the onset of viral DNA replication. The difference between these two classes of virus-specific RNA was recognized both by competition experiments and by base ratio analysis. It was calculated from the data of the saturation experiment with late RNA that at least 76% of the SV40 genome is represented in this RNA, whereas the competition experiments provide a clear indication that a smaller fraction (approximately 37%) of the SV40 genome is represented in the early RNA. The hybridization experiments to determine, by "exhaustion", the relative amounts of early and late RNA, show that the amount of late RNA is approximately 40 times greater. This result is consistent with the view that progeny viral DNA acts as a template for RNA synthesis, and it will be of interest to examine whether the dependence of late RNA synthesis on virus DNA synthesis requires a specific gene product (in addition to that controlling virus DNA synthesis), as is suggested by experiments on the control of late functions in the development of bacteriophage lambda (Dove, 1966).

The results with virus-free transformed cells have shown that the sensitivity for detection of virus-specific nucleotide sequences in RNA from transformed cells can be considerably enhanced by eluting the labelled RNA bound to the virus DNA and rehybridizing this eluted RNA with virus DNA. Using this double hybridization procedure, the amount of [^{32}P]RNA, from SV40 transformed cells, bound to SV40 DNA was 30 to 100 times greater than the amount bound to polyoma DNA, *E. coli* DNA, or blank filters.

The competition experiments between virus-specific RNA from transformed cells and the late RNA from virus-yielding cells have indicated that only 31 to 37% of the SV40 genome is represented in the transformed cell RNA. Furthermore, the base ratio analysis has shown that the early and transformed-cell virus-specific RNA's are richer in purines and poorer in uridine than the late RNA, and that the transformed cell RNA appears to be richer in adenine and poorer in guanine than the early RNA. Competition experiments to examine the relationship between the transformed-cell virus-specific RNA and the early virus-specific RNA have not, so far, been performed, since no saturation of even very small amounts of SV40 DNA was obtained with the amounts of early and transformed-cell RNA that were available.

TABLE 4

Base composition of SV40-specific [^{32}P]RNA from transformed 3T3 cells and virus yielding BS-C-1 cells

Type of virus-specific [^{32}P]RNA	Method of preparing the [^{32}P]mono-nucleotides	No. of experi-ments	Moles/100 moles nucleotides: cytidylic acid	adenylic acid	guanylic acid	uridylic acid	guanylic + cytidylic acids	Purines/ pyrimidines
1 Late RNA from infected BS-C-1 cells	A	3	19·2	28·8	23·5	27·3	42·7	1·09
	B	2	18·8	30·1	24·3	26·8	43·1	1·19
2 Early RNA from infected BS-C-1 cells	A	2	17·2	34·8	27·8	20·3	45·0	1·67
	B	3	18·9	32·2	30·6	18·3	49·5	1·69
3 From transformed 3T3 cells	A	2	16·5	41·8	21·8	19·8	38·3	1·75
	B	2	14·4	42·4	21·4	21·8	35·8	1·76

[^{32}P]RNA extracted from SV40-infected BS-C-1 cells labelled from 2 to 72 hr p.i. (= late) and from 2 to 22 hr p.i. (= early) and [^{32}P]RNA extracted from SV40-transformed 3T3 cells were incubated with filters containing 20 μg SV40 DNA (which was sufficient to exhaust all the RNA available for hybridization). After hybridization and RNase treatment, the bound [^{32}P]RNA was eluted from the filter, and the base composition determined as described under Materials and Methods. The data are the average values obtained from the indicated number of experiments (the maximum deviation was 3%).

In method A, the [^{32}P]RNA eluted from the filters was first precipitated with 5% perchloric acid and then hydrolysed with alkali; in method B, the eluted [^{32}P]RNA was hydrolysed to the [^{32}P]mononucleotides without prior precipitation.

It has been reported that in some lines of cells transformed by SV40, there can be an induction of virus synthesis (Sabin & Koch, 1963; Gerber, 1966; Black, 1966), which is apparently mediated by cell fusion (Gerber, 1966; Koprowski, Jensen & Steplewski, 1967; Watkins & Dulbecco, 1967). It will thus be of interest to determine if there is transcription of the same part of the virus genome in both inducible and non-inducible lines of SV40 transformed cells, and whether the non-inducible lines contain only part of the virus genome. Whether the virus DNA is incorporated into specific chromosomes, or whether it functions as an extra chromosomal entity, remains to be determined.

The skilful technical assistance of Mr Eliyahu Nava is gratefully acknowledged. This work was supported by grant DRG 860-B from the Damon Runyon Memorial Fund for Cancer Research, and by a grant from The Jane Coffin Childs Memorial Fund for Medical Research.

REFERENCES

Benjamin, T. L. (1966). *J. Mol. Biol.* **16**, 359.
Black, P. H. (1966). *J. Nat. Cancer Inst.* **37**, 487.
Black, P. H., Crawford, E. M. & Crawford, L. V. (1964). *Virology*, **24**, 381.
Denhardt, D. T. (1966). *Biochem. Biophys. Res. Comm.* **23**, 641.
Dove, W. F. (1966). *J. Mol. Biol.* **19**, 187.
Dulbecco, R. (1966). *Perspectives Biol. Med.* **9**, 298.
Dulbecco, R., Hartwell, L. H. & Vogt, M. (1965). *Proc. Nat. Acad. Sci., Wash.* **53**, 403.
Fujinaga, K. & Green, M. (1966). *Proc. Nat. Acad. Sci., Wash.* **55**, 1567.
Fujinaga, K. & Green, M. (1967). *Proc. Nat. Acad. Sci., Wash.* **57**, 806.
Gerber, P. (1966). *Virology*, **28**, 501.
Gershon, D., Sachs, L. & Winocour, E. (1966). *Proc. Nat. Acad. Sci., Wash.* **56**, 918.
Gillespie, D. & Spiegelman, S. (1965). *J. Mol. Biol.* **12**, 829.
Kaye, A. M. & Winocour, E. (1967). *J. Mol. Biol.* **24**, 475.
Koprowski, H., Jensen, F. C. & Steplewski, Z. (1967). *Proc. Nat. Acad. Sci., Wash.* **58**, 127.
Mandell, J. D. & Hershey, A. D. (1960). *Analyt. Biochem.* **1**, 78.
Reich, P. R., Black, P. H. & Weissmann, S. M. (1966). *Proc. Nat. Acad. Sci., Wash.* **56**, 78.
Richards, O. C. (1967). *Proc. Nat. Acad. Sci., Wash.* **57**, 156.
Sabin, A. B. & Koch, M. A. (1963). *Proc. Nat. Acad. Sci., Wash.* **50**, 407.
Sachs, L. (1965). *Nature*, **207**, 1272.
Sachs, L. (1967). In *Current Topics in Developmental Biology*, vol. 2, p. 129, ed. by A. Monroy & A. A. Moscona. New York: Academic Press.
Sebring, E. D. & Salzman, N. O. (1964). *Analyt. Biochem.* **8**, 126.
Todaro, G. J., Green, H. & Goldberg, B. D. (1964). *Proc. Nat. Acad. Sci., Wash.* **51**, 66.
Warnaar, S. O. & Cohen, J. A. (1966). *Biochem. Biophys. Res. Comm.* **24**, 554.
Watkins, J. F. & Dulbecco, R. (1967). *Proc. Nat. Acid. Sci., Wash.* **58**, 1396
Weil, R. (1961). *Virology*, **14**, 46.
Winocour, E. (1967*a*). *Virology*, **31**, 15.
Winocour, E. (1967*b*). In *The Molecular Biology of Viruses*, ed. by J. S. Colter and W. Paranchych, p. 577. New York: Academic Press.

REGULATION OF TRANSCRIPTION OF THE SV40 DNA IN PRODUCTIVELY INFECTED AND IN TRANSFORMED CELLS

KINICHIRO ODA AND RENATO DULBECCO

The expression of genes of the oncogenic virus SV40 is regulated. In productive infection, some genes are expressed before viral DNA replication begins ("early" genes); others are expressed after it has begun ("late" genes).[1] In transformed cells, which contain the complete viral genome,[2] some genes are expressed; others are not. Whether either type of regulation is at the level of transcription of viral genes is the question raised in the work to be reported here. The basic technique was hybridization of RNA synthesized in infected or transformed cells to viral DNA, and hybridization competition experiments. Results partly similar to those reported here were recently published by Aloni *et al.*[3]

Materials and Methods.—Cells, media, and viruses: The following cells were used: African green monkey kidney cells (AGMK) and human embryonic kidney (HEK) purchased from Flow Laboratories; the permanent BSC-1 line of AGMK cells; and the SV3T3 line (SV40-transformed 3T3) obtained from Dr. Howard Green. Medium was reinforced Eagle's medium with 10% fetal bovine serum for AGMK cells, 10% fetal bovine serum and 10% tryptose phosphate for BSC-1 cells, and 10% calf serum for SV3T3 and HEK cells. SV40 virus was propagated in BSC-1 cells. $E46^+$ virus (no. 1237), obtained from Dr. W. P. Rowe, was propagated in AGMK cells. SV40 was titrated by plaque formation in BSC-1 cells; $E46^+$ on HEK cells.

Labeling of DNA was carried out with H^3-thymidine (1.9 c/mmole); that of RNA with H^3-uridine (24.7 c/mmole, together with 2×10^{-5} *M* of the unlabeled thymidine).

Cell fractionation into nuclei, membranes, and cytoplasmic fraction was carried out essentially according to Attardi and Attardi.[4]

Preparation of viral and cellular DNA: SV40 DNA was extracted from virus with water-saturated phenol, pH 8.0, containing 0.2 *M* Na trichloroacetate and 0.01 *M* ethylenediaminetetraacetate (EDTA), and then with chloroform-isoamyl alcohol (24:1). It was sedimented through a self-forming gradient of CsCl (initial $\rho = 1.5$) in 0.01 *M* tris-(hydroxymethyl)aminomethane-HCl (Tris-HCl), pH 8.0, and 0.002 *M* EDTA. Component I was collected, dialyzed, and concentrated by evaporation at room temperature in a dialysis tubing to *ca.* 100 μg/ml, and stored at 4°C.

$E46^+$ virus was incubated for 5 hr at 37°C with 500 μg papain, 0.002 *M* EDTA, and 0.3 *M* 2-mercaptoethanol; 1% sodium dodecyl sulfate (SDS) was then added at 37°; after 30 min, DNA was extracted as above.

Cellular DNA was prepared from nuclei according to Marmur,[5] omitting RNase treatment.

Preparation of RNA: Cells or nuclei were suspended in 0.05 *M* Na acetate, pH 5.1 (*ca.* 2×10^7 cells/ml), and made 1% with SDS. The mixture was heated at 60° for 10 min and, after the addition of $^1/_5$ vol 5 *M* Na perchlorate, was extracted two or three times with 2 vol of chloroform-isoamyl alcohol. Nucleic acids were precipitated with 2 vol ethanol, resuspended in 0.05 *M* Tris, pH 8.0, and 0.01 *M* $MgCl_2$ at the initial concentration. After 20 min at 37° with 20 μg/ml of electrophoretically purified pancreatic DNase I (Worthington), RNA was re-extracted two to three times with chloroform-isoamyl alcohol, precipitated with ethanol, and stored at −20°.

RNA was extracted from cytoplasm or polysome fraction according to Oda and Joklik.[6]

DNA-RNA hybridization was carried out essentially according to Gillespie and Spiegelman.[7] SV40 DNA was converted to Component II by controlled DNase treatment, and brought to pH 12.5 with 1 N NaOH at 0°. After 10 min the pH was adjusted to 7.0 with 1 N HCl. The DNA solution was heated at 100°C for 5 min, rapidly cooled, and diluted with 6 × standard saline citrate (SSC) to 0.1–1 μg/ml; it was slowly filtered through presoaked Millipore membrane filters. When a membrane had less than 1 μg DNA, 5 μg of denatured *E. coli* DNA were added. Hybridization was carried out in 1 ml of 6 × SSC, 0.1% SDS at 66°C for 18–24 hr. To some mixtures with more than 10^6 cpm input RNA, 1 mg of yeast RNA was added in order to decrease background, and the edge of the membrane filter where DNA was not immobilized was cut off after it was washed with 6 × SSC. The addition of yeast RNA and SDS did not alter the efficiency of hybridization. Counts fixed by labeled RNA with the same amount of *E. coli* DNA as SV40 were 0.003–0.001% of input RNA. This background was subtracted from the experimental counts. In competition experiments, labeled RNA was always below saturation.

Sucrose density gradient centrifugation: Polyribosomes were separated by centrifuging the cytoplasmic fraction into 15–30% w/w sucrose gradient in RSB (RSB = 10^{-2} M tris pH 7.4, 10^{-2} M KCl, and 1.5×10^{-3} M $MgCl_2$). RNA's were separated in 15–30% w/w sucrose gradient in 0.5% SDS, 0.005 M Tris, pH 7.4, and 0.1 M NaCl. The gradients were collected through a Gilford automatic absorbance recorder.

Results.—Time course of synthesis and localization of virus-specific RNA in productively infected BSC-1 cells: Figure 1 shows that most of the virus-specific RNA was synthesized after the onset of DNA synthesis, which in these cells is known to be entirely viral.[8] It appeared last in the membranes. Although the membranes had the lowest total counts (Fig. 1*a*), they had the highest proportion of counts (Fig. 1*b*), probably owing to the slow labeling of membrane-associated ribosomes.[9]

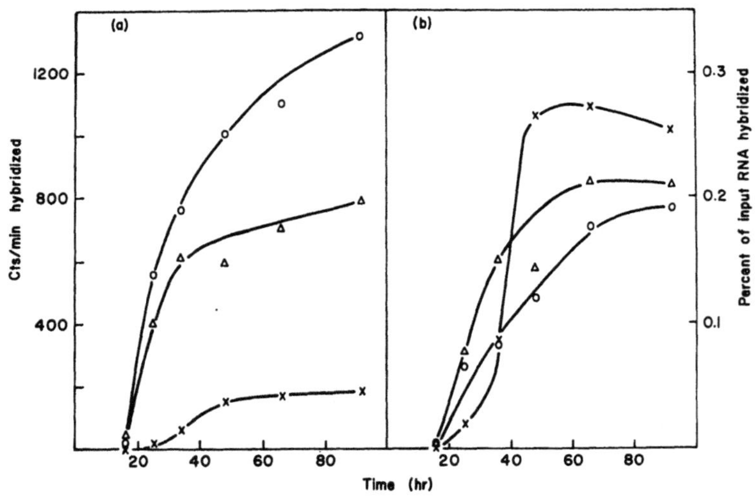

FIG. 1.—Time course of synthesis of SV40-specific RNA in infected BSC-1 cells. Confluent monolayers of BSC-1 cells were infected with SV40 at a multiplicity of 100. Other parallel cultures were mock-infected (controls). At the times indicated, H^3-uridine (10 μc/ml) was added to ten 100-mm cultures. After 3 hr, the cells were washed and fractionated into nuclei, membrane, and cytoplasm, and the RNA was extracted from each fraction. The RNA of the nuclear and cytoplasmic fractions were each dissolved in 2.2 ml of SSC/10; that of the membrane fraction in 1.3 ml SSC/10. Aliquots of 0.4 ml were hybridized with 1 μg SV40 DNA. In parallel cultures labeled with H^3-thymidine, viral DNA synthesis was found to begin between 16 and 20 hr.

O, Nuclear fraction; Δ, cytoplasmic fraction; ×, membrane fraction.

Virus-specific RNA in the membrane fraction is not an accidental contamination from either nuclei or free polysomes. In fact, cytoplasmic fractions from cells labeled with H^3-thymidine before infection, followed by excess cold thymidine at infection, contained 1–3 per cent of all counts; and centrifugation of a labeled membrane fraction in a sucrose gradient yielded no counts at the level of free polysomes.

To see whether the virus-specific RNA's of membranes and of polysomes are different, they were compared in hybridization competition experiments. Labeled membrane RNA (Fig. 2*a*) and labeled free polysome RNA (Fig. 2*b*) were

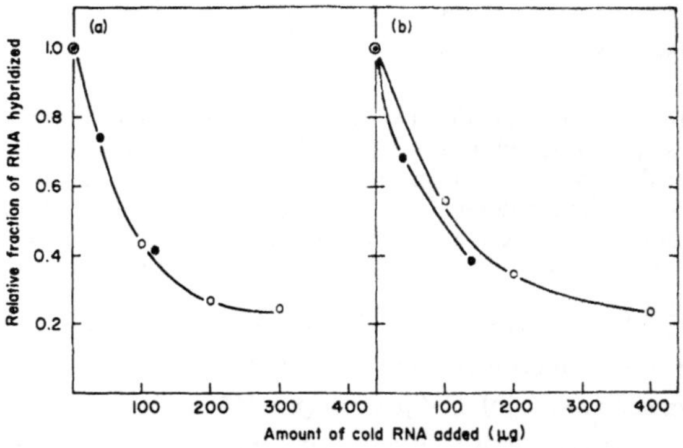

FIG. 2.—Hybridization competition between free-polysome and membrane-associated SV40-specific RNA. BSC-1 cultures were infected, labeled, and fractionated as indicated in Fig. 1. Label was added 70 hr after infection. The polysome fraction was obtained after centrifuging the cytoplasmic fraction in a 15–30% sucrose density gradient. Labeled RNA's were derived from forty 100-mm cultures; cold RNA's from 90 cultures. RNA's used for each hybridization mixture were in (*a*) 36 μg of labeled membrane RNA (2.8×10^3 cpm/μg); in (*b*) 85 μg of labeled free-polysome RNA (1.1×10^3 cpm/μg). Competing cold RNA was added in increasing amounts to the hybridization mixtures. O, Cold free polysome RNA; ●, cold membrane RNA. Cpm hybridized without cold RNA were 602 in (*a*) and 1445 in (*b*).

each competed by either cold membrane or free polysome RNA. Although in no case was there enough cold RNA to reach a plateau of hybridized counts, the identity of the points obtained suggests that membrane-associated and free polysome RNA are identical.

Characteristics of virus-specific RNA synthesized in productive infection: The distribution of sedimentation rates of virus-specific RNA is given in Figure 3. Early RNA, synthesized within 20 hours of infection, was much less than late RNA, synthesized after 90 hours; both were heterogeneous. The heterogeneity was not caused by breakdown during fractionation, since there was no breakdown of 28*S* or 16*S* ribosomal RNA in the same gradient. The heterogeneity may therefore reflect the size heterogeneity of the messengers.

Competition experiments between early and late RNA, given in Figures 4 and 5, show that cold late RNA competes about 95 per cent against labeled late RNA (Fig. 4*b*) and about 90 per cent against labeled early RNA (Fig. 4*a*). Early unlabeled RNA, however, competes only about 40 per cent against late labeled RNA (Fig. 5*c*). In the competition of cold late toward labeled early RNA, a definite plateau was not reached, and this is attributed to the low concentration of both labeled and competing RNA. It seems likely that the curve would reach essentially complete competition if larger amounts of competing RNA were employed. We conclude that, although late RNA contains all sequences present in

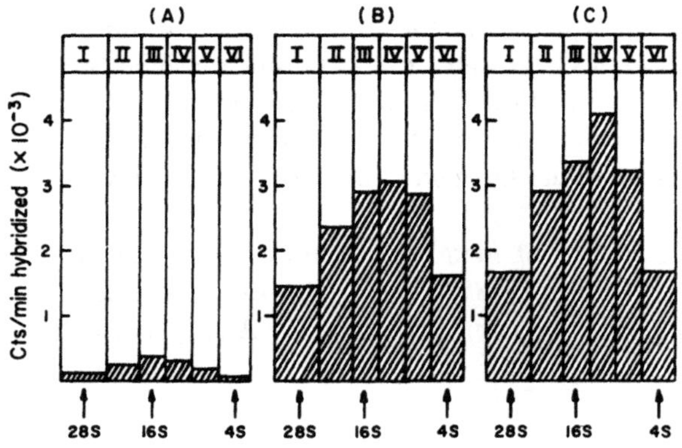

FIG. 3.—Size distribution of SV40-specific RNA in infected BSC-1 cells. BSC-1 cultures were infected as in Fig. 1. Three batches, 20 cultures (100 mm) each, were labeled with H^3-uridine as follows: (*A*) 30 μc/ml, between 10 and 22 hr after infection; (*B*) 25 μc/ml, between 50 and 56 hr; and (*C*) 25 μc/ml, between 93 and 99 hr. The RNA extracted from each batch was sedimented in a SDS-sucrose gradient. Six fraction pools were collected as follows: I = 7 fractions; II = 5 fractions; III, IV, and V = 4 fractions; VI = 5 fractions. Each fraction represents a constant cut of the gradient. The RNA in each fraction was precipitated with ethanol after adding 0.5 μg of yeast RNA as carrier.

After dissolving each pool in 1.4 ml of SSC/10, 0.3 ml of each pool for the early RNA (*A*), or 0.2 ml for the middle (*B*) and late (*C*) RNA were hybridized to 2 μg of SV40 DNA. The cpm shown are those hybridized by each pool divided by the number of fractions in the pool. The arrows indicate the positions of the peak of the two rRNA's and of tRNA.

early RNA, 90 per cent is the maximal degree of competition observable in practice under the conditions employed.

Characteristics of SV40-specific RNA in cells infected by E46⁺ virus: E46⁺ virus is a mixture of hybrid virus, which contains recombinant adeno-SV40 DNA enclosed in adeno-7 capsids,[10] and adenovirus-7. The proportion of SV40 DNA incorporated in the hybrid molecule is about 0.40.[11] The SV40 component of the hybrid DNA expresses early SV40 functions, both in productively infected and in transformed cells.[12] Cold E46⁺ RNA competes about 90 per cent with the hybridization of labeled early RNA (Fig. 5*a*). This is probably maximal competition, as discussed above. Thus, the early SV40 genes are all contained in the SV40 DNA expressed in E46⁺ infection. Since, as shown below, early SV40 RNA and E46⁺ RNA with SV40 specificity behave identically toward SV3T3 RNA, the two virus-specific RNA's may be identical.

There is somewhat more viral RNA in E46⁺-infected AGMK cells than in early

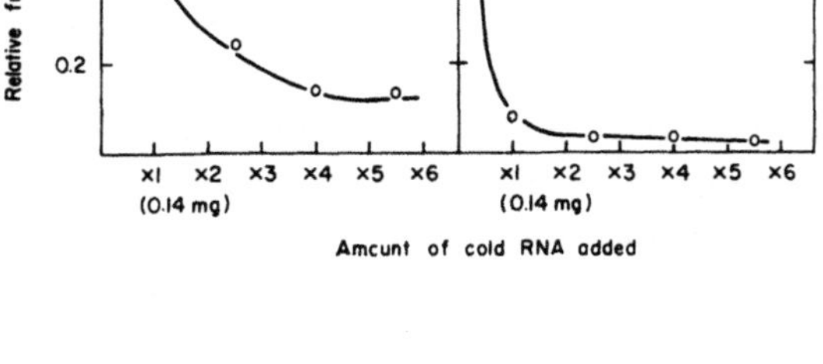

FIG. 4.—Hybridization competition of labeled early and late RNA with cold late RNA. BSC-1 cultures were infected as in Fig. 1, and labeled with H^3-uridine as follows: (*a*) 30 μc/ml, between 6 and 18 hr; (b) 20 μc/ml, between 93 and 99 hr. Hybridization competition was carried out in two steps: increasing amounts of cold late RNA were first hybridized to 0.1 μg of SV40 DNA; then the labeled RNA was added ((*a*) 93 μg, 6.4×10^4 cpm/μg; (*b*) 8 μg, 9×10^3 cpm/μg). Incubation under hybridization conditions was continued for an additional 8 hr. Cpm hybridized without competing RNA were 259 in (*a*) and 1366 in (*b*).

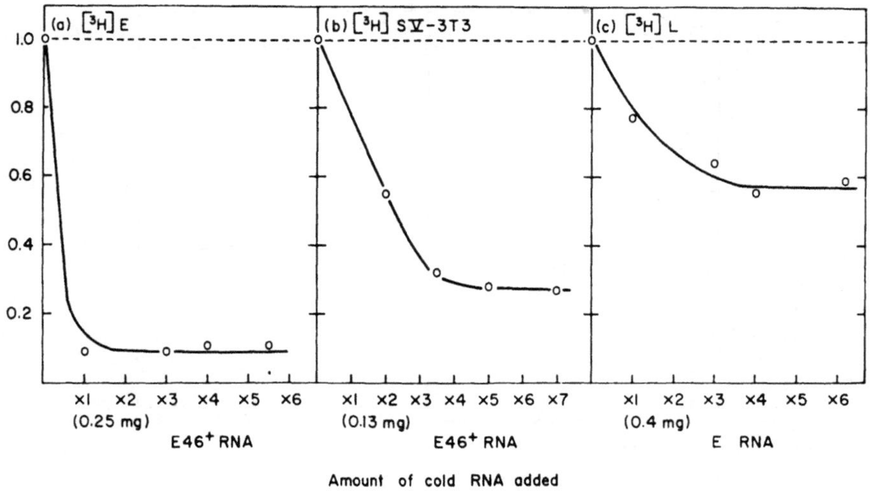

FIG. 5.—Hybridization competition of labeled early, late, and SV3T3 RNA with cold E46$^+$ or early RNA. Cold E46$^+$ RNA was prepared from African green monkey kidney cells primary cultures 60–72 hr after infection with E46$^+$ virus at a multiplicity of 4. Cold early RNA was prepared from BSC-1 cells, 18 hr after infection, as in Fig. 1. Labeled RNA's were employed as follows: (*a*) early RNA, 93 μg per hybridization mixture (6.4×10^4 cpm/μg); (*b*) SV3T3 RNA, 153 μg (4.6×10^4 cpm/μg); (*c*) late RNA (9×10^3 cpm/μg). Experimental procedure as in Fig. 4, with 0.1 μg SV40 DNA per filter. Cpm hybridized without competing RNA's: (*a*) 307, (*b*) 267, and (*c*) 314.

SV40 infection. Therefore, E46$^+$ is a convenient source of early SV40 RNA. In E46$^+$ infection, contrary to SV40 infection, there is a slight increase in the rate of synthesis of SV40 RNA after the viral DNA replicates. The characteristics of SV40 RNA in E46$^+$ infection will be described more fully elsewhere.

Characteristics of virus-specific RNA in transformed cells: Hybridization competition experiments of this RNA with the various RNA's defined above are reported in Figures 5 and 6. Unlabeled E46$^+$ competes less efficiently against labeled SV3T3 cell RNA (Fig. 5*b*) than against early RNA (Fig. 5*a*), thus showing that viral DNA sequences not expressed in early lytic infection are expressed in the transformed cells. Cold SV3T3 RNA competes about 90 per cent against labeled SV3T3 RNA (maximal competition) (Fig. 6*d*), but only about 60 per cent against labeled early SV40 or E46$^+$ RNA (Fig. 6*a* and *b*), and about 40 per cent against labeled late RNA (Fig. 6*c*). Thus both early SV40 and E46$^+$ RNA contain sequences of approximately the same length, not present in the virus-specific RNA of SV3T3 cells. Late RNA contains an even longer length of sequences not present in the RNA of SV3T3 cells.

Discussion.—The competition hybridization experiments gave the following results: (*a*) Early viral RNA transcribes about 40 per cent of the late genes; in contrast, late RNA transcribes early genes. (*b*) Early RNA is essentially identical to the SV40 RNA synthesized in AGMK cells infected by E46$^+$ virus. (*c*) The SV40 sequences transcribed in SV3T3 cells are about one-third those of late RNA. (*d*) Although early and SV3T3 RNA's transcribe approximately the same length of viral DNA, they are not identical. They overlap in part, but each

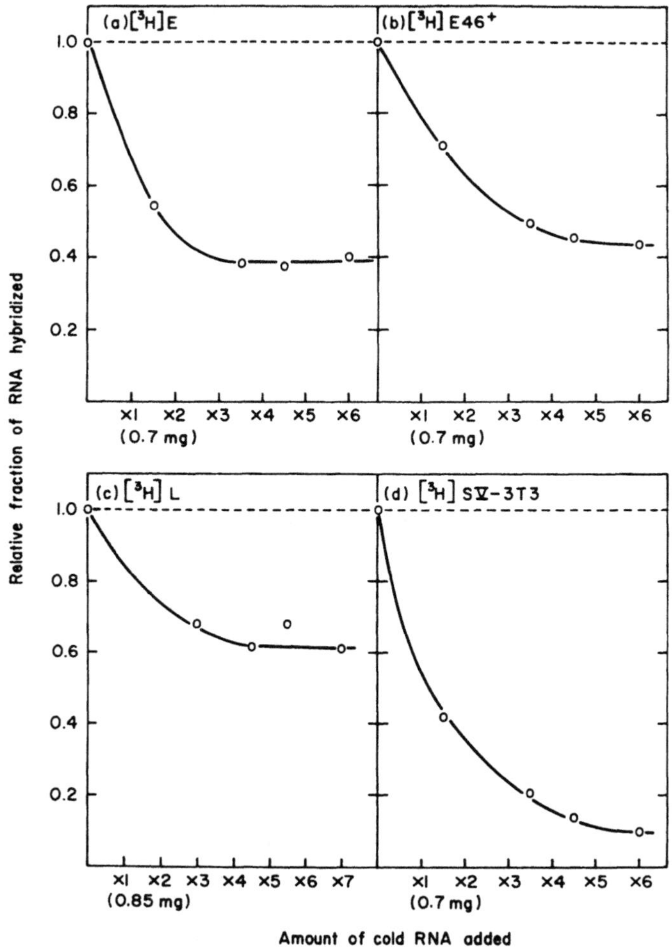

FIG. 6.—Hybridization competition of labeled early, E46$^+$, late, and SV3T3 RNA with cold SV3T3 RNA. The various RNA's were prepared as in Figs. 4 and 5. Labeled RNA's were employed as follows: (*a*) early RNA, 93 μg per hybridization mixture (6.4 $\times 10^4$ cpm/μg); (*b*) E46$^+$ RNA, 10 μg (2.8 $\times 10^4$ cpm/μg); (*c*) late RNA, 12 μg (9 $\times$ 10^3 cpm/μg); (*d*) SV3T3 RNA, 218 μg (5 $\times$ 10 cpm/μg). Experimental procedure as in Fig. 4 with 0.1 μg SV40 DNA per filter. Cpm hybridized without competing RNA (*a*) 389, (*b*) 631, (*c*) 2128, and (*d*) 390.

mRNA contains sequences absent in the other RNA. Results (*a*) and (*c*) are similar to those reported by Aloni *et al.*[3]

For a quantitative analysis, the data must be corrected for maximal degree of competition obtainable in practice, which is 95 per cent for labeled late RNA and 90 per cent for labeled early or SV3T3 RNA's. Another correction is for unspecific competition (e.g., with 3T3 or BSC-1 RNA); this amounts to about 15 per cent. The corrected data are given in Table 1 and represented diagrammatically in Figure 7.

The main conclusion from these results is that lack of expression of some viral

TABLE 1. *Results of hybridization competition experiments.*

Labeled RNA	Cold RNA	Maximal inhibition of hybridization observed	Corrected competition
Late	SV3T3	0.38	0.287
$E46^+$	SV3T3	0.60	0.600
Early	SV3T3	0.60	0.600
SV3T3	$E46^+$	0.70	0.735
Late	Early	0.42	0.337

Corrected competition was obtained by subtracting 0.15 from column 3 (unspecific competition) and normalizing the figures to either 0.80 (lines 1 and 5) or 0.75 (lines 2–4), as discussed in the text. The length of the overlap region between early RNA and SV3T3 RNA is calculated as 0.202 of the total genome from lines 3 and 5 (0.600×0.337), and as 0.210 from lines 4 and 1 (0.735×0.287).

genes, either in the early period of productive infection or in transformed cells, is caused by failure of transcription. The sequences transcribed in the two cases are not identical, pointing to two mechanisms by which transcription is blocked. The early block of the transcription of late genes may be a property of the virus, since it is tied up to the replication of the viral DNA; in contrast, the block of transcription in transformed cells is likely to be a property of the host cells. It may be the main factor making the 3T3 cells nonpermissive for SV40 multiplication. Since infectious virus is released after the SV3T3 cells are fused to BSC-1 cells,[2] the fusion must activate transcription.

The differences in the viral DNA sequences transcribed in early productive infection and in transformed cells are intriguing. On the basis of previous results, one difference could be an early gene that participates in the specification of the virus-specific thymidine kinase and is not expressed in SV3T3 cells.[13] The inference that a late gene is transcribed in the transformed cells is of special interest, but there is no hint as to the nature of this gene.

The study of the viral RNA bound to membranes has not supported the hypothesis that it may represent a special class devoted to the synthesis of a membrane protein.[4] Since viral RNA appears late in membranes (see Fig. 1) and competes well with free-polysome RNA, it may be a sample of free-polysome RNA secondarily associated with membrane-bound ribosomes. Some viral RNA could also be demonstrated in the membrane fraction of transformed SV3T3 cells. The function of this RNA is still an open question.

The identity of the early SV40 RNA with the RNA synthesized in cells productively infected by the $E46^+$ virus could arise because, by accident, the hybrid virus contains exactly the early SV40 genes. Or late genes are present but are

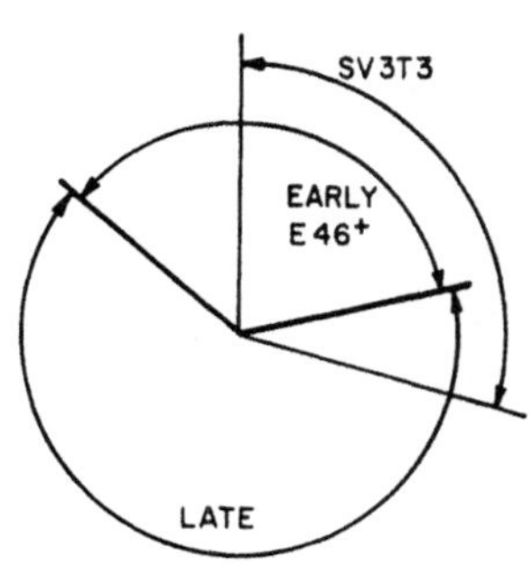

FIG. 7.—Diagram representing the proportions of SV40 DNA transcribed under the different conditions examined in the text.

not transcribed in these cells, owing to a control mechanism similar to that limiting the early transcription in productive SV40 infection; in E46$^+$ infection, however, transcription would remain limited, even after the hybrid viral DNA has replicated.

This work was supported by a grant from the National Institutes of Health (grant no. CA 07592).

[1] Rapp, F., J. L. Melnick, and T. Kitahara, *Science*, **147**, 625 (1965).
[2] Koprowski, H., F. C. Jensen, and Z. Steplewski, these PROCEEDINGS, **58**, 127 (1967); Watkins, J. F., and R. Dulbecco, these PROCEEDINGS, **58**, 1396 (1967).
[3] Aloni, Y., E. Winocour, and L. Sachs, *J. Mol. Biol.*, **31**, 415 (1968).
[4] Attardi, B., and G. Attardi, these PROCEEDINGS, **58**, 1051 (1967).
[5] Marmur, J., *J. Mol. Biol.*, **3**, 208 (1961).
[6] Oda, K., and W. K. Joklik, *J. Mol. Biol.*, **27**, 395 (1967).
[7] Gillespie, D., and S. Spiegelman, *J. Mol. Biol.*, **12**, 829 (1965).
[8] Gershon, D., L. Sachs, and E. Winocour, these PROCEEDINGS, **56**, 918 (1966).
[9] Blobel, G., and V. R. Potter, *J. Mol. Biol.*, **26**, 293 (1967).
[10] Baum, S. G., P. R. Reich, C. J. Hybner, W. P. Rowe, and S. M. Weissman, these PROCEEDINGS, **56**, 1509 (1966).
[11] Oda, K., and R. Dulbecco, in preparation.
[12] Rowe, W. P., and S. G. Baum, these PROCEEDINGS, **52**, 1340 (1964); Rapp, F., J. L. Melnick, J. S. Butel, and T. Kitahara, these PROCEEDINGS, **52**, 1348 (1964); Black, P. H., and G. Todaro, these PROCEEDINGS, **54**, 374 (1965).
[13] Hatanaka, M., and R. Dulbecco, these PROCEEDINGS, **58**, 1888 (1967).

THE TRANSCRIPTION OF THE SV40 GENOME IN PRODUCTIVELY INFECTED AND TRANSFORMED CELLS

G. Sauer and J. R. Kidwai

A particular feature of cells which have been transformed by DNA-containing tumor viruses is the restricted expression of the viral genome. These cells do not display any of the viral functions which occur late during a productive cycle of infection—such as extensive replication of the viral DNA, production of viral coat protein, and maturation of infectious viral progeny. They do contain, however, tumor antigen which is synthesized early after infection. Hence, it appears that in transformed cells only "early" but no "late" products are being synthesized.

With regard to the mechanism of viral oncogenesis, it is of particular importance to determine whether the block, which prevents the expression of "late" viral functions in transformed cells, occurs at the level of transcription or at the level of translation.

It has been established by various experimental approaches[1–4] that the viral DNA, in some form, persists in transformed cells. These results, however, do not indicate the extent to which the viral genome is actually transcribed.

SV40 specific messenger RNA (mRNA) sequences synthesized in SV40-transformed cells were first reported by Aloni *et al.*[5] According to their results, about one third of the viral genome is transcribed in SV40-transformed 3T3 cells. In a recent publication, Oda and Dulbecco[6] reported results similar to those of Aloni *et al.*[5] and to those presented in this paper.

It was not known, however, whether the same sequences of the viral genome were being copied in SV40-transformed cells of different origins. Using the DNA-RNA hybridization technique, we have examined the characteristics of SV40 mRNA in virus-free 3T3 and African green monkey kidney (GMK) cells transformed by SV40. It will be shown that in transformed 3T3 cells only a section of the viral genome is transcribed (40% homology with late mRNA). On the other hand, in SV40-transformed GMK cells a far larger portion of the SV40 DNA is being copied (80% homology with late mRNA). This difference may be related to particular properties of the host cell system, the 3T3 cells being non-permissive for the productive cycle of infection, whereas GMK cells permit the growth of SV40. In the latter cells transformation can be achieved only under conditions of low input multiplicities.[7, 8]

Thus, the common denominator for various SV40-transformed cell species is the regulation of viral functions at the level of transcription, even though the sections of the SV40 genome transcribed may be variable.

Materials and Methods.—Virus and cells: The SV40 strain Rh 911,[9] which had been plaque-purified twice, was used in these experiments. The SV40 pool titered 5×10^7 plaque-forming units (PFU) per ml in primary GMK cells.

CV-1 cells, a permanent line of monkey kidney origin,[10] were productively infected with 20 PFU per cell. Only confluent cultures were used. The adsorption period was 2 hr.

The following SV40-transformed cells were investigated: (1) 3T3 cells transformed by SV40;[11] (2) SV40-transformed GMK cells, designated GMK–EVa,[3] which yield infectious SV40 upon fusion with indicator cells; and (3) a clone derived from these cells called GMK EVa-Cl 2A-1, which upon fusion proved to be non-virus-yielding.[3] The above-mentioned cells could not be successfully superinfected with SV40, and all were free of infectious SV40. This was demonstrated by freezing and thawing 2×10^6 cells and assaying the cell lysates for plaque production on GMK cells. All plaque assays proved to be negative.

The cultures were maintained in Earle's balanced salt solution, with twice the concentration of amino acids and vitamins, and in Eagle's basal medium supplemented with 10% calf serum.

Preparation of SV40 DNA: The virus was purified in saturated KBr according to the method described by Yoshiike and Axelrod.[12] For extraction of the DNA the suspension of purified virus was made 1 *M* with respect to NaCl and shaken at room temperature (15 min) with an equal volume of 1 part of 2 *M* sodium trichloroacetate containing 0.01 *M* ethylenediaminetetraacetate (EDTA), pH 7.1, and 9 parts of 80% phenol. The aqueous phase was extracted once more with 80% phenol. After removal of the remaining phenol from the aqueous phase with ether (the ether was removed by bubbling N_2 through the solution), the DNA was precipitated with ethanol (2.5 × the volume). The precipitate was suspended in 0.1 × SSC (SSC = 0.15 *M* NaCl, 0.015 *M* sodium citrate), denatured (5 min at 100°C, then cooled rapidly in ice water), and layered on a methylalbumin-kieselguhr column.[13] The viral DNA was then eluted by increasing concentrations of NaCl. The fractions containing the viral DNA were pooled and the DNA was precipitated with ethanol.

Preparation of ^3H-labeled RNA: The cells were labeled for different periods of time (see *Results*) with [^{3}H]-uridine (spec. act., 27 c/m*M;* The Radiochemical Centre, Amersham). At the end of the labeling period, the cells were washed extensively with buffered salt solution, harvested by scraping, and suspended in 0.13 *M* NaCl, 0.05 tris (hydroxymethyl)aminomethane (Tris-HCl), pH 7.8. The procedure of extraction was similar to that described by Benjamin.[2] The cell suspension was adjusted to 1% sodium dodecyl sulfate (SDS) and 4×10^{-3} *M* $MgCl_2$ and shaken at 60°C for 5 min with an equal volume of 80% phenol which contained 0.1% 8-hydroxyquinoline. The mixture was then chilled in ice water and shaken again for 3 min at 60°C. After being cooled, the mixture was centrifuged and the aqueous phase was extracted twice with phenol at room temperature. Traces of phenol were removed by three successive extractions with ether, and the ether was removed by bubbling N_2 through the solution. The final aqueous phase was made 1 *M* for NaCl, and the RNA was precipitated by adding 2.5 times the volume of ethanol. The RNA was then dissolved in 0.1 × SSC containing 0.005 *M* $MgCl_2$ and treated with 20 μg/ml of DNase (RNase-free) for 45 min at 37°C. The mixture was then extracted again with phenol at room temperature. The aqueous phase was made 1 *M* for NaCl and mixed with 2.5 times the volume of ethanol. The precipitated RNA was resuspended in 0.1 × SSC and passed through a Sephadex G–100 column. The leading fractions which contained the bulk of the material (as determined by spectrophotometry) were pooled, precipitated, and redissolved in 2 × SSC. This material, after determination of the specific activity, was used in the hybridization experiments.

DNA-RNA hybridization: The DNA-RNA hybridization was carried out according to Benjamin's modification[2] of the technique described by Gillespie and Spiegelman.[14] The SV40 DNA in 0.1 × SSC was denatured (10 min at 100°C, followed by rapid cooling in an ice bath), adjusted to 0.5 *M* KCl, 0.01 Tris-HCl, pH 7.25, and passed slowly under suction on nitrocellulose membrane filters which had been presoaked in the same buffer. The filters containing the DNA were dried at room temperature and then heated at 80°C for 4 hr. The hybridization assay was carried out in Tricarb scintillation vials. The labeled RNA (in 2 × SSC) was added to the membrane filters and incubated for 24 hr at 60°C, the total volume of the reaction mixture being 2 ml. After this period of time the hybrid formation had reached a plateau. The filters were washed extensively with 2 × SSC, incubated with 20 μg/ml RNase (which had been heated for 15 min at 85°C) in 1 × SSC for 1 hr at 37°C. The filters were then rinsed in 0.5 *M* KCl, washed from each side with the same buffer, and dried. The radioactivity was then determined in the Tricarb liquid scintillation spectrometer.

Results.—SV40 mRNA in productively infected cells: The experiments described here were concerned with the functional status of the SV40 genome during the productive cycle of infection in CV-1 cells. The growth cycle of SV40 can be divided into two major periods with regard to the onset of viral DNA replication. Prior to the onset of viral DNA replication, T-antigen is formed, whereas late functions (viral coat protein synthesis and virus maturation) can be detected not earlier than 24–27 hours postinfection, i.e., after the start of viral DNA synthesis. Thus, SV40-specific RNA that is synthesized prior to 24 hours after infection will be referred to as "early." The SV40 mRNA that is transcribed later will be referred to as "late."

To demonstrate the specificity of the hybridization reaction, SV40 DNA was incubated with increasing amounts of RNA extracted 48 hours postinfection from productively infected CV-1 cells or with RNA from noninfected CV-1 cells (Table 1). It can be seen from the results that no homology exists between the viral DNA and the cellular RNA.

TABLE 1. *Hybridization of SV40 DNA with SV40-specific RNA and with RNA from uninfected CV-1 cells.*

[^{3}H]RNA from	Cpm assayed ($\times 10^4$)	Cpm bound to filter
SV40-infected CV-1 cells	2.25	212
	4.5	396
	9.0	627
Uninfected CV-1 cells	1.75	24
	3.5	23
	9.0	26

CV-1 cells were infected with SV40 at a multiplicity of 20 PFU/ml. The cells were labeled with 20 μc/ml [^{3}H]-uridine for 30 min at 47 hr postinfection. Uninfected CV-1 cells were labeled similarly and the RNA extracted from infected and uninfected cells was tested for its ability to form a hybrid complex with 0.6 μg of SV40 DNA immobilized on filter pads. The specific activity of late and CV-1 RNA was 1850 cpm/μg RNA and 1500 cpm/μg RNA, respectively. The values indicated in the table are corrected for the nonspecific background bound to the blank filters.

The relative amounts of SV40 mRNA present in RNA preparations obtained early and late after infection were determined by exhausting the extraneous material from the RNA which is hybridizable with immobilized SV40 DNA. The results presented in Table 2 indicate that the amount of SV40-specific RNA which is produced 48 hours postinfection is about 25 times greater than the amount synthesized early after infection.

The increasing rate of SV40 mRNA synthesis after the onset of viral DNA replication is also reflected in the slopes of the saturation curves presented in Figure 1. The level of saturation is attained more rapidly with late than with early RNA preparations. Assuming that all RNA species synthesized early and late postinfection have the same specific activities, it appears that about four

TABLE 2. *Relative amounts of SV40-specific RNA in productively infected CV-1 cells at different times after infection.*

Type of [^{3}H]RNA	Cpm assayed	Cpm Bound to Filter: 1st hybridization	Cpm Bound to Filter: 2nd hybridization	Cpm Bound to Filter: 3rd hybridization	mRNA (% of input)
Early	16.74×10^5	950	54	20	0.06
Late	1.76×10^5	2470	36	21	1.42

CV-1 cells were infected with SV40 at a multiplicity of 20 PFU per cell. The cells were labeled with 20 μc/ml [^{3}H]-uridine for 30 min at 17 and 47 hr postinfection, respectively. The RNA was extracted as described under *Materials and Methods* and hybridized for 24 hr at 60°C with 0.5 μg of SV40 DNA immobilized on a filter. In order to exhaust the RNA from material hybridizable with SV40 DNA, the filter was replaced by a new filter also containing 0.5 μg of immobilized SV40 DNA. This procedure was repeated once more. The values indicated in the table are corrected for the nonspecific background bound to the blank filters.

times more late SV40 mRNA than early SV40 is bound to the SV40 DNA at saturation level.

To detect whether early SV40 mRNA shares homologous sequences with late SV40 mRNA, a competition experiment was completed (Fig. 2). Two controls were included in this experiment. Since large amounts of unlabeled "cold" RNA extracted early after infection are required for the competition, it might be argued that high concentrations of RNA interfere nonspecifically with the hybridization reaction. Therefore, heterologous "cold" HeLa RNA was added in increasing amounts to the hybridization reaction between late [^{3}H]SV40 RNA and SV40 DNA (the late SV40 RNA being added in all cases at saturation level). It can be seen, however, that even large amounts of HeLa RNA do not interfere with the hybridization reaction between late [^{3}H]SV40 RNA and SV40 DNA. It is also shown that unlabeled late SV40 RNA successfully competes with late [^{3}H]SV40 RNA, because the binding of late [^{3}H]SV40 mRNA to SV40 DNA is completely inhibited by the addition of increasing amounts of late cold SV40 mRNA.

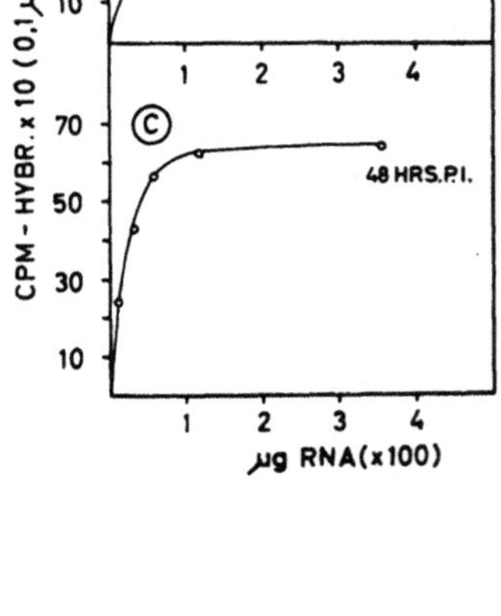

FIG. 1.—Saturation of SV40 DNA with early and late [^{3}H]SV40 mRNA from productively infected CV-1 cells. CV-1 cells were infected with SV40 at a multiplicity of 20 PFU/ml. At 15, 26 and 48 hr postinfection, respectively, the cells were labeled for 30 min with 20 μc/ml of [^{3}H]-uridine. The RNA was extracted and hybridized, in increasing amounts, with 0.1 μg of SV40 DNA immobilized on filters. The specific activities were: 15 hr postinfection, 8500 cpm μg; 26 hr postinfection, 7300 cpm/μg; and 48 hr postinfection, 7800 cpm/μg.

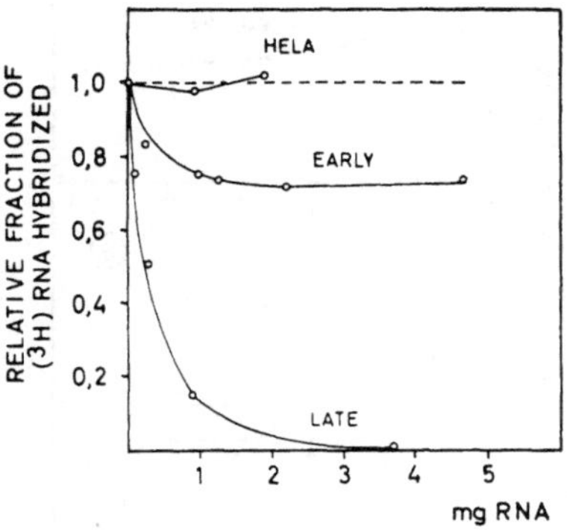

FIG. 2.—Hybridization competition between ³H-labeled late, unlabeled late, and unlabeled early SV40 mRNA. Late [³H]RNA was extracted 51 hr postinfection from CV-1 cells infected with 20 PFU/ml of SV40. The cells had been labeled with 20 μc/ml [³H]-uridine from 48 to 51 hr. The specific activity of this preparation was 5440 cpm/μg RNA.

Filters containing 0.05 μg of immobilized SV40 DNA were incubated with 240 μg of late RNA (this is a saturating amount), and with increasing amounts of (1) unlabeled HeLa RNA, (2) unlabeled early RNA which had been extracted 17 hr postinfection from SV40-infected CV-1 cells, and (3) unlabeled late RNA which had been extracted 51 hr postinfection. The broken line indicates the level of hybridization of late [³H]RNA without any cold RNA.

The competition between cold early and late [³H]SV40 mRNA revealed that 28 per cent of the sequences present in late SV40 mRNA are transcribed prior to the onset of viral DNA replication. Also, it is evident from the results of the competition experiment that sequences of the viral genome, other than early, are being copied late. However, early SV40 mRNA is present also in preparations of late SV40 RNA. Thus, the data clearly demonstrate the existence of at least two different species of SV40 mRNA.

SV40 mRNA in transformed cells: To examine the homologies shared between RNA extracted from transformed cells and late SV40 mRNA, competition experiments were performed. First, saturation plateaus were established for late [³H]SV40 mRNA obtained from productively infected cells. Late [³H]SV40 mRNA was then added to each hybridization experiment at saturation level, and unlabeled RNA from transformed cells was added in increasing amounts to the hybridization reaction mixture.

As shown in Figure 3(*A*), the RNA from SV40-transformed 3T3 cells inhibits the hybridization of late [³H]RNA by 40 per cent. Thus, the sequences of the viral genome transcribed in transformed 3T3 cells are smaller than those transcribed late during a productive cycle of infection. 3T3 cells are nonpermissive for the growth of SV40, even after infection with high multiplicities.[15] Therefore, the transcription of the viral genome must be curtailed by deficiencies of the host cell.

If host cell factors were playing a prominent role in the regulation of the transcription of the viral DNA, then SV40 mRNA sequences other than those detected in 3T3 cells might be present in transformed GMK cells which are originally permissive. The RNA extracted from SV40-transformed GMK cells (GMK-EVa) which did not contain any infectious virus was subjected to a hybridization competition with late [³H]SV40 mRNA. As can be seen in Figure 3(*B*), the hybridization of late SV40 mRNA is inhibited by 80 per cent. This experiment shows that considerably more nucleotide sequences are represented in SV40-transformed GMK cells than in transformed 3T3 cells.

Previous reports[3, 16] demonstrated that, upon fusion with GMK cells, both GMK-EVa and SV40-transformed 3T3 cells revealed infectious SV40 although a

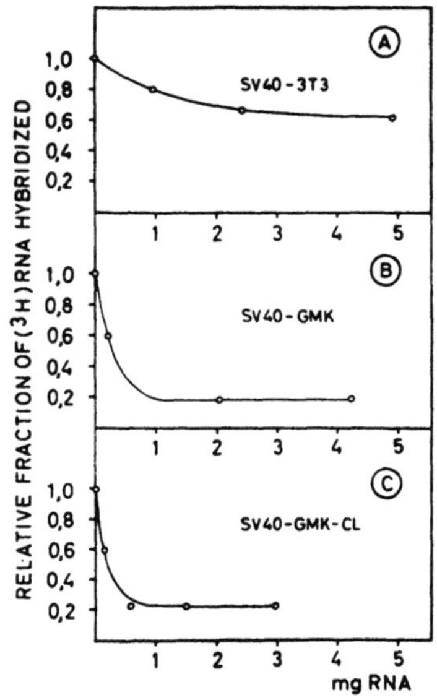

FIG. 3.—Hybridization competition between late [³H]SV40 mRNA and unlabeled RNA from 3T3 and African green monkey kidney cells transformed by SV40. Filters containing 0.05 μg of immobilized SV40 DNA were incubated with 240 μg of late [³H]RNA (for details see legend to Fig. 2) and with increasing amounts of (*A*) unlabeled RNA from SV40-transformed 3T3 cells, (*B*) unlabeled RNA from SV40-transformed GMK-EVa cells, and (*C*) unlabeled RNA from SV40-transformed EVaCl 2A-1 cells.

cell clone derived from GMK-EVa proved to be non-virus-yielding in fusion experiments.[3] In the latter case, mutations of the SV40 genome might account for the failure to initiate virus production after fusion. Such mutations or deletions could conceivably be reflected in the base sequences transcribed in the non-yielding cell line. That this is not the case is revealed by a comparison of the curves shown in Figure 3(*B*) (GMK-EVa) and Figure 3(*C*) (clone 2A-1 derived from GMK-EVa). In either case the hybridization of late SV40 mRNA is inhibited by 80 per cent.

So far, homologies between SV40 mRNA from transformed cells and late SV40 mRNA have been studied, but the results of these studies do not indicate the extent to which early sequences are transcribed in transformed cells.

Since only small quantities of SV40 mRNA are produced early postinfection (Table 2), it was difficult to obtain the amounts of early [³H]SV40 mRNA required for saturation of the SV40 DNA. In preliminary hybridization competition experiments (unpublished), the early [³H]RNA had to be used below saturation level (about 50% of the amount required for saturation). In these experiments the binding of early [³H]RNA to the SV40 DNA appears to be almost completely inhibited by unlabeled RNA extracted from the SV40-transformed GMK-EVa Cl 2A-1 cells. Therefore, most, if not all, early sequences seem to be present in these transformed cells.

Discussion.—The aim of these experiments was to determine whether the late functions of the SV40 genome are blocked in transformed cells at the level of transcription and, if so, whether the same sections of the viral DNA are copied in various cell species transformed by SV40.

Generally, in all SV40-transformed cell lines fewer SV40 mRNA sequences are represented than during a productive cycle of infection. This is true for both SV40-transformed 3T3 and GMK cells. Thus, the regulation of the viral genome functions apparently takes place at the level of transcription.

The data presented here have shown that the sections of the viral genome copied in SV40-transformed GMK cells are twice as large as in 3T3 cells. This observation may shed some light on the mechanisms which control the transcription of the SV40 genome. With regard to the 3T3 cells, it is clear that particular properties of the host cells must be responsible for the limited transcription of the SV40 DNA, since virus multiplication does not occur even after infection with high multiplicities.

The problems posed by GMK cells are much more difficult to explain. In this case the permissiveness is multiplicity-dependent. Large numbers of virus particles per cell cause lysis, whereas transformation of at least fractions of the cell culture can be achieved by very low multiplicities of infection.[7, 8]

The implications of these facts may be discussed in the light of two different hypotheses: (*a*) For reasons presently unknown, certain sections of the SV40 genome cannot be transcribed in GMK cells when only one or very few complete genomes are present in one cell. Such sequences would be transcribed, however, upon infection with high multiplicities. (*b*) The presence of mutant SV40 genomes in SV40 pools which are able to code for early but not for all late viral functions has been demonstrated previously.[17] Assuming that all base sequences of the SV40 DNA are transcribed in GMK cells (regardless of the number of viral genomes initially present per cell), it seems likely that only those cells which have been infected with mutant viral genomes would be transformed. If this hypothesis were correct, no infectious SV40 should be induced by fusion experiments. This applies for the cell clone derived from GMK-EVa in which no SV40 is inducible by fusion experiments.[3] Conversely, the parental cultures (GMK-EVa) which had not been subjected to any cloning procedure yielded infectious virus upon fusion.[3] This, however, may be the result of complementation processes between different mutant SV40 genomes which, although present in different cells, participated after fusion in the formation of one syncytium. Further experiments will be required to decide the correct hypothesis.

With regard to the mechanism of viral oncogenesis, the restricted transcription of the SV40 genome appears to be the rule, since the complete transcription would kill the host cell. Furthermore, the sections of the SV40 DNA copied in transformed cells are larger than those transcribed early during a productive cycle, although no late functions become apparent. The biological significance of this fact remains to be elucidated. It can be concluded from these observations that there are several different SV40 mRNA species. Further studies would be greatly facilitated by the characterization of their biophysical properties.

Summary.—SV40 mRNA has been studied by means of the DNA-RNA hybridization technique in productively infected monkey cells and in several transformed (virus-free) cell species of monkey and mouse origin.

Prior to the onset of viral DNA replication, during a productive cycle, 28 per cent of the viral genome is transcribed (early mRNA). This is followed by the synthesis of late SV40 mRNA which also includes the early mRNA species. In competition experiments, RNA from transformed 3T3 cells inhibits the hybridization of late [^{3}H] SV40 mRNA by 40 per cent. Thus, more sequences of the SV40 genome are transcribed than early during a productive cycle of infection. The SV40 mRNA in transformed monkey kidney cells has also been examined for homologies with late [^{3}H]SV40 mRNA, with results indicating that about 80 per cent of the SV40 DNA sequences present in late SV40 mRNA are transcribed in transformed monkey kidney cells. Similar results have been obtained in the case of a clone derived from SV40-transformed GMK cells, which proved to be non-virus-yielding upon fusion.

Therefore, depending on the cell species, different portions of the viral genome are copied in the transformed cells. The common feature shared by all transformed cell species is that fewer sections of the genome are transcribed than those which are operative late during a productive cycle of infection.

The authors are gratefully indebted to Dr. H. Koprowski and Mr. F. Jensen, of The Wistar Institute, Philadelphia, who kindly supplied the transformed cells.

[1] Axelrod, D., E. T. Bolton, and K. Habel, *Science*, **146**, 1466 (1964).
[2] Benjamin, T. L., *J. Mol. Biol.*, **16**, 359 (1966).
[3] Koprowski, H., F. C. Jensen, and Z. Steplewski, these PROCEEDINGS, **58**, 127 (1967).
[4] Westphal, H., and R. Dulbecco, these PROCEEDINGS, **59**, 1158 (1968).
[5] Aloni, Y., E. Winocour, and L. Sachs, *J. Mol. Biol.*, **31**, 415 (1968).
[6] Oda, K., and R. Dulbecco, these PROCEEDINGS, **60**, 525 (1968).
[7] Fernandes, M. V., and P. S. Moorhead, *Texas Rep. Biol. Med.*, **23**, 242 (1965).
[8] Jensen, F. C., personal communication.
[9] Girardi, A. J., these PROCEEDINGS, **54**, 445 (1965).
[10] Jensen, F. C., A. J. Girardi, R. V. Gilden, and H. Koprowski, these PROCEEDINGS, **52**, 53 (1964).
[11] Todaro, G. J., H. Green, and B. D. Goldberg, these PROCEEDINGS, **51**, 66 (1964).
[12] Yoshiike, K., and D. Axelrod, personal communication.
[13] Mandell, J. D., and A. D. Hershey, *Anal. Biochem.*, **1**, 78 (1960).
[14] Gillespie, D., and S. Spiegelman, *J. Mol. Biol.*, **12**, 829 (1965).
[15] Todaro, G. J., and H. Green, these PROCEEDINGS, **55**, 302 (1966).
[16] Watkins, J. F., and R. Dulbecco, these PROCEEDINGS, **58**, 1396 (1967).
[17] Sauer, G , H. Koprowski, and V. Defendi, these PROCEEDINGS, **58**, 599 (1967).

Strand Orientation of Simian Virus 40 Transcription in Productively Infected Cells

Dona M. Lindstrom and Renato Dulbecco

Strand orientation of transcription in BSC-1 cells infected by simian virus 40 (SV40) was investigated by annealing RNAs extracted from infected cells with asymmetric complementary RNA from SV40 DNA synthesized *in vitro* by RNA polymerase from *Escherichia coli.* The results suggest that the early viral RNA sequences (those made before replication of viral DNA) and the late RNA sequences (those synthesized concurrently with replication of viral DNA) are transcribed from opposite strands of the SV40 DNA. The RNA synthesized *in vitro* is of the same polarity as the early *in vivo* RNA and is complementary to the late *in vivo* RNA. Knowledge of strand selection in lytic infection is a step towards understanding the regulation of transcription of this viral DNA.

The small DNA virus, simian virus 40 (SV40), grows lytically in permissive African green monkey cells and usually transforms nonpermissive cells, such as mouse or hamster, in which a complete replicative cycle is absent. This difference of interaction with host cells often correlates with a difference in transcription of the viral genome. In the lytic cycle about one-third to one-half of the genome is transcribed early in infection, before replication of viral DNA begins, and comprises the class of early RNA(s) (1–5). After DNA replication has begun, the remainder of the genome (1, 6) is transcribed such that late in infection both early and late RNA sequences are synthesized. In certain transformed cells, transcription is limited to about 40% of the genome (1, 2). Variable amounts of the genome are transcribed in other transformed lines (6). Most of the sequences transcribed in transformed cells are also transcribed early in lytic infection (2, 4).

Abbreviations: cRNA, complementary RNA; SSC, standard saline citrate, 0.15 M sodium chloride–0.015 M sodium citrate; SV40, simian virus 40.

To understand how the various modes of transcription might be regulated, we have investigated which strand of viral DNA serves as template for RNA synthesis *in vivo*. This was made possible by the observation that, *in vitro*, *Escherichia coli* RNA polymerase preferentially transcribes only one DNA strand, often in its entirety (7). We compared viral RNA made early and late in infection with that made *in vitro*. The results show that early sequences are of the same polarity as the *in vitro* RNA, whereas the late sequences are complementary to the *in vitro* RNA. We conclude that early and late viral RNA species are transcribed from different strands of the SV40 DNA.

MATERIALS AND METHODS

Cells and Virus. BSC-1 cells from E. Winocour were grown in Dulbecco's modification of Eagle's medium (8) containing 10% fetal bovine serum (Microbiological Associates), 500 units/ml of penicillin (Charles Pfizer & Co.), and 100 μg/ml of streptomycin (Charles Pfizer & Co.). Small plaque SV40 was propagated at low multiplicity (<0.01 plaque forming units/cell) in BSC-1 cells as described by Hatanaka and Dulbecco (9).

Preparation of SV40 DNA. Confluent cultures of BSC-1 cells in 100-mm plastic petri dishes were infected with 0.4 ml of a stock preparation (2.3×10^8 plaque forming units/ml; input multiplicity of infection = 0.5 plaque forming units/cell). After 8 days, SV40 DNA was selectively extracted by the procedure of Hirt (10). The supernatant was extracted three times with redistilled phenol [saturated with 1 M Tris (pH 7.4)], once with $CHCl_3$–isoamyl alcohol (24:1 v/v), and precipitated with two volumes of 95% ethanol at −20°. The precipitate was dissolved in 0.1 M Tris (pH 7.4)–1 mM EDTA, and the DNA was centrifuged in an ethidium bromide (400 μg/ml)–CsCl (density = 1.580 g/cm^3) gradient at 43,000 rpm for 48 hr in a Spinco type 50 rotor. The lower band containing the closed twisted circular form I DNA was extracted three times with an equal volume of isopropanol: H_2O (9:1 v/v), and extensively dialyzed against 0.01 M Tris (pH 7.4)–1 mM EDTA. Form I DNA was then sedimented in a 5–20% (w/v) neutral sucrose gradient containing standard saline citrate–0.1 M Tris (pH 7.4)–1 mM EDTA for

17.5 hr at 25,000 rpm in a Spinco SW27 rotor at 4°. The 21S fraction was precipitated overnight with two volumes of ethanol at −20°. DNA concentration was determined by the procedure of Burton (11) or by absorbence at 260 nm.

Preparation of In Vivo RNA. Late RNA was prepared by infection of confluent monolayers of BSC-1 cells with 0.4 ml of a stock preparation (2×10^5 plaque forming units/ml; input multiplicity of infection = 20 plaque forming units/cell). 48 hr after infection, the cell layers were washed and RNA was extracted by one of two alternate procedures. (*a*) Cells were lysed with 0.6% sodium dodecyl sulfate, 0.01 M Tris (pH 7.4), 1 mM EDTA, and NaCl added to 0.5 M. The lysate was allowed to stand at 4° overnight and was then centrifuged at 12,000 × *g* for 1 hr. The supernatant was extracted three times with redistilled phenol [saturated with 1 M Tris (pH 7.4)], once with $CHCl_3$–isoamyl alcohol, and precipitated with ethanol. The precipitate was dissolved in 0.01 M NaAc (pH 5.3)-0.01 M $MgCl_2$, and electrophoretically purified DNase (Worthington Biochemical), which had been previously treated with iodoacetic acid (12), was added at a concentration of 20 μg/ml for 1 hr at 37°. DNase was removed by phenol and $CHCl_3$ extraction and the RNA was precipitated with ethanol. (*b*) Cells were lysed in 1% sodium dodecyl sulfate–0.01 M NaAc (pH 5.3) and extracted three times with redistilled phenol [saturated with 1.0 M NaAc (pH 5.3)], once with $CHCl_3$–isoamyl alcohol and precipitated with ethanol. The precipitate was treated with DNase as above. After removal of DNase, the RNA was selectively precipitated in 2 M NaCl for 18 hr at 4°, and again treated with DNase. The RNA was precipitated with ethanol and stored at −20° after removal of the DNase. RNA concentration was determined by orcinol reaction (13) or by absorbence at 260 nm. In final RNA preparations, less than 0.5 μg DNA/100 μg RNA could be detected by the Burton assay for DNA.

RNA from uninfected BSC-1 cells was extracted by procedure (*a*).

Early RNA was prepared from infected BSC-1 cells as above, except 12.5 μg/ml of cytosine arabinoside hydrochloride (Sigma) was added to the media at the time of infection. 25 hr after infection, RNA was isolated as described in procedure (*a*).

Preparation of In Vitro Complementary RNA (cRNA). RNA complementary to SV40 DNA was synthesized in a volume of 200 μl containing 5.2 μg of SV40 form I DNA, 15 μg *E. coli* RNA polymerase [purified according to Burgess (14) on glycerol gradients], 0.04 M Tris (pH 7.9), 0.15 M KCl, 0.01 M $MgCl_2$, 0.1 mM dithiothreitol, 0.06 mM [8-^{3}H]ATP (20.5 Ci/mmol; Schwartz BioResearch), 0.055 mM [5-^{3}H]CTP (22 Ci/mmol; New England Nuclear), 0.067 mM [5,6-^{3}H]UTP (36.5 Ci/mmol; New England Nuclear), 0.125 mM GTP (P. L. Biochemicals), 1 mg/ml phosphoenolpyruvate (Sigma), and 9 μg/ml pyruvate kinase (Calbiochem). After incubation for 90 min at 37°, the reaction mixture was applied to a 2.0-ml Sephadex G-50 column equilibrated with SSC–0.1% sodium dodecyl sulfate–0.01 M Tris (pH 7.4)–1 mM EDTA. The cRNA was pooled, extracted twice with phenol and once with $CHCl_3$–isoamyl alcohol, and precipitated with ethanol with 50 μg of yeast RNA. The yield was 9.0 μg of cRNA having a specific radioactivity of 1.2×10^8 dpm/μg, as calculated from the specific radioactivity of the incorporated nucleotides.

RESULTS AND DISCUSSION

Analysis of complementary RNA (cRNA)

Using SV40 form I DNA as template, *E. coli* RNA polymerase synthesizes cRNA that is highly asymmetric and of heterogenous size. When this cRNA is self-annealed, between 80 and 95% remains RNase sensitive depending on the particular cRNA preparation. In agreement with previous data (7), this suggests that between 90 and 97.5% of the cRNA is transcribed from one DNA strand. The size heterogeneity of the cRNA is shown by centrifugation in a dimethylsulfoxide gradient (15) in which cRNA forms a broad band with a median sedimentation constant about 21 S. Fractions consisting of different sizes of cRNA were pooled as indicated in Fig. 1. These fractions retained their original S values when resedimented. As determined by self-annealing (Fig. 2), the different size fractions were of approximately equal asymmetry.

28S cRNA is the length of a complete single strand of SV40 DNA (4). If the RNA polymerase transcribes sequentially, 28S cRNA should contain all the sequences in the template DNA strand in nearly equal quantities. For this reason, the 28S size fraction of cRNA was used in the experiments described below.

Annealing of late *in vivo* RNA with asymmetric cRNA

Unlabeled late RNA was extracted from BSC-1 cells infected with SV40 48 hr after infection. When asymmetric 28S cRNA

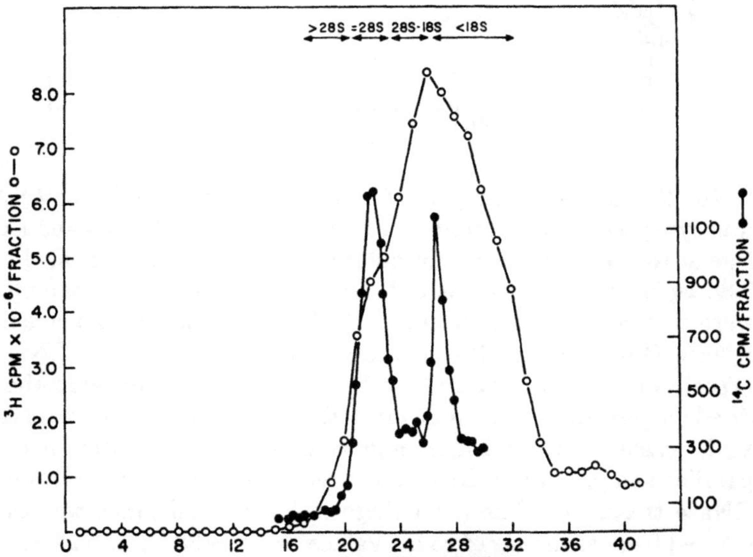

FIG. 1. Size fractionation of cRNA. 330 μl containing 4.5 μg cRNA in 45 μl of H_2O–180 μl of dimethylsulfoxide (Me_2SO) (Matheson, Coleman & Bell, spectroquality)–105 μl of dimethylformamide (Mallinckrodt Chemical) was layered on a 11.0 ml 0–10% sucrose gradient in 99% Me_2SO, containing 1% (v/v) 0.1 M EDTA in H_2O. The cRNA (○——○) was centrifuged for 19 hr at 34,000 rpm. in a Spinco SW41 rotor at 29°. Fractions were collected, and 1-μl aliquots were dried on 24-mm Whatman GF/C glass fiber discs and counted in a toluene-based scintillation fluid. Fractions greater than 28 S, equal to 28 S, 28 S–18 S, and less than 18 S, were pooled as indicated in the figure, adjusted to 0.1 M NaCl and precipitated with ethanol with 50 μg/ml of carrier yeast RNA. Fractions were treated once with DNase as described in *Methods*, precipitated with ethanol, and stored at −20°. Marker ribosomal RNA (●——●), prepared from BSC-1 cells labeled for 72 hr with media containing 2 μCi/ml [^{14}C]uridine was extracted by procedure (*a*) of *Methods*, except DNase treatment was omitted. Marker RNA was sedimented in a parallel gradient as above. Fractions were collected on glass fiber discs and counted.

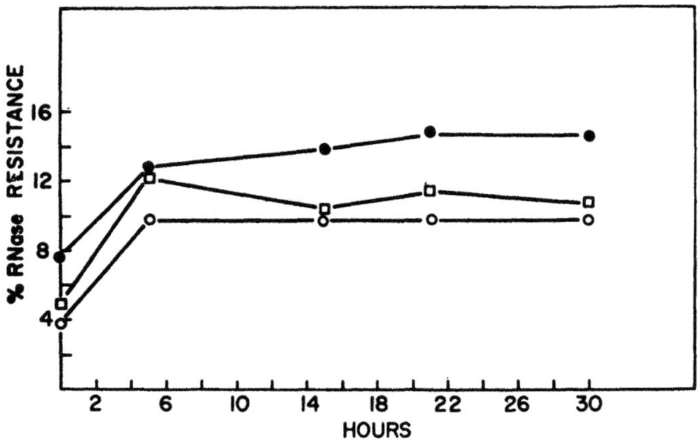

FIG. 2. Self-annealing of different size classes of cRNA. cRNA was synthesized and fractionated as described in *Methods* and the legend to Fig. 1. Fractions of different sizes were self-annealed at 0.88 μg/ml in 2 × SSC at 65°. At the indicated times, samples were removed, incubated in 2 × SSC with 20 μg/ml pancreatic RNase (Worthington Biochemical) and 15 units/ml T1 RNase (Worthington Biochemical) for 1 hr at 37°. Samples were then dried on glass fiber discs, precipitated with cold 5% trichloroacetic acid, washed with ethanol, and counted. At each time-point, parallel samples were dried and precipitated as above without RNase treatment. The percentage of RNase resistance was calculated from each pair of count values. ●, 18 S–28 S; □, >28 S; ○, = 28 S.

was annealed with this unlabeled late RNA, 55–60% of the input cRNA became resistant to RNase digestion, whereas the RNase resistance of the cRNA was not increased by RNA from uninfected BSC-1 cells (Fig. 3). That the RNase-resistant material consists of double-stranded RNA is shown by two results. First, previous treatment of the late RNA with 0.3 N KOH for 18 hr at 25° abolishes its ability to protect [^{3}H]cRNA from RNase digestion; second, the RNase-resistant RNA, formed by the annealing of late RNA to [^{3}H]cRNA, bands at the density predicted for double-stranded RNA in Cs_2SO_4 equilibrium gradients. The incomplete protection of cRNA against RNase digestion by late RNA is not due to the unavailability of some cRNA sequences. In fact, the addition of denatured SV40 DNA to the annealing mixture after the late RNA has annealed to 55–60% of the [^{3}H]cRNA, protected more than 90% of the [^{3}H]cRNA.

These results suggest that sequences in late RNA are transcribed from the DNA strand opposite to that serving as template for cRNA. Since the true late sequences (those which are synthesized after viral DNA replication begins) comprise 60% of the genome (1–5), the amount of protection of [^{3}H]cRNA by late RNA is consistent with annealing of only the true late sequences with cRNA.

Other size classes of [^{3}H]cRNA were annealed to late RNA and the results were the same as those obtained with 28S cRNA.

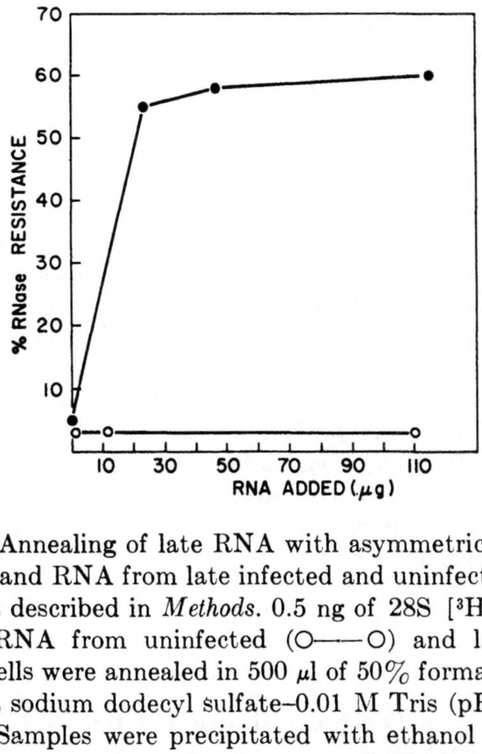

FIG. 3. Annealing of late RNA with asymmetric 28S cRNA. [^{3}H]cRNA and RNA from late infected and uninfected cells was prepared as described in *Methods*. 0.5 ng of 28S [^{3}H]cRNA and unlabeled RNA from uninfected (○——○) and late infected (●——●) cells were annealed in 500 μl of 50% formamide–0.9 M NaCl–0.1% sodium dodecyl sulfate–0.01 M Tris (pH 7.4) for 72 hr at 37°. Samples were precipitated with ethanol with carrier yeast RNA, dissolved in 200 μl of 2 × SSC, digested with 20 μg/ml pancreatic RNase and 15 units/ml T1 RNase for 1 hr at 37°. Samples were applied to glass fiber discs, dried, precipitated with cold 5% trichloroacetic acid, washed with ethanol, and counted. Aliquots of each sample were removed before RNase treatment and precipitated as above to determine total counts present before RNase treatment. RNase resistance was calculated from each pair of count values. 100% of the cRNA remained precipitable by ethanol after the 72-hr annealing.

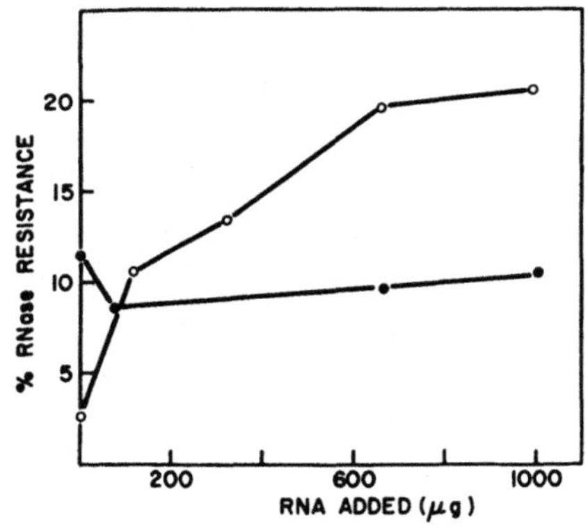

FIG. 4. Annealing of early RNA with asymmetric 28S cRNA and with symmetric 28S cRNA. (●——●) 0.5 ng of 28S asymmetric cRNA and the indicated amount of early RNA were annealed and analyzed as in the legend to Fig. 3. (○——○) 0.5 ng of symmetric 28S cRNA, which had been denatured by boiling for 8 min in water, and the indicated amount of early RNA were annealed and analyzed as above. Symmetric cRNA self-annealed to the extent of 3% in the absence of added early RNA as shown in the figure. Symmetric 28S cRNA was prepared by RNase digestion of self-annealed 28S cRNA. The cRNA used for the preparation of symmetric cRNA was synthesized and fractionated according to size as described in *Methods* and the legend to Fig. 1, except the reaction volume was 112 μl containing 4.1 μg of SV40 form I DNA, 7.9 μg of *E. coli* RNA polymerase, 0.114 mM [8-^{3}H]ATP (20.5 Ci/mmol), 0.083 mM [5-^{3}H]CTP (28.5 Ci/mmol), 0.128 mM [5,6-^{3}H] UTP (36.5 Ci/mmol), and 0.24 mM GTP. 28S RNA was pooled, treated with DNase, and self-annealed for 21 hr at 1.7 μg/ml in 6 × SSC at 65°. RNase digestion was performed as in the legend to Fig. 2, and the RNases were removed by two phenol extractions and one $CHCl_3$–isoamyl alcohol extraction.

Annealing of early RNA with asymmetric and symmetric cRNA

Early RNA was prepared from BSC-1 cells infected with SV40 that were treated from the time of infection with 12.5 μg/ml of cytosine arabinoside to inhibit replication of SV40 DNA. Under these conditions, over 99% of tritiated thymidine incorporation into DNA is inhibited and only early RNA sequences should be transcribed (4).

Annealing of unlabeled early RNA with 28S asymmetric [^{3}H]cRNA (Fig. 4) did not increase the RNase resistance of the [^{3}H]cRNA above the 10% value produced by self-annealing cRNA. However, the same early RNA protected 20% of the sequences in *symmetric* 28S cRNA, which should contain sequences from both strands of the DNA in equal proportion. Symmetric 28S was prepared by RNase digestion of self-annealed 28S cRNA and was denatured by boiling before being annealed with early RNA. The annealing with symmetric cRNA suggests that the failure of early RNA to anneal with asymmetric cRNA was not due to the absence of viral sequences in the *in vivo* RNA, but because early RNA and asymmetric cRNA are *not* complementary and are, therefore, transcribed from the same DNA strand. Protection of 20% of the symmetric 28S cRNA is equivalent to 40% protection of one strand. According to independent evidence (1–5), this is the fraction of the genome transcribed early in infection.

These results suggest that the early and late viral RNAs are transcribed *in vivo* from opposite strands of the SV40 DNA and that the cRNA, synthesized *in vitro* by *E. coli* RNA polymerase, is transcribed from the same strand as the early RNA and is complementary to the late RNA. This model of transcription is in agreement with the data of G. Khoury and M. Martin (16), who independently demonstrated transcription from both strands of the SV40 DNA in late lytic infection (when both early and true late sequences are synthesized) and from one strand during *in vitro* cRNA synthesis with the *E. coli* RNA polymerase.

Experiments in which lytic RNA is hybridized to the separated strands of the SV40 DNA are necessary to confirm the present model of strand orientation of transcription.

This work was supported by a grant from the National Institutes of Health (no. 5 RO 1 CA07592).

1. Aloni, Y., Winocour, E. & Sachs, L. (1968) "Characterization of the simian virus 40-specific RNA in virus-yielding and transformed cells," *J. Mol. Biol.* **31**, 415–429.
2. Oda, K. & Dulbecco, R. (1968) "Regulation of transcription of the SV40 DNA in productively infected and in transformed cells," *Proc. Nat. Acad. Sci. USA* **60**, 525–532.
3. Sauer, G. & Kidwai, J. R. (1968) "The transcription of the SV40 genome in productively infected and transformed cells," *Proc. Nat. Acad. Sci. USA* **61**, 1256–1263.

4. Sauer, G. (1971) "Apparent differences in transcriptional control in cells productively infected and transformed by SV40," *Nature* **231**, 135–138.
5. Oxman, M. N., Levine, A. S., Crumpacker, C. S., Levin, M. J., Henry, P. H. & Lewis, A. M., Jr. (1971) "Studies of nondefective adenovirus 2-simian virus 40 hybrid viruses. IV. Characterization of the simian virus 40 ribonucleic acid species induced by wild-type simian virus 40 and by the nondefective hybrid virus, Ad2^{+}ND$_{1}$," *J. Virol.* **8**, 215–224.
6. Martin, M. A. & Axelrod, D. (1969) "SV40 gene activity during lytic infection and in a series of SV40 transformed mouse cells," *Proc. Nat. Acad. Sci. USA* **64**, 1203–1210.
7. Westphal, H. (1970) "SV40 DNA strand selection by *Escherichia coli* RNA polymerase," *J. Mol. Biol.* **50**, 407–420.
8. Vogt, M. & Dulbecco, R. (1963) "Steps in the neoplastic transformation of hamster embryo cells by polyoma virus," *Proc. Nat. Acad. Sci. USA* **49**, 171–179.
9. Hatanaka, M. & Dulbecco, R. (1966) "Induction of DNA synthesis by SV40," *Proc. Nat. Acad. Sci. USA* **56**, 736–740.
10. Hirt, B. (1967) "Selective extraction of polyoma DNA from infected mouse cell cultures," *J. Mol. Biol.* **26**, 365–369.
11. Burton, K. (1956) "A study of the conditions and mechanism of the diphenylamine reaction for the colorimetric estimation of the deoxyribonucleic acid," *Biochem. J.* **62**, 315–323.
12. Zimmerman, S. B. & Sandeen, G. (1966) "The ribonuclease activity of crystallized pancreatic deoxyribonuclease," *Anal. Biochem.* **14**, 269–277.
13. Schneider, W. C. (1957) in *Methods in Enzymology*, eds. Colowick, S. P. & Kaplan, N. O. (Academic Press, Inc., New York), pp. 680–684.
14. Burgess, R. R. (1969) "A new method for the large scale purification of *Escherichia coli* deoxyribonucleic acid-dependent ribonucleic acid polymerase," *J. Biol. Chem.* **244**, 6160–6167.
15. Strauss, J. H., Jr., Kelly, R. B. & Sinsheimer, R. L. (1968) "Denaturation of RNA with dimethyl sulfoxide," *Biopolymers* **6**, 793–807.
16. Khoury, G. & Martin, M. (1972) "A comparison of SV40 DNA transcription *in vivo* and *in vitro*," *Nature*, in press.

A Map of Simian Virus 40 Transcription Sites Expressed in Productively Infected Cells

George Khoury, Malcolm A. Martin,
Theresa N. H. Lee, Kathleen J. Danna and Daniel Nathans

The topographical location of "early" and "late" sequences on the physical map of the simian virus 40 genome was determined by reacting unlabeled RNA from monkey cells productively infected with simian virus 40 with the separated strands of each of the 11 DNA fragments formed by digesting supercoiled virus DNA with the restriction endonuclease of *Hemophilus influenzae*. Stable species of early RNA are complementary to the minus strands of the contiguous fragments A, H, I and B, while the late RNA is transcribed predominantly from the plus strands of fragments A, C, D, E, K, F, J, G and B, which also form a continuous set on the physical map. This result is in agreement with previous findings, which indicated that transcription before DNA synthesis is confined to the minus DNA strand, while late transcription includes sequences located predominantly on the plus strand.

The direction of DNA transcription on the plus and minus template strands was ascertained by first preparing a population of linear simian virus 40 DNA molecules oriented by cleavage with the *Escherichia coli* R_1 restriction endonuclease and then digesting them with *E. coli* exonuclease III. DNA reassociation studies using the separated strands of these "half-molecules" and specific simian virus 40 DNA fragments indicated that transcription proceeds from A → H → I → B on the minus DNA strand, and in the opposite direction on the plus DNA strand.

1. Introduction

Transcription of the simian virus 40 genome in productively-infected cells occurs in two recognizable stages. Before the onset of virus DNA synthesis a portion of the genome is transcribed into stable "early" RNA (Aloni *et al.*, 1968; Oda & Dulbecco, 1968; Sauer & Kidwai, 1968; Carp *et al.*, 1969). Following the onset of virus DNA synthesis, early RNA synthesis continues and is accompanied by the appearance of a new species of RNA; together, these RNA species are complementary to the equivalent of one strand of SV40† DNA (Aloni *et al.*, 1968; Martin & Axelrod, 1969). Recently it has been found that early SV40 RNA is transcribed from 30 to 40% of

† Abbreviation used: SV40, simian virus 40.

one strand of the virus DNA (designated the minus strand), while the specifically late SV40 RNA (i.e. RNA found only after the onset of virus DNA synthesis) is transcribed from 60 to 70% of the plus strand (Khoury & Martin, 1972; Lindstrom & Dulbecco, 1972; Khoury *et al.*, 1972; Sambrook *et al.*, 1972). Although these results have clearly identified the early and late template strands and have indicated what fraction of each strand is transcribed into stable virus RNA, they do not indicate the topology of early and late regions of the SV40 genome.

To localize the early and late template regions we have made use of specific fragments of SV40 DNA generated by bacterial restriction endonucleases. Of most use have been the 11 fragments produced by *Hemophilus influenzae* restriction endonuclease (Danna & Nathans, 1971), which vary in size from 22·5% to 4% of the length of the SV40 genome. The order of these fragments in the SV40 DNA molecule is now known (Danna & Nathans, 1972), as is their relation to the cleavage sites of other restriction enzymes, thus providing a physical map of the SV40 genome (Danna *et al.*, 1973). We have used these enzyme cleavage products to determine the regions of the plus and minus strands of SV40 DNA that are transcribed into stable RNA in productively infected cells, and also to determine the direction of early and late transcription. In a subsequent communication, we shall present a similar analysis of SV40 transcription in virus transformed cells.

2. Materials and Methods

(a) *Preparation of virus and virus DNA*

Small plaque SV40 (from strain 776, obtained from K. Takemoto) was plaque purified and a stock prepared in the BSC-1 line of African green monkey kidney cells, as described previously (Danna & Nathans, 1971). ^{32}P-labeled SV40 DNA I was obtained from virus purified by isopycnic centrifugation in CsCl as described previously (Yoshiike, 1968; Trilling & Axelrod, 1970). The specific activities of the labeled DNA preparations varied from 1 to 4×10^5 cts/min per μg. ^{14}C-labeled SV40 DNA I was extracted from infected monkey cells by differential salt precipitation (Hirt, 1967) and purified by equilibrium density centrifugation in CsCl/ethidium bromide (Radloff *et al.*, 1967) followed by sedimentation through a neutral sucrose gradient; the specific activity was 3×10^4 cts/min per μg.

(b) *Digestion of SV40 DNA with* H. influenzae *restriction endonuclease*

To obtain specific fragments of SV40 DNA, [^{32}P] or [^{14}C]DNA I was incubated with *H. influenzae* restriction endonuclease in a mixture containing 25 to 70 μg of DNA, 0·03 unit of enzyme (Smith & Wilcox, 1970), 50 mM-NaCl, and 6 mM each of $MgCl_2$ and Tris·HCl (pH 7·5), in a volume of 0·10 ml, as described previously (Danna & Nathans, 1972). After incubation at 37°C for 2 h, the reaction was complete as determined by electrophoresis and autoradiography of a portion of the reaction mixture (see below).

(c) *Isolation of DNA fragments*

The 11 fragments of SV40 DNA present in an *H. influenzae* enzyme digest (fragments A to K) were separated by electrophoresis in 4% polyacrylamide slab gels measuring 15 cm × 40 cm × 0·16 cm, as described elsewhere (Danna & Nathans, 1972). ^{32}P-labeled fragments were detected by autoradiography of the wet gel slab and DNA was eluted from segments of gel containing individual fragments with 0·1 times SSC (SSC is 0·15 M-NaCl, 0·015 M-sodium citrate). To localize ^{14}C-labeled fragments, the ^{14}C-labeled digest was mixed with a small amount of ^{32}P-labeled digest and the above procedure followed. The ^{32}P was then allowed to decay to undetectable levels before use of the ^{14}C-labeled fragments. Individual ^{14}C-labeled fragments were then tested for purity by electrophoresis and autoradiography of dried gel slabs; all fragments were pure by this criterion.

(d) *Cleavage of SV40 DNA with* E. coli *R_I restriction endonuclease*

E. coli R_I restriction endonuclease (Yoshimori, 1971) generously supplied by H. Boyer, was incubated with ^{32}P-labeled SV40 DNA I as previously described (Danna *et al.*, 1973). The product had a sedimentation rate of 14·5 S in a 5% to 20% neutral sucrose gradient, corresponding to full length linear molecules (Morrow & Berg, 1972).

(e) *Exonuclease III digestion of R_I linear DNA*

The ^{32}P-labeled linear DNA product of the *E. coli* R_I cleavage was incubated at 35°C with 10 units of *E. coli* exonuclease III (kindly provided by H. Gerry) in a reaction mixture containing 5 μg of DNA, 5 mM-$MgCl_2$, 5 mM-β-mercaptoethanol, 50 μg bovine serum albumin/ml, and 50 mM-Tris·HCl (pH 7·4) in a volume of 0·21 ml. Portions of 5 μl of the reaction mixture were removed at various times and the percentage of DNA soluble in cold 5% trichloroacetic acid determined. The values observed were $< 0{\cdot}1\%$, 24%, 45% and 49% at 0, 20, 40 and 60 min, respectively. The reaction was stopped at 60 min by addition of an equal vol. of water-saturated phenol and the aqueous layer plus two 0·2-ml washes of the phenol layer were extracted 5 times with ether. The ether was removed by a stream of nitrogen and the solution was then dialyzed against 0·05 M-NaCl. The exonuclease-treated DNA sedimented in an alkaline sucrose gradient as a broad band with an S value of 13·5 (expected value for half-length strands is 11·5 S), suggesting some variability in the extent of exonuclease digestion of individual molecules. Since exonuclease III digests duplex DNA from the 3′ end, this preparation will be referred to as 5′ "half molecules".

(f) *Preparation of RNA*

"Early lytic" RNA was prepared from BSC-1 cells 24 to 36 h after infection with SV40 (multiplicity of 40 to 80 plaque-forming units/cell) in the presence of 20 μg arabinosyl cytosine/ml as described previously (Khoury *et al.*, 1972). Late lytic RNA was purified from BSC-1 cells 40 to 60 h after infection using the same multiplicity of virus. SV40 complementary RNA was prepared *in vitro* with *E. coli* RNA polymerase and SV40 DNA I as described previously (Khoury & Martin, 1972).

(g) *Separation and purification of SV40 DNA strands*

The ^{32}P-labeled fragments of SV40 DNA, obtained by digestion with the *H. influenzae* restriction endonuclease, were dialyzed against 10 mM-Tris·HCl (pH 7·5), and 10 mM-NaCl. Reaction mixtures containing approximately 0·05 to 0·15 μg of a specific, heat-denatured DNA fragment, 10 μg of complementary RNA, 10 mM-NaCl, 10 mM-Tris·HCl in 1·5 ml were incubated at 60° for 1 h. The plus and minus strands of each of the 11 fragments were then separated on hydroxyapatite and were purified by methods described previously (Khoury *et al.*, 1972).

(h) *DNA-RNA hybridization*

Small amounts (1 to 5 ng) of the plus or minus strand of each specific SV40 DNA fragment were incubated with early (0·2 to 0·5 mg) or late (0·1 to 0·3 mg) lytic RNA in 1·0 M-NaCl, 80 mM-phosphate buffer and 0·2% sodium dodecyl sulfate in 0·15 to 0·25 ml at 68°C for 18 to 24 h. DNA–RNA hybrid formation was monitored on hydroxyapatite (Kohne, 1969). DNA–RNA hybrids formed with fragments A or B, and lytic RNA, were also examined by treatment with the single-strand specific nuclease, S_1, as has been described in a previous publication (Khoury *et al.*, 1973). The reaction mixtures for S_1 nuclease analysis were diluted to 3 ml to give a final concentration of 300 mM-NaCl, 10^{-2} mM-$ZnSO_4$, 30 mM-sodium acetate buffer (pH 4·0) and 30 μg denatured salmon sperm DNA/ml, and divided into 3 equal portions, each containing about 350 cts/min of ^{32}P-labeled DNA. 2 of the samples were treated with an excess of the S_1 enzyme and the third served as a control. The fraction of the DNA resistant to the single strand specific enzyme was determined from the average of the nuclease resistant radioactivity in the 2 treated samples compared to the untreated sample.

(i) *DNA–DNA hybridization*

DNA–DNA hybridization was carried out with ^{14}C-labeled denatured fragments of SV40 DNA and the separated strands of the ^{32}P-labeled 5′ half-molecules produced by sequential cleavage of DNA I with *E. coli* R_I endonuclease and exonuclease III. The strands of 5′ half molecules were separated as described above for the separation of the strands of DNA fragments. These separated strands will be referred to as the plus and minus 5′ half strands. For hybridization, approximately 1 to 4 ng of either the plus or minus 5′ half strand was incubated with approximately 20 to 40 ng of a specific heat-denatured ^{14}C-labeled DNA fragment in 1·2 M-NaCl, 100 mM-phosphate buffer, 0·3% sodium dodecyl sulfate for 24 h at 68°C. The extent of DNA reassociation was then analyzed on hydroxyapatite (Britten & Kohne, 1968).

(j) *Detection of radioactivity*

After the addition of 0·2 mg of carrier yeast RNA, samples were adjusted to 5% trichloroacetic acid and collected on nitrocellulose membrane filters (B-6, Carl Schleicher and Schuell Co., Keene, N. H.). The filters were washed with 10 ml of cold 0·01 N-HCl, dried, and counted in a toluene-based scintillation fluid. A background of 20 cts/min was subtracted from each result.

3. Results

(a) *Separation of fragment strands*

The plus and minus strands of each of the 11 fragments (A to K), produced by digestion of SV40 DNA I with the *H. influenzae* restriction endonuclease, were separated by incubating denatured fragments with an excess of SV40 complementary RNA followed by chromatography on hydroxyapatite, as described previously for unfractionated virus DNA (Khoury *et al.*, 1972). Less than 4% of the purified separated strands of each fragment "self-associated" when incubated under the conditions used for DNA–RNA hybridization. The strand which reacted with SV40 complementary RNA has been designated the minus strand, and the other, the plus strand.

(b) *Hybridization of the plus and minus strands of DNA fragments with RNA from SV40-infected BSC-1 cells*

As shown previously, 30 to 40% of the minus strand of SV40 DNA is expressed early in a productive infection and 60 to 70% of the plus strand is expressed late in infection (Lindstrom & Dulbecco, 1972; Khoury *et al.*, 1972; Sambrook *et al.*, 1972). In order to determine which of the 11 DNA fragments contained the sequences complementary to early RNA and to late RNA, the individual strands of each DNA fragment were incubated with early or late lytic RNA. The percentage of each [^{32}P] DNA fragment which hybridized to added RNA was then determined by hydroxyapatite chromatography. The results of these experiments, shown in Table 1, indicated a complex pattern of transcription.

TABLE 1

Hybridization of early or late lytic RNA to the separated strands of unique SV40 DNA fragments

DNA fragment	[^{32}P]DNA in hybrid molecules (%) (−) DNA strand Early RNA	Late RNA	(+) DNA strand Early RNA	Late RNA
A	87	91	9	83
B	87	85	12	87
C	9	46	15	90
D	5	14	19	88
E	8	12	16	90
F	3	7	21	91
G	10	14	16	91
H	39	51	6	35
I	23	30	7	46
J	10	10	14	85
K	5	7	12	87

RNA from SV40-infected BSC-1 cells (early, 0·2 to 0·5 mg; late, 0·1 to 0·3 mg) was hybridized under the conditions described in Materials and Methods to the plus or minus strand of each SV40 DNA fragment produced by cleavage with the *H. influenzae* restriction endonuclease. Samples containing 250 to 400 cts/min were analyzed by hydroxyapatite chromatography to determine the percentage of [^{32}P]DNA in hybrid molecules. In the absence of RNA, this value was less than 4% for either strand of the 11 fragments. Similarly, low levels of hybridization (<2%) were found when uninfected monkey cell RNA was incubated with either strand of a control fragment (E).

Focusing first on the reaction between early RNA and the minus strands of DNA fragments, which are derived from the early template strand (column 1 of Table 1) we see that only fragments A, B, H and I hybridized with early RNA to a significant extent. At the level of RNA used, fragments A and B were essentially saturated, whereas H and I were not. Therefore, we conclude that at least a portion of the SV40 genome corresponding to each of the A, B, H and I fragments is transcribed into early SV40 RNA, and that stable "early" transcripts from segments A and B are probably more abundant than those from H and I.

Considering next the reaction between late RNA and the plus strands of the DNA fragments, which are derived from the specifically late template strand (column 4 of Table 1), we see that all fragments hybridized with late RNA. At the level of RNA used, fragments A, B, C, D, E, F, G, J and K were essentially saturated, whereas fragments H and I were not. We conclude from these results that at least a portion of the SV40 genome corresponding to each of the fragments is transcribed into late RNA and that stable late transcripts from A, B, C, D, E, F, G, J and K are probably more abundant than those from H and I.

The reaction between the minus strands of DNA fragments and late RNA (column 2 of Table 1) shows the expected hybridization with fragments A, B, H and I since early SV40 RNA species are known to be present during the late phase of productive infection (Aloni *et al.*, 1968; Oda & Dulbecco, 1968; Sauer & Kidwai, 1968; Carp *et al.*, 1969). In addition, the minus strand of fragment C is unique in that it hybridizes with late SV40 RNA but not with early RNA. The possible significance of this last result will be considered in the Discussion.

Also shown in Table 1 (column 3) is the reaction between early RNA and the plus strands of DNA fragments. As shown in the Table, the plus strands of several fragments showed a low level of hybridization with early RNA. The significance of the low levels of reaction between plus (late template) strands of some of the other fragments and early RNA is not clear (see Discussion).

(c) *Hybridization of fragments H and I with different concentrations of RNA*

The minus strands of fragments A and B and the plus strands of all fragments except H and I were saturated by the amounts of late lytic RNA used in this study. The low levels of hybridization of both the plus and minus strands of fragments H and I with late RNA (Table 1) suggested that stable transcripts from these regions of SV40 DNA may be less abundant than transcripts from other regions of the genome. To explore this possibility further, the hybridization reaction with these fragments was done with different concentrations of late RNA. As shown in Table 2, the extent

TABLE 2

Effect of RNA concentration on hybridization experiments with DNA fragments H and I

DNA fragment	Strand	RNA concentration† (mg/ml)	Duplex molecules (%)
H	+	2·4	16
		3·2	35
H	−	2·4	39
		3·2	51
I	+	2·4	27
		3·2	46
I	−	2·4	24
		3·2	30
J	+	1·6	85
J	−	1·6	10
K	+	1·2	87
K	−	1·2	7

Under conditions described in Materials and Methods, 1 to 4 ng (approx. 300 cts/min) of either strand of a DNA fragment was incubated with the indicated concentrations of late lytic RNA. The percentage of radioactivity in duplex molecules was analyzed by hydroxyapatite chromatography.

† Total cellular RNA obtained late in the productive cycle of SV40 infection of BSC-1 cells.

of hybridization of each strand of fragment H and of fragment I was dependent on the concentration of late RNA. This was the case even at RNA levels which were at least two to three times higher than that needed to saturate the plus strand of the other DNA fragments, including fragments J and K which are smaller than H and I. These data, together with those presented in Table 1, thus suggest that stable transcripts of both strands of fragments H and I are present in lower relative concentrations than RNA species transcribed from SV40 DNA sequences corresponding to the other nine DNA fragments. Alternatively, the results obtained with fragments H and I could point to a previously unsuspected heterogeneity of these fragments (see Discussion).

(d) *Estimation of the proportion of each strand of fragments A and B which hybridize with lytic RNA*

The results presented in Table 1 indicate that more than 85% of the plus and minus strands of fragments A and B reacted with late lytic RNA. In this experiment, the DNA–RNA reaction was monitored by hydroxyapatite chromatography which fails to discriminate between complete and partial DNA–RNA hybrids. For this reason we decided to use the S_1 single-strand specific nuclease assay to determine the fraction of the minus and plus strands which form DNA–RNA hybrids. Since early SV40 RNA sequences are represented in late lytic RNA, and since there is more SV40-specific RNA late after infection, late lytic RNA was used in these studies to insure that the DNA–RNA reactions had reached saturation.

The separated strands of ^{32}P-labeled fragments A and B were incubated with saturating amounts of late lytic RNA and then exposed to the S_1 nuclease as described in Materials and Methods. The proportion of either DNA strand of these fragments which formed hybrid structures with RNA was determined as the percentage of [^{32}P]DNA resistant to the S_1 enzyme. This value provided an estimate of the extent of transcription on each strand of fragment A or B (Table 3). Approximately 33% of the plus strand and 60% of the minus strand of fragment A appear to be transcribed, while 51% of the plus strand and 40% of the minus strand of fragment B are expressed. Since each of these sets adds up to nearly 100%, one interpretation of these results is that the minus and plus strand templates do not overlap.

TABLE 3

Extent of hybridization of lytic RNA to a specific DNA fragment

DNA fragment	Strand	Resistance of hybrid to S_1 enzyme (%)
A	+	33
A	−	60
B	+	51
B	−	40

Approximately 5 ng of the plus or minus strand of fragments A and B were incubated with 0·8 to 1·0 mg of late lytic RNA in 0·75 ml as described in Materials and Methods. After 24 h incubation, the sample was diluted to 3 ml and divided into thirds; 2 of these were treated with an excess of S_1 enzyme, while the third served as a control. The percentage resistance to S_1 nuclease is the average of the radioactive counts in the 2 treated samples compared to the control. The values in this Table represent the average of 2 of these experiments and differed by less than 10%

(e) *Map positions of early and late template regions*

Danna *et al.* (1973) have constructed a physical map of the SV40 genome based on the sites of cleavage of SV40 DNA by restriction endonuclease from *H. influenzae*, *H. parainfluenzae*, and *E. coli* (R_I) (Fig. 1). We can therefore relate the results of hybridization of DNA fragments with SV40 lytic RNA to the map position of individual fragments.

Early RNA hybridized to the minus strands of fragments A, H, I and B, which form a contiguous group in the map (Fig. 1). Since only 60% of the minus strand of A and

40% of the minus strand of B are expressed, the early template region appears to extend from near the middle of A to near the middle of B, i.e. about 30% of the length of the virus genome. This interpretation assumes that the DNA sequences transcribed into stable early RNA are entirely contiguous. Late lytic RNA, on the other hand, hybridized almost completely with the plus strands of fragments A, C, D, E, K, F, J, G and B. As seen in the cleavage map shown in Figure 1, these fragments also form a contiguous group. Since about 33% of the plus strand of A and about 51% of the plus strand of B are expressed, the late template region forms a segment encompassing about 70% of the length of the virus genome (Fig. 1). As noted above, very high concentrations of late lytic RNA hybridize with the plus strands of fragments H and I but the reaction is incomplete. This suggests that a minor species of late lytic RNA is transcribed from parts of the sequences corresponding to fragments H and I (see Discussion).

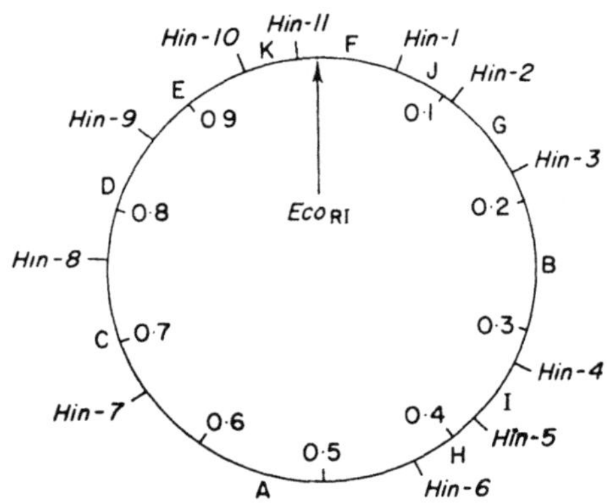

FIG. 1. A cleavage map of the SV40 genome (Danna *et al.*, 1973). The *E. coli* R_I endonuclease site (*Eco* R_I) is the zero point and map distances are given as fractional length of SV40 DNA in the direction F → J → G– – – –. *Hin-1, 2, 3*, etc. refer to the *H. influenzae* endonuclease cleavage sites. The map positions of these are: *Hin-1*, 0·060; *Hin-2*, 0·105; *Hin-3*, 0·175; *Hin-4*, 0·325; *Hin-5*, 0·375; *Hin-6*, 0·430; *Hin-7*, 0·655; *Hin-8*, 0·760; *Hin-9*, 0·860; *Hin-10*, 0·945; *Hin-11*, 0·985.

(f) *Direction of transcription*

Although the results just presented indicate the position of early and late template regions on the SV40 map, they do not indicate the direction of transcription of the early and late genes, i.e. clockwise or counter-clockwise transcription as the map is drawn. To determine the direction of transcription, it is sufficient to establish the 5′ to 3′ direction of the minus and plus strands of SV40 DNA relative to the cleavage map, since RNA is transcribed in the 3′ to 5′ direction of the template strand, and the minus and plus strands are known to be templates for early and late RNA, respectively.

In order to relate the 5′ to 3′ direction of the minus and plus strands of SV40 DNA to the cleavage map, we have carried out the experiment diagrammed in Figure 2. ^{32}P-labeled SV40 DNA I was first cleaved with the *E. coli* R_I restriction endonuclease to obtain unique, full length linear molecules. Since the R_I enzyme cleaves within fragment F (Danna *et al.*, 1973), these molecules have the map order F_1 J G B I H A

C D E K F_2 (see Fig. 2). In the next step, the linear molecules were digested with *E. coli* exonuclease III, which removed the 3′ halves of each strand, leaving 5′ half molecules. The individual minus and plus 5′ half strands were then isolated by annealing with SV40 complementary RNA. Finally, we determined which 5′ half strand (plus or minus) contains fragment J and fragment G sequences, and which contains fragment K and fragment E sequences by annealing each half strand with denatured ^{14}C-labeled fragments J, G, K or E. The sequences corresponding to these four fragments are nearest to sequence F, which is present at each end of the R_I linear molecules (Fig. 2).

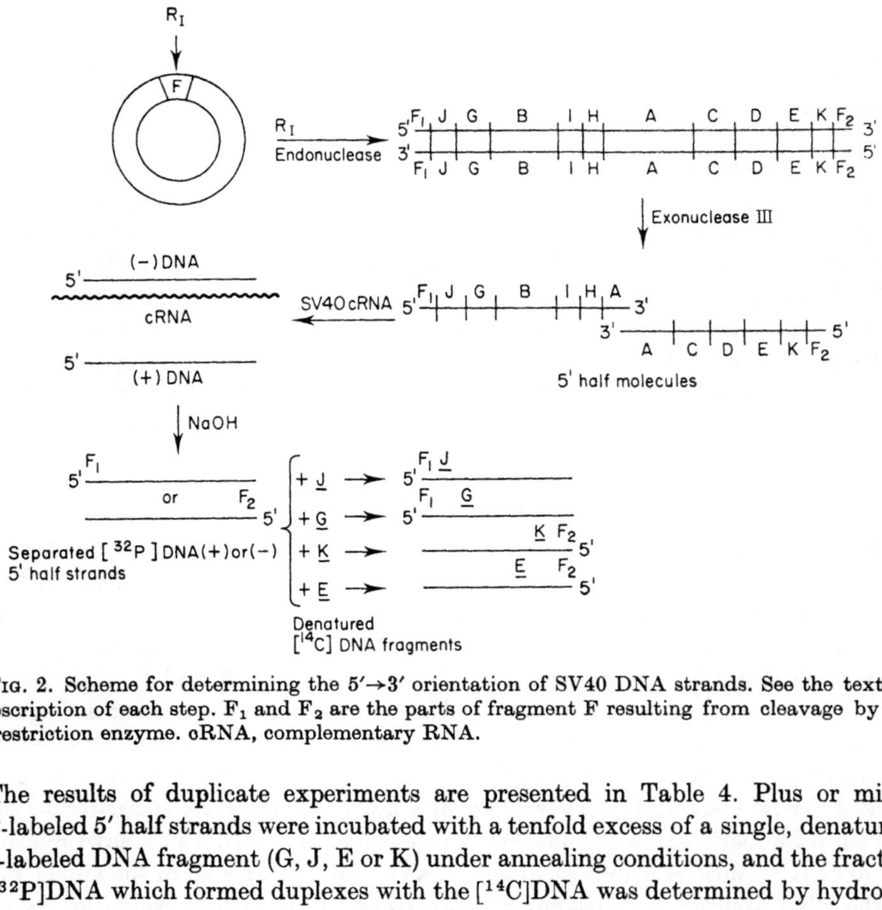

FIG. 2. Scheme for determining the 5′→3′ orientation of SV40 DNA strands. See the text for a description of each step. F_1 and F_2 are the parts of fragment F resulting from cleavage by the R_1 restriction enzyme. cRNA, complementary RNA.

The results of duplicate experiments are presented in Table 4. Plus or minus ^{32}P-labeled 5′ half strands were incubated with a tenfold excess of a single, denatured, ^{14}C-labeled DNA fragment (G, J, E or K) under annealing conditions, and the fraction of [^{32}P]DNA which formed duplexes with the [^{14}C]DNA was determined by hydroxyapatite chromatography. The results indicate that fragments E and K hybridize preferentially with the *plus* 5′ half strand, while fragments G and J hybridize preferentially with the *minus* 5′ half strand of SV40 DNA. In each case, more than 85% of the ^{14}C-labeled DNA fragments reassociated. Theoretically one would expect that nearly all of the ^{32}P-labeled DNA would anneal in the presence of an excess of the appropriate [^{14}C]DNA fragment if the half-strands had remained intact. However, when the separated ^{32}P-labeled 5′ half strands used in this experiment were examined in alkaline sucrose, they sedimented in the 5·4 to 10·7 S region of the gradient which corresponds to a single-stranded molecule 1·7 to 6·1 × 10^5 daltons in size, thus explaining why no more than 40% of the [^{32}P]DNA reassociated. The data, nevertheless, are

TABLE 4

Hybridization of the separated strands of ^{32}P-labeled SV40 DNA half-molecules with specific ^{14}C-labeled SV40 DNA fragments

Denatured [^{14}C]DNA fragment	[^{32}P]DNA in hybrid molecules (%)	
	(+) DNA strand	(−) DNA strand
G	6	40
J	11	35
E	40	20
K	30	15

Hybridization reactions were done as described in Materials and Methods. The percentage of [^{32}P]DNA in hybrid molecules was analyzed on hydroxyapatite as previously described. Values represent the average of 2 experiments and differed by less than 10%.

internally consistent, i.e. adjacent fragments E and K preferentially bind to one 5′ half strand (plus), while fragments G and J bind preferentially to the other (minus) 5′ half strand. Since the minus strand is the template for early SV40 RNA, we can conclude that the orientation of the early template strand is 5′ F J G B I H A . . . . . 3′, i.e. 5′—3′ clockwise on the cleavage map shown in Figure 1, and the orientation of the plus strand (late template strand) is 5′ F K E D C A . . . . 3′, i.e. 5′—3′ counterclockwise on the cleavage map. It follows, therefore, that transcription of early genes proceeds from A to B counter-clockwise on the minus DNA strand and transcription of late genes proceeds from A to B in a clockwise direction on the plus DNA strand.

4. Discussion

The major conclusions of this study are presented in the map shown in Figure 3. We have confirmed previous results indicating that early in productive infection, transcription of the SV40 genome occurs on the minus strand template. Later in infection, additional transcription occurs on the plus strand template. The early region of SV40 DNA includes segments of fragments A, B, H and I, while the late region of the virus genome encompasses polynucleotide sequences present in fragments C, D, E, K, F, J, G and parts of A and B. In addition, there is evidence for late transcription of at least part of the sequences corresponding to the plus strands of fragments H and I. These transcripts, however, are at a much reduced level compared to those from the other parts of the genome. Within fragments A and B, which have relatively abundant RNA species complementary to each strand, nearly all of the stable virus-specific RNA appears to be asymmetrically transcribed, since in each case we were able to account for over 90% of the transcribing activity by summing the hybridization to each DNA strand. Although we cannot make any statement about the precise location of transcribed regions on the plus and minus strands of fragments A and B, the simplest explanation of this result is that the parts of A and B nearest H and I are transcribed from the minus strand template and the parts nearest C and G, respectively, are transcribed from the plus strand template. The early region would then encompass sequences present between 0·26 to 0·57 map unit and the late region would comprise the rest of the genome (Fig. 3).

The direction of transcription of early and late SV40 RNA has been established by

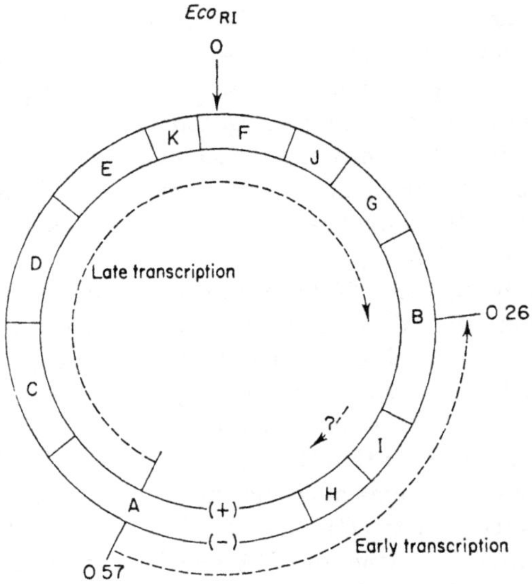

FIG. 3. Diagrammatic representation of transcription, early and late in the SV40 lytic cycle as it relates to the physical map of the viral genome.

determining the 5′ to 3′ orientation of the minus and plus strands of SV40 DNA, which serve as the early and late templates, respectively. As indicated in Figure 3, early transcription occurs in a counter-clockwise direction from A to B and late transcription occurs in a clockwise direction from A to B. On the assumption that the abundant species of stable SV40 RNA that we have used in these studies is the immediate transcription product, we can localize initiation sites for both early and late transcription near the middle of fragment A (at 0·57 map unit) and termination sites for both early and late transcription near the middle of fragment B (at 0·26 map unit). Alternatively, if large regions of the DNA are symmetrically transcribed (Aloni, 1972) with subsequent degradation of specific RNA sequences, the true initiation and termination sites may be at different locations. Furthermore, we have not excluded the possibility that other initiation and termination sites may be present within both the early and late regions. In this regard, experiments carried out *in vitro* with *E. coli* RNA polymerase and an SV40 DNA I template suggest that transcription with this enzyme begins outside fragment A (Westphal *et al.*, 1973; Zain, Dhar, Weissman, Lebowitz & Lewis, unpublished observations). It will be of considerable interest to determine what special nucleotide sequences may exist at the transition sites within fragments A and B, and whether these sites are related to the specific cleavage of SV40 component I DNA by the single-strand specific nuclease S_1 (J. Morrow & P. Berg, personal communication).

The map position for the early region of the SV40 genome shown in Figure 3 is entirely consistent with observations made with adeno-SV40 hybrid viruses by Kelly & Lewis (personal communication) who found that the entire early region of SV40 is included in the DNA of a hybrid virus containing 43% of the SV40 genome. They have localized within the early segment those regions which are required for expression of U antigen, T antigen, and the SV40 tumor-specific transplantation antigen. Recently

Patch *et al.* (1972) found that the SV40 DNA segment of $AD2^{+}ND_1$ contains parts of both early and late information and, therefore, has a probable transcriptional initiation or termination site. A comparison of the map of the SV40 segment of the adenovirus hybrid with the cleavage map of the SV40 genome (Morrow & Berg, 1972; T. Kelly, personal communication; P. Lebowitz, personal communication) suggests that this site corresponds to the termination point within fragment B. Therefore, it is likely that transcription of SV40 DNA sequences in $AD2^{+}ND_1$ infected cells begins within the adenovirus portion of the DNA.

The incomplete hybridization of *both* the minus and plus strands of fragments H and I with the RNA from infected monkey cells (Table 1) suggests that transcription of SV40 DNA may be more complex than initially thought. As pointed out above, fragments H and I, which are physically located in the middle of the early region of the viral genome, were the only two that failed to react completely with amounts of lytic RNA known to be saturating for the other nine *H. influenzae* DNA fragments. Although heterogeneity of these two fragments could explain this finding, there is at present no other evidence for heterogeneity (Danna *et al.*, 1973). The results presented in Table 2 suggest that the relative abundance of RNA complementary to fragment H or I DNA is much lower than the other, stable lytic RNA species encountered thus far. It is not clear at present whether the low concentrations of virus specific RNA complementary to DNA fragments H and I reflect a diminished rate of synthesis or the degradation of specific polyribonucleotide sequences transcribed from this region of the virus genome. Either of these possibilities would reduce the concentrations of specific RNA sequences and explain the results presented in Table 2. The reaction of a significant proportion of the *plus* strands of fragments H and I with late lytic RNA (Tables 1 and 2) suggests the existence of late DNA sequences (or early information on the plus DNA strand) within the early segment of the virus DNA. As such, this class of RNA may have an important role with respect to the regulation of SV40 DNA transcription.

Low levels of reaction (6 to 21%) were also observed between early lytic RNA and the plus strands of the 11 SV40 DNA fragments (Table 1). Previous reports have indicated that early RNA is complementary to the minus strand of virus DNA (Lindstrom & Dulbecco, 1972; Khoury *et al.*, 1972; Sambrook *et al.*, 1972). The early lytic RNA used in the experiment shown in Table 1 was prepared from monkey cells infected in the presence of arabinosyl cytosine. Incomplete inhibition of SV40 DNA synthesis at a time during productive infection (24 to 36 h) when late RNA is known to be present may account for the observed low levels of reaction between early RNA and the plus strands of the various fragments. Alternatively, this result could reflect the incomplete post-transcriptional degradation of RNA complementary to the plus strand of SV40 DNA. Several reports have appeared recently which suggest that integration of virus genetic information into chromosomal DNA may occur during productive infection by SV40 (Lavi & Winocour, 1972; Tai *et al.*, 1972; Hirai & Defendi, 1972). Whether the RNA transcribed from such integrated SV40 DNA (Jaenisch, 1972; Rozenblatt & Winocour, 1972) is related to those RNA sequences that hybridize incompletely with the DNA fragments remains to be determined.

The partial hybridization of late, but not early RNA, to the minus strand of fragment C (Table 1) is worthy of comment. Since fragment C contains the origin of DNA replication (Nathans & Danna, 1972; Danna & Nathans, 1972), it is possible that this RNA represents a primer for initiation of DNA replication (Wickner *et al.*, 1972). Since SV40 DNA replication proceeds bidirectionally from fragment C (Danna &

Nathans, 1972; Fareed *et al.*, 1972), one might expect primers for each strand at the initiation site.

It has been shown that the pattern of transcription in SV40 transformed cells is significantly different from that observed during lytic infection (Khoury *et al.*, 1973). The use of SV40 DNA fragments produced by restriction endonucleases should permit a topographic analysis of these differences.

This work was supported by grants from the National Institutes of Health and the Whitehall Foundation.

REFERENCES

Aloni, Y. (1972). *Proc. Nat. Acad. Sci.*, U.S.A. **69**, 2404–2409.

Aloni, Y., Winocour, E. & Sachs, L. (1968). *J. Mol. Biol.* **31**, 415–429.

Britten, R. J. & Kohne, D. E. (1968). *Science*, **161**, 529–540.

Carp, R. I., Sauer, G. & Sokol, F. (1969). *Virology*, **37**, 214–226.

Danna, K. J. & Nathans, D. (1971). *Proc. Nat. Acad. Sci., U.S.A.* **68**, 2913–2917.

Danna, K. J. & Nathans, D. (1972). *Proc. Nat. Acad. Sci., U.S.A.* **69**, 3097–3100.

Danna, K. J., Sack, G. & Nathans, D. (1973). *J. Mol. Biol.* **78**, 363–376.

Fareed, G. C., Garon, C. F. & Salzman, N. P. (1972). *J. Virol.* **10**, 484–491.

Hirai, K. & Defendi, V. (1972). *J. Virol.* **9**, 705–707.

Hirt, B. (1967). *J. Mol. Biol.* **26**, 365–369.

Jaenisch, R. (1972). *Nature New Biol.* **235**, 46–47.

Khoury, G. & Martin, M. A. (1972). *Nature New Biol.* **238**, 4–6.

Khoury, G., Byrne, J. C. & Martin, M. A. (1972). *Proc. Nat. Acad. Sci., U.S.A.* **69**, 1925–1928.

Khoury, G., Byrne, J. C., Takemoto, K. K. & Martin, M. A. (1973). *J. Virol.* **11**, 54–60.

Kohne, D. E. (1969). *Carnegie Inst. of Wash. Yearbook*, **67**, 310–320.

Lavi, S. & Winocour, E. (1972). *J. Virol.* **9**, 309–316.

Lindstrom, D. M. & Dulbecco, R. (1972). *Proc. Nat. Acad. Sci., U.S.A.* **69**, 1517–1520.

Martin, M. A. & Axelrod, D. (1969). *Proc. Nat. Acad. Sci., U.S.A.* **64**, 1203–1210.

Morrow, J. F. & Berg, P. (1972). *Proc. Nat. Acad. Sci., U.S.A.* **69**, 3365–3369.

Nathans, D. & Danna, K. J. (1972). *Nature New Biol.* **236**, 200–202.

Oda, K. & Dulbecco, R. (1968). *Proc. Nat. Acad. Sci., U.S.A.* **60**, 525–532.

Patch, C., Lewis, A. M. & Levine, A. S. (1972). *Proc. Nat. Acad. Sci., U.S.A.* **69**, 3375–3379.

Radloff, R., Bauer, W. & Vinograd, J. (1967). *Proc. Nat. Acad. Sci., U.S.A.* **57**, 1514–1521.

Rozenblatt, S. & Winocour, E. (1972). *Virology*, **50**, 558–566.

Sambrook, J., Sharp, P. A. & Keller, W. (1972). *J. Mol. Biol.* **70**, 57–71.

Sauer, G. & Kidwai, J. R. (1968). *Proc. Nat. Acad. Sci., U.S.A.* **61**, 1256–1263.

Smith, H. O. & Wilcox, K. (1970). *J. Mol. Biol.* **51**, 379–391.

Tai, H. T., Smith, A. C., Sharp, P. A. & Vinograd, J. (1972). *J. Virol.* **9**, 317–325.

Trilling, D. M. & Axelrod, D. (1970). *Science*, **168**, 268–271.

Westphal, H., Delius, H. & Mulder, C. (1973). In *Lepetit Colloquia on Biology and Medicine*, Vol. **4**, North Holland Publishing Co., Amsterdam and London.

Wickner, W., Brutlag, D., Schekman, R. & Kornberg, A. (1972). *Proc. Nat. Acad. Sci., U.S.A.* **69**, 965–969.

Yoshiike, K. (1968). *Virology*, **34**, 391–401.

Yoshimori, R. N. (1971). Ph.D. Thesis, University of California, San Francisco Medical Center.

Presence of Cell and Virus Specific Sequences in the Same Molecules of Nuclear RNA from Virus Transformed Cells

Randolph Wall and James E. Darnell

Hybridization studies show that nuclear RNA in transformed cells contains both host and virus specific sequences. This suggests virus-specific mRNA could be formed by selective cleavage of high molecular weight nuclear RNA.

CELLS transformed by the oncogenic DNA virus, SV40, contain viral DNA covalently integrated into their chromosomes[1] but there is conflicting evidence concerning the precise amount of viral DNA in transformed cells. Some studies have indicated the equivalent of only one viral genome per transformed cell[2], while other studies have indicated as many as sixty equivalents of viral DNA per transformed cell[3]. The integrated viral DNA is transcribed because RNA-DNA hybridization studies have shown that virus specific RNA exists in SV40 transformed cells[4–6]. This RNA presumably directs the formation of several characteristic virus specific proteins found in transformed cells including T-antigen and transplantation specific tumour antigen[7].

We have examined the transcription of SV40 DNA in transformed cells to help to delineate the nuclear events involved in the production of polysomal mRNA in mammalian cells. It has been reported that virus specific sequences are present in large HnRNA ($>4\times10^6$ daltons) and in smaller polysomal mRNA ($<1.5\times10^6$ daltons) from SV40 transformed mouse cells (SV3T3 cells)[8]. Because the molecular weight of double stranded SV40 DNA is 3×10^6 daltons[7], the large virus specific HnRNA molecules were considerably longer than a single strand of SV40 DNA. Such large virus specific HnRNA molecules are not aggregates containing small SV40 specific RNA molecules, for treatment of transformed cell nuclear RNA with dimethyl sulphoxide (which destroys nucleic acid interaction) failed to change the sedimentation pattern of virus specific RNA[8]. This suggests that virus specific mRNA molecules in the polysomes of transformed cells arise in the nucleus through the selective cleavage of larger SV40 specific HnRNA molecules. It is therefore of interest to determine whether the large virus specific HnRNA molecules contain covalently linked cellular RNA sequences or are composed entirely of viral sequences which could have arisen through the uninterrupted transcription of multiple serially integrated SV40 DNA molecules.

Our experiments suggest that large SV40 specific HnRNA molecules contain covalently linked sequences which hybridize specifically to cellular DNA from the parental untransformed cell line (3T3), whereas smaller polysomal SV40 specific mRNA molecules are primarily composed of sequences which bind to SV40 DNA.

Experimental Procedures

3T3, SV3T3 and BSC-1 cells were grown as monolayers in large roller bottles in Eagle's minimal essential medium[9] with 10% foetal calf serum and were labelled at about 4×10^6 cells/ml. for 3 h with uridine-5-^{3}H (New England Nuclear, 25 Ci/mmol) at 1 mCi/ml.

Polysomes were isolated by sucrose gradient centrifugation of cytoplasmic extracts prepared in an isotonic buffer (0.14 M NaCl, 0.01 M Tris (*p*H 8.4), 0.0015 M $MgCl_2$) containing 0.5% NP40, a non-ionic detergent[8]. Polysomal RNA or total nuclear RNA in 0.05 M acetate, 0.01 M EDTA and 0.5% SDS was extracted twice with phenol at 45° C, precipitated twice with 2 M LiCl at 4° C, and then sedimented through sucrose gradients according to Lindberg and Darnell[8].

High titre SV40 lysates (small plaque variety of SV40, obtained from Dr K. Takemoto) were made in roller cultures of monkey kidney (BSC-1) cells[10]. Purified SV40 was prepared and type 1 viral DNA was isolated free of cellular DNA contamination as previously reported[8]. After brief DNAase treatment to break the closed circles[11], heat denatured SV40 DNA was bound to nitrocellulose filters according to Gillespie and Spiegelman[12]. Untransformed 3T3 cell DNA, prepared from NP40 washed nuclei, was denatured by boiling and then bound to 'Millipore' filters[12]. SV40 DNA filters to be used in formamide hybridization were presoaked for 12–15 h at 45° C in Denhardt's solution[15] (0.2% Ficoll (average molecular weight 40,000), 0.2% polyvinylpyrrolidone (average molecular weight 360,000), and 0.2% bovine albumin) dis-

Table 1 Selection of SV40 Specific RNA

RNA sample	Input c.p.m.	c.p.m. bound to* 20γ SV40 DNA	Blank
HnRNA SV3T3	18×10^6	6,800	1,160
HnRNA SV3T3 alkali degraded†	16×10^6	2,650	290
HnRNA 3T3	18×10^6	1,430	1,020
Polysomal RNA SV3T3	40×10^6	9,570	800
Polysomal RNA 3T3	30×10^6	800	400

* RNA bound in 8 h incubation at 45° C in 30% formamide hybridization solution. No RNAase treatment was used. Results presented represent single preparations of each type of RNA. Similar results were obtained with at least three preparations of each type of RNA.

† Treated as described in ref. 14; RNA reduced to <28.5.

solved in 0.3 M NaCl, 0.01 M EDTA, 0.01 M TES (N-Tris (Hydroxymethyl)) methyl-2-aminoethane sulphonic acid) (*p*H 7.4). Drained presoaked filters were not rinsed before addition of RNA solutions for hybridization.

Selections of unbroken virus specific RNA molecules on SV40 DNA were performed in 30% formamide (V/V) in 0.3 M NaCl, 0.01 M EDTA, 0.01 M TES (*p*H 2.4), at 45° C for 8 h without RNAase digestion of the resultant hybrids[14,16]. Formamide to be used in hybridization solutions was partially purified of breakdown products by equilibration against N_2 followed by ether extractions. Before elution DNA-RNA hybrid filters were washed three times at 45° C in 30% formamide hybridization solution and then rinsed four times at 25° C with 0.001 M EDTA. More than 95% of the bound RNA molecules were then released by two 15 min elutions at 75° C in 0.001 M EDTA. The samples were adjusted to 0.1 M, NaCl, 0.01 M TES and 0.001 M MgCl, digested with DNAase (20γ/ml. for 20 min), boiled, and brought to 0.3 M NaCl, 0.01 M EDTA and 0.1% SDS. Rehybridization to fresh SV40 or 3T3 DNA filters was carried out at 65° C with one DNA filter of each type and one blank filter in each vial. These hybrid filters were washed three times at 65° C and scored for bound RNAase resistant hybrid using pancreatic and T1 RNAase[13].

Selection of SV40 Specific RNA

Hybridization conditions were sought in which no molecular breakage occurred during incubation and specific hybrid formation could be achieved without the necessity of RNAase treatment (that is low temperature incubation in formamide)[14,16]. The results in Table 1 demonstrate that the hybridization of polysomal RNA or HnRNA from SV3T3 cells to SV40 DNA filters was some six to twelve times better than to blank filters. Moreover, when equivalent amounts of radioactive HnRNA or polysomal RNA from 3T3 cells were exposed to SV40 DNA, the amount of radioactivity bound was only slightly greater than that bound to blank filters. The selection of virus specific RNA was thus both dependent on SV40 DNA on the filters and specific for RNA from transformed cells. In the course of developing the conditions for the SV40 specific RNA selection experiments, a number of smaller hybridization reactions were carried out with various types of RNA containing virus specific sequences. Hybridization with these various types of RNA to SV40 DNA was determined both before and after RNAase treatment. Table 2 shows that RNA from cells productively infected with SV40 or RNA made in an *in vitro* reaction employing

Table 2 SV40 Specific Hybridization of Various RNA Samples

RNA sample	Input c.p.m.	c.p.m. bound to 1γ SV40 RNA − RNAase	c.p.m. bound to 1γ SV40 RNA + RNAase	Bound RNA resistant to RNAase (%)
HnRNA SV3T3	2×10^6	620	94	15
Polysomal RNA SV3T3	2×10^5	226	97	43
Total cell RNA, SV40 infected	1.2×10^5	481	337	70
cRNA SV40	650	400	290	72

* Hybrid formed after 18 h incubation at 45° C in 30% formamide hybridization solution. Hybrid counts have been corrected for radioactivity bound to blank filters which did not exceed 20 c.p.m. cRNA prepared by transcribed γ of SV40 DNA with *E. coli* RNA polymerase prepared according to the method of Burgess (gift from Dr R. Sederoff).

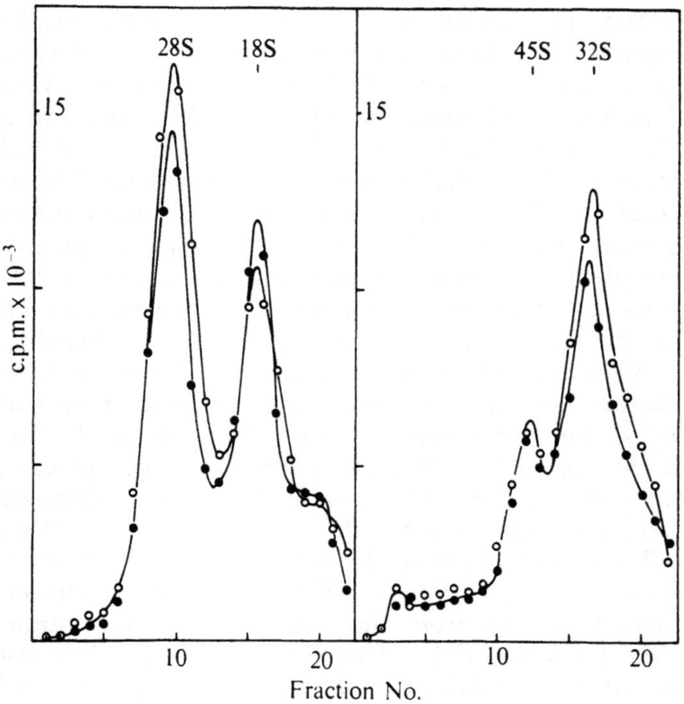

Fig. 1 Sucrose gradient sedimentation profiles of cytoplasmic polyribosomal RNA (left panel) or HnRNA (right panel), after 45 s incubation at 45° C in 30% formamide hybridization solution (○); after 8 h incubation in formamide hybridization conditions (●). The sedimentation pattern of the originally isolated RNA appeared identical to the 45 s sample and is not shown.

SV40 DNA and *E. coli* DNA-dependent RNA polymerase[18] both formed hybrids which were largely resistant to RNAase. Also almost half the RNA from the polyribosomes of transformed cells which bound to SV40 DNA was resistant to RNAase digestion. By contrast, only 15% of the HnRNA of transformed cells which was bound to the SV40 DNA filters resisted RNAase digestion. These results suggested that the high molecular weight HnRNA contained regions which were not viral specific. If this were so, then degradation of the HnRNA should separate the viral specific RNA from the non-viral portion of the poly nucleotide chain. This was found to be the case. When total HnRNA was reduced in size to less than 28S (Table 1) by alkali treatment[17] the amount of HnRNA bound in formamide without RNAase treatment was decreased by a factor of three.

We then investigated whether the undegraded HnRNA carrying virus specific regions might also contain host specific sequences. Undegraded HnRNA and polysomal RNA from SV3T3 and untransformed 3T3 cells were hybridized to 20γ SV40 DNA filters for 8 h in 30% formamide hybridization solution at 45° C. Earlier hybridization experiments had indicated that this DNA concentration and incubation time removed more than 85% of the total virus specific content of a sample of the starting preparation of transformed cell RNA. The extent of covert breaks introduced in the various RNA samples during preparation was assessed by removing a sample of the hybridization reaction after 45 s incubation at 45° C and comparing its sedimentation profile to an untreated aliquot of the same sample. Estimation of RNA breaks introduced during hybridization was determined by sucrose gradient centrifugation of a sample from the hybridization vial after incubation for 8 h at 45° C. The HnRNA and polysomal RNA were maintained essentially intact during these procedures as seen in Fig. 1. SV40 selected RNA molecules were then recovered from the washed filters by elution in 0.001 M EDTA at 75° C. These molecules were then tested for their viral and cellular content by rehybridization to SV40 DNA or by hybridization to DNA from 3T3 cells, the non-transformed parental cell-type.

A substantial fraction of the SV3T3 polysomal RNA which was initially selected on SV40 DNA rehybridized to viral

Table 3 Hybridization of SV40 Selected RNA to 3T3 or SV40 DNA

RNA sample	Input c.p.m.	Hybridized to* SV40 DNA		20γ 3T3 DNA	
		24 h	96 h	6 h	24 h
HnRNA-SV3T3, selected by SV40 DNA	6,400	172 (2.7)	220 (3.5)	60 (0.9)	116 (1.8)
HnRNA-SV3T3, unselected	36,000	0	0	168 (0.5)	380 (1.1)
Polysomal RNA-SV3T3, selected by SV40 DNA	8,000	2,620 (33)	3,220 (40)	–	–

* Hybridizations were performed at 65° C and tested for RNAase resistant hybrid as described in procedures.

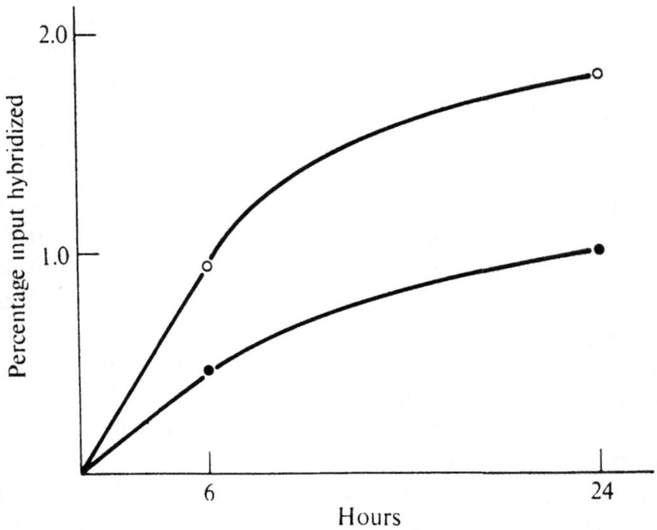

Fig. 2 Kinetics of hybrid formation of SV40 selected (○) and unselected (●) SV3T3 HnRNA on 3T3 DNA. Hybridization was performed with 20γ 3T3 DNA filters at 65° C and RNAase resistant hybrid measured as described in procedures.

DNA (Table 3). As a consequence of the small amount of radioactivity available in the SV40 selected polysomal sample, a sizable proportion of which was virus specific, it was not possible to determine if any of selected polysomal counts bound specifically to 3T3 DNA. But this finding does not necessarily preclude the possibility of a small fraction of covalently linked cellular RNA sequences in the selected polysomal RNA.

The SV40 selected HnRNA was also greatly enriched in SV40 specific sequences, because 3–4% of the input rehybridized to SV40 DNA as compared to about 0.01% RNAase resistant hybrid found with unselected material (see Table 2 and ref. 8). The fraction of selected HnRNA which hybridized to SV40, DNA was, however, far less than that of the selected polysomal molecules. Furthermore, the selected HnRNA exhibited a significant level of binding to 3T3 DNA. The total amount hybridized to cellular DNA is low both for SV40-selected and the original HnRNA molecules due to the fact that only RNA molecules arising from the so-called reiterated DNA sequences bind during a hybridization test of the type used in this work[14,16,19,20]. The rate of hybridization of the 3T3 specific fraction of selected HnRNA molecules is approximately twice as fast as that seen for unselected SV3T3 HnRNA (Fig. 2) probably because of the presence of labelled ribosomal precursor RNA in the unselected input which is absent from the selected RNA (R. W., unpublished results). The fact that the rate of binding of the SV40 selected HnRNA to 3T3 DNA is therefore about the same rate as the unselected molecules means that a very large fraction of the total HnRNA molecule bearing SV40 sequences is in fact cell specific. This is consistent with the finding that only 3–4% of the SV40 DNA selected HnRNA rehybridizes to SV40 DNA. By comparison, the viral DNA selected polysomal mRNA from transformed cells is largely composed of SV40 specific sequences since some 40% of the selected RNA hybridizes back to SV40 DNA. (SV40 cRNA made by *E. coli* polymerase which has been hybridized to SV40 DNA and recovered also gives about a 50% hybridization during a second exposure to DNA.)

It has been proposed that mRNA in eukaryotic cells derives from the regulated cleavage of HnRNA[8,21]. Although our results do not prove that high molecular weight HnRNA serves as a precursor to polysomal mRNA, they are consistent with the possibility that virus specific mRNA molecules in transformed cells may be derived through the selective cleavage of cell specific sequences from high molecular HnRNA molecules which contain both viral and cellular sequences.

[1] Sambrook, J., Westphal, H., Srinivasan, P. R., and Dulbecco, R., *Proc. US Nat. Acad. Sci.*, **60**, 1288 (1968).
[2] Gelb, L., Kohne, D., and Martin, M., *J. Mol. Biol.*, **57**, 129 (1971).
[3] Westphal, H., and Dulbecco, R., *Proc. US Nat. Acad. Sci.*, **59**, 1158 (1968).
[4] Oda, K., and Dulbecco, R., *Proc. US Nat. Acad. Sci.*, **60**, 525 (1968).
[5] Aloni, Y., Winocour, E., and Sachs, L., *J. Mol. Biol.*, **31**, 415 (1968).
[6] Martin, M., and Axelrod, D., *Proc. US Nat. Acad. Sci.*, **64**, 1203 (1969).
[7] Green, M., *Ann. Rev. Biochem.*, **39**, 701 (1970).
[8] Lindberg, U., and Darnell, J. E., *Proc. US Nat. Acad. Sci.*, **65**, 1089 (1970).
[9] Eagle, H., *Science*, **130**, 432 (1959).
[10] Carp, R. I., and Gilden, R. V., *Virology*, **28**, 150 (1966).
[11] Vinograd, J., Lebowitz, J., Radloff, R., Watson, R., and Laipis, P., *Proc. US Nat. Acad. Sci.*, **53**, 1104 (1965).
[12] Gillespie, D., and Spiegelman, S., *J. Mol. Biol.*, **12**, 829 (1965).
[13] Soeiro, R., and Darnell, J. E., *J. Mol. Biol.*, **44**, 551 (1969).
[14] Pagoulatos, G., and Darnell, J. E., *J. Mol. Biol.*, **54**, 517 (1970).
[15] Denhardt, D., *Biochem. Biophys. Res. Commun.*, **23**, 641 (1966).
[16] McConaughy, B., Laird, C., and McCarthy, B., *Biochemistry*, **8**, 3289 (1969).

[17] Darnell, J. E., and Balint, R., *J. Cell Physiol.*, **76**, 349 (1970).
[18] Westphal, H., *J. Mol. Biol.*, **50**, 407 (1970).
[19] Melli, M., and Bishop, J. O., *J. Mol. Biol.*, **40**, 117 (1969).
[20] Britten, R. J., and Kohne, D. E., *Science*, **161**, 529 (1968).
[21] Scherrer, K., Latham, H., and Darnell, J. E., *Proc. US Nat. Acad. Sci.*, **49**, 240 (1963).
[22] Penman, S., Scherrer, K., Becker, Y., and Darnell, J. E., *Proc. US Nat. Acad. Sci.*, **49**, 654 (1963).

VII
Temperature-Sensitive Mutant Studies of Cell Transformation in Polyoma Virus

INDUCTION OF CELLULAR DNA SYNTHESIS BY POLYOMA VIRUS

R. Dulbecco, L. H. Hartwell, and M. Vogt

The transformation of normal cells into tumor cells by polyoma virus is caused by the interaction of susceptible cells with the DNA of the virus. Thus, purified polyoma virus DNA has been shown to transform cells cultured *in vitro*,[1] whereas the empty protein shells of the virus do not produce this effect.[2] Consequently, a knowledge of the functions of the viral genes is basic to an understanding of the mechanisms of cell transformation. With the hope of identifying these functions, we have initiated a study of the biochemical events which occur after the cytocidal infection of mouse kidney cells by polyoma virus. This article describes the effects of virus infection upon DNA synthesis and upon the activity of enzymes involved in DNA synthesis. One of the most interesting findings was that the virus induces the synthesis of *cellular* DNA in addition to viral DNA.

Materials and Methods.—*Virus:* The polyoma virus used was of the large plaque type.[3] It was grown and purified according to Winocour.[4] The virus was assayed by plaque formation on secondary mouse embryo cultures.

Cell cultures: The cells were derived from kidneys of 10–14-day-old mice. After trypsinization, a large number of cells (0.1 ml of packed cells per 100-mm Petri dish) were plated onto plastic Petri dishes in a reinforced[5] Eagle's medium with 10% horse serum. Within 24–48 hr after plating, the cultures consisted primarily of confluent layers of epithelial cells. These dense cultures had a very low DNA-synthesizing activity from the second day on.

Infection of cultures: The cultures were infected 48–64 hr after the plating of the cells. The medium was removed and 0.4 ml of a suspension of purified virus containing 5×10^9 PFU/ml was added. The cultures were incubated for 1 hr at 37°C in a well-humidified incubator flushed with a CO_2-air mixture; after this time fresh medium was added. All control cultures were mock-infected under identical conditions but without virus.

DNA synthesis: The amount of H^3-thymidine taken up into DNA was determined from the radioactivity in acid-insoluble material (H^3-thymidine did not label RNA). The amount of P^{32}-orthophosphate taken up into DNA was determined after SDS-phenol extraction of the cultures and alkali digestion of the RNA.

SDS-phenol extraction of DNA: Cell cultures were washed 4 times with Tris-buffered saline (TD:NaCl, 0.8%; KCl, 0.038%; Na_2HPO_4, 0.01%; Sigma 7–9, 0.3%); the cells were removed in a SDS-solution containing 0.15 *M* NaCl, 1.5×10^{-3} *M* Na-citrate, 0.1 *M* Tris buffer, pH 8.0, 0.5% sodium dodecyl sulfate (Coleman, Matheson, and Bell), and 5×10^{-3} Versene. After addition of Na-trichloroacetate to a final concentration of 0.3 *M*, the mixture was extracted 3 times with 80% phenol (in 20% 0.5 *M* Tris buffer, pH 8.0). The aqueous phase was extracted with 2 vol of ether and dialyzed against SSC/10 (SSC:0.15 *M* NaCl, 0.015 *M* Na-citrate).

Chromatography of DNA: The methylated albumin column was prepared according to the procedure of Mandell and Hershey.[6] Fractions were collected and precipitated with 5% trichloroacetic acid after the addition of 2 mg of bovine serum albumin.

The DNA-DNA hybridization: The DNA-DNA hybridization was carried out according to the procedure of McCarthy and Bolton,[7] except that final concentration of agar was 4%. The DNA for the agar column was isolated from baby mouse kidneys by SDS-phenol extraction; it was treated with RNAse, re-extracted, and collected by ethanol precipitation.

The concentration of DNA in the agar was 0.31 mg per gram of DNA-agar. Hybridization was carried out by incubating 0.3 mg of DNA-agar with 0.3 ml of heat-denatured DNA in 2 × SSC for 16 hr at 60°C. The DNA had been sonicated for 5 min in an MSE ultrasonic disintegrator. The mixture was then placed in a jacketed column and washed 10 times with 10-ml portions of 2 × SSC at 60°C in order to remove the unhybridized DNA. The hybridized DNA was subsequently eluted at 75°C with 5 washes of SSC/100 (10 ml each). To each fraction, 0.2 mg of herring sperm DNA and 1 ml of 100% trichloroacetic acid were added in the cold to precipitate the eluted DNA. The precipitates were collected on Millipore filters and counted in a liquid scintillation counter.

The *assay for infectious DNA* was carried out by plaque formation[8] after SDS-phenol extraction.

RNA synthesis: The amount of H^3-uridine taken up into cell RNA was determined by isolating the RNA mononucleotides after alkali digestion.

Protein synthesis: The amount of H^3-leucine taken up into acid-insoluble material was determined.

Enzyme assays: Thymidine kinase: Extracts were prepared by homogenizing cells in cold distilled water containing 2-mercaptoethanol (1/10,000 v/v); Tris buffer of pH 7.9 was added to a final concentration of 10^{-2} *M*, and the extracts were centrifuged for 10 min at 12,000 × *g*. The supernatants were assayed following the method of McAuslan[9] by using H^3-thymidine as substrate. The phosphorylated derivatives were separated from thymidine by paper chromatography in an isopropanol-NH_3 solvent.[10]

dCMP-deaminase: Extracts were prepared by homogenizing cells in a buffer solution containing 0.03 *M* phosphate buffer, pH 7.3, 2 × 10^{-4} *M*, $MgCl_2$, 3 × 10^{-2} *M* NaF, and 2 × 10^{-4} *M* dCTP. The assay conditions were similar to those of Maley and Maley[11] with H^3-dCMP as the substrate. dUMP was separated from dCMP by chromatography in a solvent consisting of 85% isopropanol, 15% 0.3 *N* HCl.

DNA polymerase: The extracts prepared for the thymidine kinase assays were used. The assay method was that of Keir[12] with heat-denatured mouse kidney DNA as primer; the labeled triphosphate was H^3-dATP.

Results.—Rate of DNA synthesis: DNA synthesis in mouse kidney cultures was studied either by following the incorporation of H^3-thymidine or P^{32}-orthophosphate, or by assaying for infectious DNA.

Table 1 shows the extent of incorporation of H^3-thymidine and of P^{32} into the DNA of infected and control cells between 16 and 30 hr after virus infection. The incorporation of either label was approximately 10 times higher in the infected cells than in the control cells. This result is in agreement with previous cytological findings.[13] The amount of DNA synthesized during this period in the infected cells was estimated from a comparison with the number of counts incorporated into growing mouse embryo cells under identical conditions of labeling within a period of doubling of total DNA. It was found to correspond to about 25 per cent of the pre-existing total cellular DNA, under the assumption that the efficiency of incorporation of H^3-thymidine into DNA was equal in the two cell systems.

The time course of incorporation of H^3-thymidine into the DNA of the infected cells and the appearance of infectious DNA are shown in Figure 1. In this experiment, the label was added immediately after virus adsorption. The results show that the incorporation of H^3 into DNA is roughly the same in control and in infected cells for the first 16 hr. After this time period, however, there is a sharp increase in the rate of DNA synthesis in the infected cells. A rapid rate of DNA synthesis continues in these cells until about 40 hr after infection; at this time DNA synthesis appears to be turned off. Surprisingly, the formation of infectious DNA is not synchronous with the incorporation of H^3-thymidine into DNA.

Nature of DNA synthesized: Viral and cellular DNA can be separated on a methylated albumin column presumably because the polyoma DNA is smaller than the extracted cellular DNA and uniform in size. Figure 2 shows the elution pattern of a mixture of P^{32}-labeled viral DNA isolated from purified virus and H^3-labeled cellular DNA from an uninfected culture of mouse kidney cells. The viral DNA elutes first from the column, relatively uncontaminated by cellular DNA.

TABLE 1

	P^{32} (cpm/culture)	H^3 (cpm/culture)
Infected culture	80×10^5	90×10^4
Uninfected culture	7.2×10^5	8×10^4

Incorporation of label into the DNA of infected and uninfected cells. Sixteen hr after virus absorption 0.25 mC of $Na_2HP^{32}O_4$ (neutralized, carrier-free) in 10 ml of phosphate-free medium was added to one infected and one uninfected culture. Similarly, 5×10^{-4} mmoles of thymidine and 0.25 mC H^3-thymidine (1.9 C/mmole) in 10 ml of medium were added to one infected and one uninfected culture. DNA was purified by SDS-phenol extraction. An aliquot was incubated with NaOH (1 *N*, 16 hr at room temperature) and precipitated with carrier bovine serum albumin in 5% trichloroacetic acid. Precipitate was collected and washed on a Millipore filter (grade HA) and counted in a liquid scintillation counter.

In order to examine the nature of the DNA synthesized after viral infection, a

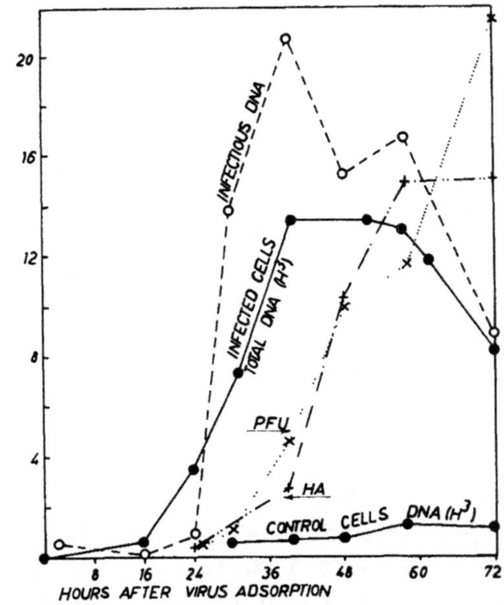

FIG. 1.—Time course of syntheses in the infected cells. Immediately after the adsorption period, 5 ml of medium containing 5×10^{-5} *M* thymidine and 25 mC of H^3-thymidine were added to each of several infected and uninfected cultures. At various times 2 infected and 2 control cultures were removed, and the amount of H^3 taken up into DNA was determined. Other cultures were infected and incubated similarly, but without adding H^3-thymidine. Pairs of these cultures were used for determining the titer of infectious DNA after SDS-phenol extraction. For determining the infectious titer (PFU) and hemagglutinating titer (HA) the cells were scraped off into the culture medium, which was then frozen-thawed three times. Every point represents the average of two determinations. The scale of the ordinate is linear, in arbitrary units. One unit corresponds to 6.7×10^2 PFU of infectious DNA, to 10^9 PFU of virus, and to 10^4 hemagglutinating units per culture.

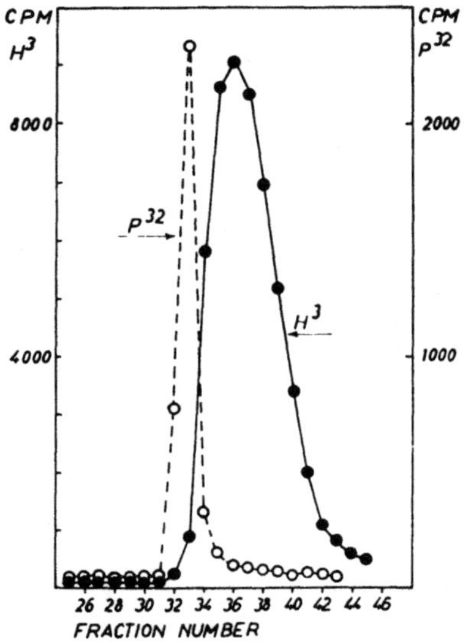

FIG. 2.—Fractions from a column of methylated albumin. The H^3-labeled infected cell DNA (● ——) is from the experiment described in the legend to Table 1. The P^{32}-labeled viral DNA (○ - - -) was phenol-extracted from a purified virus preparation. The total number of counts present in each fraction is given.

mixture consisting of DNA isolated from infected cells and DNA isolated from noninfected cells was chromatographed on a column of methylated albumin. The DNA from the infected cells was labeled by a pulse of H^3-thymidine from 16 to 30 hr after virus adsorption. The control cellular DNA was labeled with P^{32} by giving to an uninfected culture a pulse of $Na_2HP^{32}O_4$ from 16 to 30 hr after a mock infection. The elution pattern of this mixture from the column is shown in Figure 3. The first peak to elute from the infected culture corresponding to the position of viral DNA is not present in the control culture. The surprising finding is that about two thirds of the DNA synthesized in the infected culture elutes late, in a position corresponding to that of cellular DNA.

In order to test whether the late-eluting DNA is cellular, rather than a viral DNA with unusual elution properties, experiments of DNA-DNA hybridization with mouse kidney DNA were performed. A large quantity of DNA was isolated from an infected culture which had been labeled with H^3-thymidine 16–30 hr after virus adsorption; this DNA was then chromatographed on a column of methylated albumin. The fractions were tested for infectivity. The elution pattern from the column and the infectivity of each fraction are presented in Figure 4. It can be seen that infectivity is present only in the early eluting peak. Three fractions from the early peak were combined (*A*) as were three fractions from the late peak (*B*). These were exposed to intense sonication, and tested for their ability to hybridize with mouse kidney DNA by the agar-column method.[7] In the first experiment, samples from *A* and *B* with approximately the same amount of radioactivity were hybridized with mouse kidney DNA. As shown in Table 2, experiment 1, the efficiency of hybridization of the DNA from peak *B* was 10 times greater than that of the DNA from peak *A*. Fractions *A* and *B* differ in total DNA content, since peak *B* contains all the pre-existing cellular DNA. This difference was eliminated in a second experiment by adding excess cellular DNA to both samples; they were then tested for their ability to hybridize with mouse kidney DNA. As shown in Table 2, experiment 2, the DNA of sample *B* again hybridized 10 times more effectively than that of sample *A*. The maximum extent of hybridization with homologous DNA was determined by hybridizing H^3-labeled DNA from uninfected cells with the same DNA preparation of mouse kidney. The extent of hybridization was similar to that of the DNA of peak *B*, as seen in Table 2, experiment 3. This shows that the DNA of peak *B* is mostly cellular DNA.

Two points thus emerge from the study of DNA synthesis in the virus-infected cells: (1) the total rate of synthesis is increased by a factor ten over that of the control cells; (2) most of the DNA synthesized is cellular. We therefore conclude that the virus induces the synthesis of cellular DNA.

Time course of viral and cellular DNA synthesis: The times and rates of synthesis of total DNA and of infectious viral DNA in the infected cells are markedly different, as seen in Figure 1. To see whether the appearance of infectious DNA reflects the synthesis of viral DNA molecules, the following experiment was carried out. The DNA of separate cultures pulse-labeled with H^3-thymidine between 20 and 22, 22 and 24, and 30 and 32 hr after infection, was fractionated in a column of methylated albumin. The ratios of cellular to viral DNA, estimated from the elution patterns, showed that the incorporation of radioactivity into viral DNA molecules corresponds to the appearance of infectious DNA.

Rate of RNA and protein synthesis: No large differences were observed in the rates of incorporation of H^3-uridine into total cell RNA or of H^3-leucine into total cell protein of infected and noninfected cultures from 0 to 30 hr after infection.

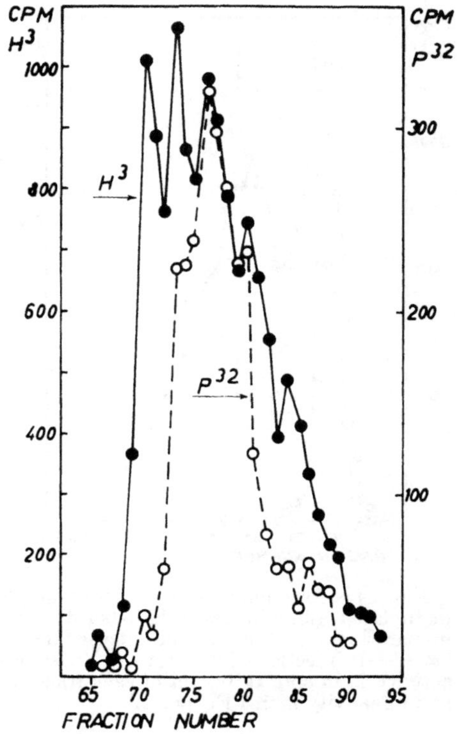

FIG. 3.—Fractions from a column of methylated albumin. The P^{32}-labeled DNA from uninfected cells (○- - -) and the H^3-labeled DNA from infected cells (● ——) are from the experiment described in the legend to Table 1. The total number of counts present in each fraction is given.

Changes in enzyme activities: The dramatic changes in the rate of DNA synthesis after infection prompted a study of the behavior of enzymes involved in DNA synthesis. The activities of the following enzymes were studied: thymidine kinase, deoxycytidine monophosphate-(dCMP-)deaminase, dCMP-kinase, deoxyadenosine monophosphate-(dAMP-)kinase, and DNA polymerase. All the enzymes studied retain the activity they have in the control cells until about 12 hr after infection. At this time, the activities of thymidine kinase, dCMP-deaminase, and DNA polymerase begin to increase (Fig. 5). Thus, the increase of the three enzyme activities follows the same time course until 28 hr, when a maximum is reached. The time at which 50 per cent of the increase is achieved is almost identical for the three enzymes. In the control cells the activities steadily decline.

The activities of dCMP and dAMP-kinase showed no changes. Thus, the pattern of changes of enzyme activities is similar to that observed in liver regeneration.[14]

Role of the capsid protein in the induction of the enzyme activities: In the infection procedure, each cell is exposed to thousands of virus particles; thus, the role of the viral protein in causing the observed effects must be evaluated. This was done by taking advantage of the separation of the virus into two bands upon equilibrium

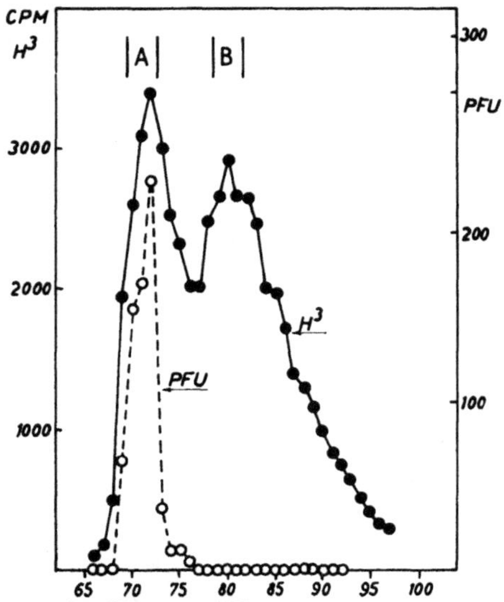

FIG. 4.—Fractions from a column of methylated albumin; same H^3-labeled DNA preparation of Fig. 3. H^3-labeled DNA (● ——); infectious DNA (○ - - -). Radioactivity is in cpm in 0.1 ml of each fraction, and infectivity is in PFU per ml.

density gradient centrifugation in CsCl: a bottom band containing complete virus particles and a top band containing primarily empty capsids. The bottom band and the top band were obtained from the same virus lysate; they had a comparable number of particles as judged from their hemagglutination titer. The infectious titer, however, was 7 times higher in the viral band; thus, the top band was contaminated with about 15 per cent infectious virus particles. Parallel cultures were infected with different amounts of the two virus preparations. Extracts of the cultures were made 24 hr after infection and assayed for thymidine kinase and dCMP-deaminase activities. The results showed that the activities of the cultures were related to the infectious and not to the hemagglutinating titers. Thus, the enzyme activities are induced by the infectious particles and not by the viral capsids.

Discussion.—The induction of cellular DNA synthesis appears to be a unique property of polyoma virus among the animal viruses studied thus far. For instance, studies with the DNA-containing vaccinia and pseudorabies viruses never showed stimulation but often inhibition of host DNA synthesis following virus infection.[15]

The property of polyoma virus to cause the enhancement of enzyme activity and DNA synthesis may well be related to the ability of polyoma virus to transform normal cells into tumor cells, for the following reasons.

The multiplication of mouse kidney cells in cultures is regulated by a control phenomenon based on population density; that is, when the cells have formed a monolayer in the Petri dish, the rate of DNA synthesis and mitosis decreases greatly, so that the cells do not pile up in a multilayered sheet. The cultures employed in these experiments had reached that stage and had, therefore, a very low rate of DNA synthesis. The infection of these cultures with polyoma virus caused the failure of the control phenomenon and induced the synthesis of new

TABLE 2

Expt.	Peak	Unhybridized DNA (cpm fractions 1–10)	Hybridized DNA (cpm fractions 10–15)	% Hybrid
1	*A*	6593	93	1.4
	B	4495	953	17.5
2	*A* + unlabeled cell DNA	7111	83	1.2
	B + unlabeled cell DNA	6396	1049	15.4
3	H^3-cell DNA	6850	1658	19.5

Hybridization of DNA from peaks *A* and *B* (Fig. 4) with mouse kidney DNA. Sample *A* contains about 1 μg of H^3-labeled DNA per ml. Sample *B* contains about 1 μg of H^3-labeled DNA plus 4 μg of unlabeled cell DNA per ml. In experiment 2, 10 μg of sonicated, heat-denatured cell DNA was added to each of the 0.3-ml aliquots of *A* and *B* before hybrid formation. In experiment 3, 10 μg H^3-labeled cell DNA was tested for hybridization.

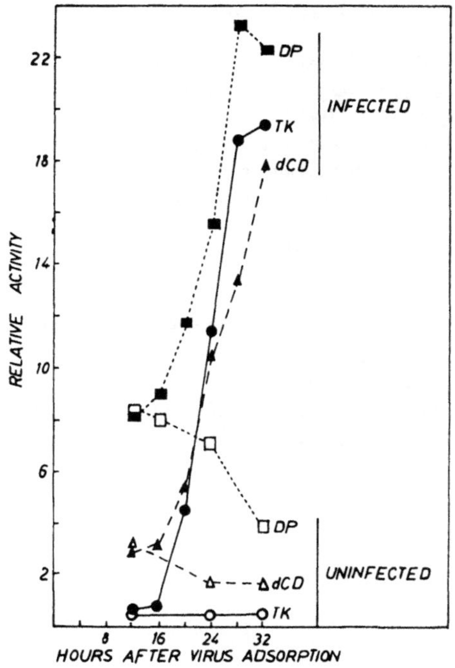

FIG. 5.—Activities of thymidine kinase (TK, ○ ——), dCMP deaminase (dCD, △ - - -), and DNA polymerase (DP, □ . . .) as a function of the time after virus adsorption. Relative activities are proportional to the counts incorporated into products and normalized for protein content of reaction mixture.

cellular DNA. These cells are destined to die as a result of the virus infection; the loss of control, however, is directly analogous to that occurring in cells which are transformed by the virus. Cells transformed *in vitro* by polyoma virus become insensitive to control by population density; they give rise to dense multilayered cultures. The function which the virus performs to release the transformed cells from control may be the same function which the virus performs to induce uncontrolled cell-DNA synthesis in the cytocidal infection of mouse kidney cells. Thus, it may be possible to study events of major importance for cell transformation in a system in which the cells are not undergoing transformation but are being killed by the virus. This system offers a distinct advantage in that nearly all of the mouse kidney cells undergo the cytocidal infection while cell transformation occurs in a small proportion of the cells in the systems so far available.

Summary.—Crowded cultures of mouse kidney cells have a very low rate of DNA synthesis, and very low activities of the enzymes involved in DNA synthesis. After infection with polyoma virus, both the enzyme activities and the rate of DNA synthesis markedly increase. It is of special interest that the DNA synthesized in the infected cells is predominantly cellular. The ability of the virus to stimulate the synthesis of cellular DNA may be related to its tumorigenic property.

The competent technical assistance of Mr. Paul Holdy and Miss Dorothy Ibalio is gratefully acknowledged.

This work was supported by grant CA-07592-01 from the National Cancer Institute, National Institutes of Health.

[1] Crawford, L., R. Dulbecco, M. Fried, L. Montagnier, and M. Stoker, these Proceedings, **52,** 148 (1964).

[2] Abel, P., and L. V. Crawford, *Virology,* **19,** 470 (1963).

[3] Vogt, M., and R. Dulbecco, *Virology,* **16,** 41 (1962).

[4] Winocour, E., *Virology,* **19,** 158 (1963).

[5] Vogt, M., and R. Dulbecco, these Proceedings, **49,** 171 (1963).

[6] Mandell, J. D., and A. D. Hershey, *Anal. Biochem.,* **1,** 66 (1960).

[7] McCarthy, B. J., and E. T. Bolton, these Proceedings, **50,** 156 (1963).

[8] Weil, R., *Virology,* **14,** 46 (1961).

[9] McAuslan, B. R., *Virology,* **20,** 162 (1963).

[10] Chargaff, E., and J. N. Davidson, in *The Nucleic Acids* (New York: Academic Press, 1955), vol. 1, p. 252.

[11] Maley, G. F., and F. Maley, *J. Biol. Chem.,* **237,** PC3311 (1962).

[12] Keir, H. M., *Biochem. J.,* **85,** 265 (1962).

[13] Minowada, J., *Gann,* **25,** 267 (1964).

[14] Davidson, J. N., *The Molecular Basis of Neoplastia* (University of Texas Press, 1962), p. 420.

[15] Kit, S., and D. R. Dubbs, *Virology,* **18,** 274 (1962); Kaplan, A. S., and T. Ben-Porat, *Virology,* **19,** 205 (1963).

Abortive Transformation by Polyoma Virus

Michael Stoker

Exposure of hamster cells to large doses of polyoma virus is known to result in permanent transformation of a small proportion of cells. Experiments with cells in suspension now suggest that a much higher proportion of infected cells undergoes a temporary change, in which at least one characteristic of transformation is acquired for a few cell divisions and then lost again. This suggests that acquisition of the transformed phenotypes does not depend upon stable integration of the viral genome.

AFTER exposure to high multiplicities of polyoma virus, a small proportion of hamster cells grow into clones of transformed cells, but most seem to be unaffected. The transformed clones can be recognized by the random orientation of the cells[1,2] and by the ability of the transformed, but not the untransformed, cells to grow in agar suspension cultures[3]. Stable lines of hamster cells, such as *BHK*21 (ref. 4) and *NIL* 2 (ref. 5), are more sensitive to transformation than are freshly isolated cell strains, but even with these a maximum of about 5 per cent of cells are transformed after exposure to virus. Transformation frequencies can be higher, however, with the 3*T*3 line of mouse cells exposed to *SV*40 virus[6,7].

During earlier studies in this laboratory on the behaviour of polyoma infected *BHK*21 cells in agar suspension cultures, it was noticed that the apparently normal cells which did not grow into recognizable transformed colonies nevertheless were larger than the uninfected controls. Size measurements suggested that these might be microclones arising from cells that had divided only a few times, but this supposition could not be confirmed because of the difficulty of removing cells from agar. Neither was it possible to find out if the supposed microclones were composed of normal cells that had grown for a few divisions, or transformed cells that had failed to continue growth.

Studies on the regulation of cell growth have recently led to the development of an alternative to the agar suspension system which permits the easier manipulation of cells and colonies*. This involves suspension culture in medium containing carboxymethyl cellulose at a concentration sufficient to immobilize the cells, contained in vessels lined with agar to prevent anchorage and spreading of the cells. Untransformed cells including *BHK*21 and *NIL* 2 cells can be maintained in this system for at least 7 days with little or no growth, but are capable of further division when exposed to a suitable stimulus, that is, anchorage to a rigid surface.

This communication describes the behaviour of polyoma infected cells in this suspension system, and shows that the virus can induce a temporary change involving at least one of the characteristics of transformation in a high proportion of cells. With the exception of a few stable transformants, these cells revert to the untransformed state on continued propagation.

Clone *C*13 of *BHK*21 cells and clone *E* isolated by Dr I. A. Macpherson from *NIL* 2 cells were propagated by standard methods, using modified Eagle's medium with 10 per cent calf serum and 10 per cent tryptose broth as growth medium. Suspension cultures* were made by adding one volume of cells to nine volumes of 1·3 per cent carboxymethyl cellulose ('Methocel', Dow type *MC*, 4,000 centipoises) in growth medium ('Methocel' medium), to give a final concentration of 1·17 per cent carboxymethyl cellulose with 2×10^5 cells/ml. Volumes of suspension (1 ml.) were cultured as " 'Methocel' suspension cultures" in 2 cm diameter holes in 0·9 per cent agar (agar-holes), or in 50 mm plastic Petri dishes lined with agar. The agar lining contained Eagle's medium and 10 per cent calf serum. The cells and colonies in 'Methocel' medium were removed with a wide bore pipette, and subsequently freed from 'Methocel' by dilution with growth medium.

The suspension of polyoma virus was an extract of infected mouse embryo cells treated with receptor destroying enzyme[8] containing 5×10^9 P.F.U. ($8{\cdot}6 \times 10^{11}$ particles)/ml. of a cloned small plaque variant of the Toronto strain. Controls were similarly made from uninfected cultures of mouse embryo cells.

Heavy (Band I) and light (Band II) fractions of polyoma virus were made by two cycles of centrifugation in caesium chloride equilibrium density gradients[9]. The virus in each fraction was spun out of the diluted caesium chloride, resuspended in water, and sterilized by filtration through 'Millipore' membranes. Both fractions contained 4×10^{12} particles/ml. More than 98 per cent of the Band I particles and less than 2 per cent of the Band II particles were "full", as judged by lack of penetration of phospho-

* Unpublished observations of Stoker, O'Neill, Berryman and Waxman.

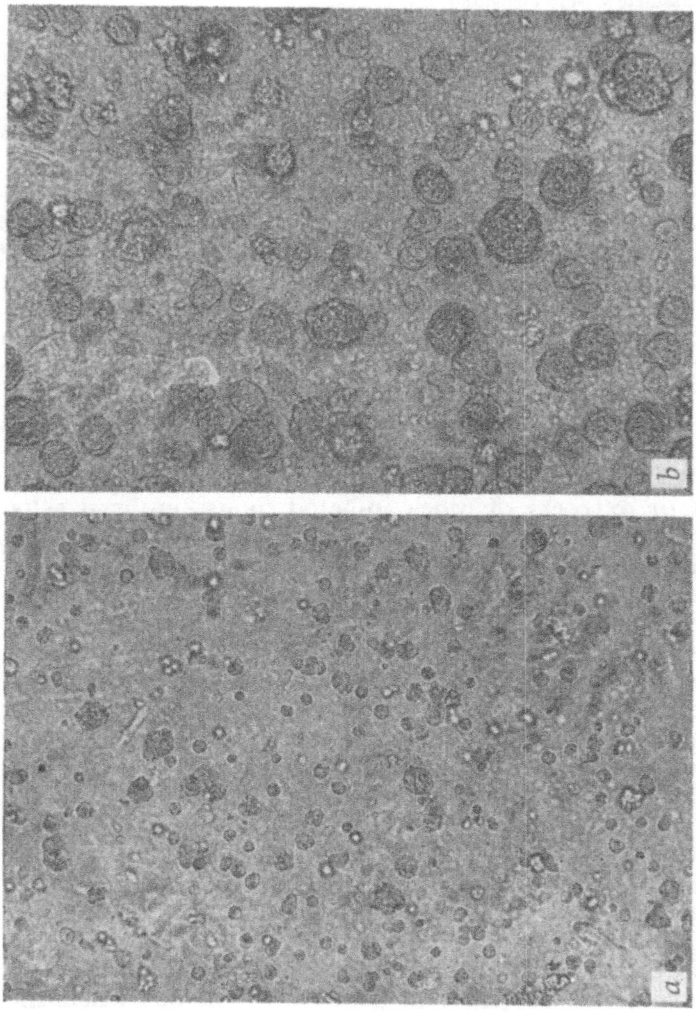

Fig. 1. *BHK*21 cells after 7 days in 'Methocel' suspension culture. (*a*) Uninfected, showing single cells and a few small colonies. (*b*) Cells exposed to polyoma virus before culture (4×10^{10} particles/2×10^5 cells) showing development of colonies. ($\times 65$.)

Table 1. GROWTH OF *BHK*21 CELLS IN 'METHOCEL' CULTURES AFTER EXPOSURE TO DENSITY GRADIENT FRACTIONATED POLYOMA VIRUS

	Particles $\times 10^8$	Plaque forming units $\times 10^6$	Size distribution of 100 cells and colonies Diameter in scale units* 1–5	6–10	11–15	16–20	21–25	26+
Polyoma								
Band I	380	160	10	58	18	10	3	1
(dense)	38	16	17	65	9	3	5	1
Band II	410	2	14	82	2	2	0	0
(light)	41	0·2	30	66	4	0	0	0
No virus	—	—	15	80	3	2	0	0

2×10^5 cells were exposed to virus dose indicated. After incubation for 6 days, diameters were measured.

* One scale unit = 3·9μ.

tungstate: Band I suspensions contained $1{\cdot}6 \times 10^{10}$ P.F.U./ml. and Band II suspensions 2×10^8 P.F.U./ml.

Virus suspensions, diluted when necessary in either *tris* or phosphate buffered saline, containing 0·5 per cent calf serum were added to cells in 'Methocel' medium by gentle stirring and pipetting (0·05 ml. of virus suspension to 1 ml. of 'Methocel' medium containing 2×10^5 cells). The excess virus was not usually removed, but in some experiments 0·1 ml. of virus suspension was allowed to adsorb for 1 h to monolayer cultures of cells, and excess virus was removed by washing. After incubation for a further 4 h in fluid medium the cells were suspended and added to 'Methocel' medium. Controls were exposed to buffer or extracts of mock infected mouse embryo cells.

Direct measurements of the size of cells and colonies in 'Methocel' were made *in situ* with a Watson image shearing eyepiece and expressed in scale units. At the magnification used, one unit equals 3·9μ.

Most uninfected *BHK*21 cells in 'Methocel' remained as single cells or pairs*. About 25 per cent increased in diameter during 7 days as a result of one or two divisions, but less than 5 per cent reached a diameter corresponding to three divisions (see calibration curve in Fig. 2). Addition of undiluted polyoma virus suspension either directly ($4{\cdot}3 \times 10^{10}$ particles, $2{\cdot}5 \times 10^8$ P.F.U./2×10^5 cells) or by earlier adsorption to monolayers ($8{\cdot}6 \times 10^{10}$ particles, 5×10^8 P.F.U./2×10^6 cells) led to growth of 25 per cent–50 per cent of the cells into spherical colonies (see Fig. 1*a* and *b*).

Uninfected *NIL* 2 clone *E* cells also failed to grow in 'Methocel', but 39 per cent grew into colonies when the cells were exposed to polyoma virus ($4{\cdot}3 \times 10^{10}$ particles/2×10^5 cells). Subsequent experiments were all carried out with *BHK*21 cells.

The colonies which developed were not cell aggregates because there was no reduction in the total number of single cells plus colonies. In one experiment, 38 per cent

* Unpublished observations of Stoker, O'Neill, Berryman and Waxman.

of the infected cells and 8 per cent of the control cells had grown into colonies estimated to contain two or more cells by 48 h. At this time the mean count of total cells and colonies/field was 132 ($S.E. \pm 7{\cdot}5$) for the controls, and 131 ($S.E. \pm 6{\cdot}8$) for the infected cultures.

Exposure of cells to the control, virus-free, extract of mouse embryo cells also resulted in some limited growth which exceeded that of the controls exposed to buffer alone. The role of the virus was therefore investigated by using density gradient purified virus, and by comparing the Band I ("full") and Band II ("empty") fractions which are known to differ in DNA content, plaque forming and transforming activity[9,10]. The DNA containing Band I particles caused the cells to grow into colonies as shown by the increase in diameter (Table 1). Similar numbers of "empty" Band II particles had little effect. Comparable results were obtained with another, separately purified, batch of virus. Subsequent experiments were all carried out with density gradient purified Band I fractions.

The number of cells/colony can be deduced from the diameter, and thus the volume in comparison with the volume of single cells. The volume of isolated single cells is variable, however, and that of cells in spherical colonies may not be comparable. Direct measurements of cell numbers in colonies were therefore made by isolating single colonies from infected cultures in microdrops and subsequently exposing them to 0·05 per cent trypsin. In these conditions, the cells separated and could be counted by microscopy. Fig. 2 shows the relationship of cell

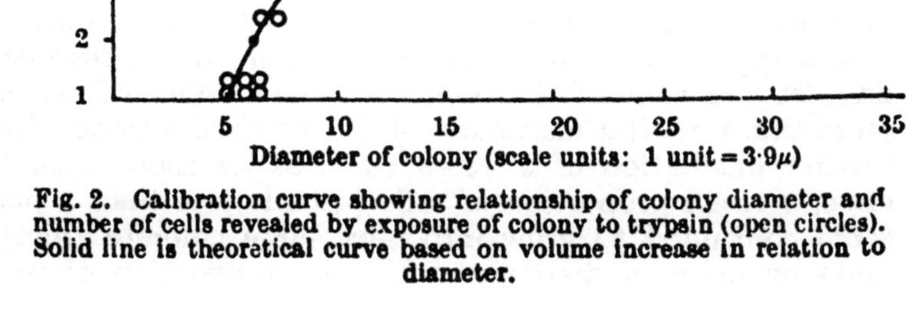

Fig. 2. Calibration curve showing relationship of colony diameter and number of cells revealed by exposure of colony to trypsin (open circles). Solid line is theoretical curve based on volume increase in relation to diameter.

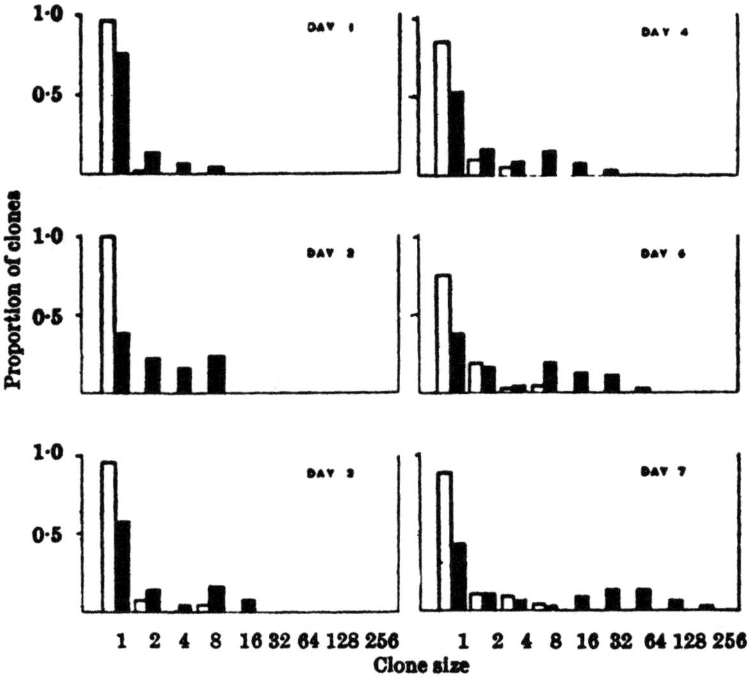

Fig. 3. Growth of *BHK*21 cells in 'Methocel' on successive days after exposure to polyoma virus (4×10^{10} particles/2×10^{5} cells). Clone size calculated from colony diameter using data shown in Fig. 2. Open columns, uninfected; closed columns, infected.

numbers to colony diameter. This fits a theoretical curve based on volume assuming a single cell diameter of 5 units (19·6μ). From this curve, colony sizes were divided into classes corresponding to doubling of cell numbers and thus mean cell divisions. (Diameters of up to six units were classified as single cells, but this is larger than the diameter of cells freshly suspended in 'Methocel' medium* so that the distinction between one, two and ever four cells is considered to be arbitrary.)

Using this calibration curve the frequency distribution of clones of different sizes was determined on successive days after addition of virus ($3{\cdot}5 \times 10^{10}$ particles/2×10^{5} cells) compared with uninfected controls. Fig. 3 shows that more than half the infected cells underwent one, two or three divisions 48 h after addition of virus. Larger clones appeared on succeeding days until, by the seventh day, 38 per cent of the population had divided four or more times, with a maximum of five or six divisions. On further incubation of infected cultures for more than 7 days, a small proportion of cells (usually less than 1 per cent) continued to grow into very large colonies thirty units or more in diameter. A small proportion of uninfected cells grew during the first week, but did not increase in size on further incubation.

The fact that about half the cells failed to grow, even at high ratios of virus : cell, could have been caused by non-viable cells which are unable to grow in 'Methocel', however stimulated. The plating efficiency of the normal cells on glass was 89 per cent when recovered after 1 day in 'Methocel' (unpublished observations), and remained high even after 7 days in suspension, but this does not necessarily signify that growth can occur in suspension. Established transformed cells grow with high efficiency of plating in 'Methocel' medium (80–90 per cent), but two transformed clones recently isolated from 'Methocel' cultures grew with lower efficiency of plating (64 per cent and 58 per cent, respectively (see Table 2)). It is not possible at this stage to judge the role of cell competence in limiting the number which are stimulated at high virus dosage.

Characteristics of Clones

Colonies of various sizes (and single cells) were removed from 'Methocel' at various times after infection by dilution of the 'Methocel' culture in fluid medium. They were then allowed to attach to glass coverslips or plastic Petri dishes in fluid medium. The colonies were stable and were not disintegrated by pipetting. Soon after plating, the colonies attached to the rigid surface, and within 1 or 2 h the cells began to spread and emigrate. By 24 h, clones of up to thirty-two cells were all attached and

Table 2. CHARACTERISTICS OF SELECTED CLONES ARISING IN SUSPENSION CULTURE FOLLOWING POLYOMA VIRUS INFECTION

Clone	Polyoma: T antigen*	Polyoma: Plaque forming units†	Transplantability‡	Percentage growing in suspension: No virus§	Percentage growing in suspension: With polyoma¶	Transformability
MC 9 (orientated)	–	–	0/4	8	70	+‖
10 ,,	–	–	0/4	8	76	
16 ,,	–	–	0/4	4	68	
24 ,,	–		0/4	8	40	
8 (random)	+	–	4/4	64		
15 ,,	+			58		

* With 1/20 dilution of immune serum from tumour bearing hamsters, 10 per cent cell suspensions, and 2 units of complement. +, complete fixation; –, no fixation.

† Number of plaques with homogenate of 4×10^6 cells.

‡ Number of hamsters with tumours/number inoculated, 6 weeks after subcutaneous injection of 10^6 cells.

§ Percentage of cells undergoing three divisions or more in 'Methocel' suspension cultures.

¶ Percentage of cells undergoing three divisions or more in 'Methocel' suspension cultures after exposure to polyoma virus, as described in text.

‖ Three of forty-seven clones that appeared in 'Methocel' were fully transformed.

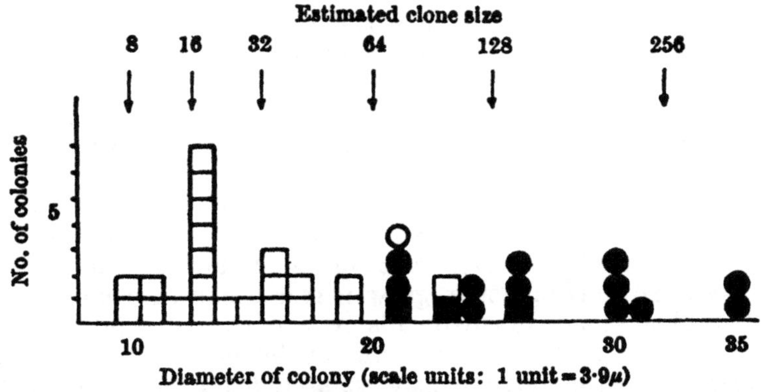

Fig. 4. Identification of selected clones of different sizes appearing in 'Methocel' suspension cultures of *BHK*21 cells exposed to polyoma virus. Each square or circle represents one clone. Squares, after 7 days; circles, after 13 days in 'Methocel'. Open, clones reverting to normal; closed, transformed clones.

spread, and larger clones were surrounded by a wide halo of elongated cells. Clones arising from single cells by division after attachment could be distinguished by their size.

When the clones were examined after growth on glass for more than 2 or 3 days, most showed parallel orientation and were indistinguishable from normal, untransformed colonies of *BHK*21 cells. A few clones, however, gave rise to typical transformed colonies in which the cells were randomly arranged. From two similar experiments, forty-seven individual clones were isolated at random and cultured separately. Of these, three failed to grow sufficiently and could not be identified, five were transformed and thirty-nine seemed to be normal in cell arrangement. Because about 33 per cent of the infected cells developed into colonies, it can be calculated that approximately 3 per cent of the original cells became fully transformed. This is higher than the proportion which grow into very large colonies (shown earlier), but is comparable with the frequency of transformation determined by other methods.

When recovered from 'Methocel' 6 or 7 days after infection, the fully transformed clones were generally larger than the clones which seemed to be normal; after longer incubation in 'Methocel' nearly all large clones (sixty-four cells or more) which developed gave rise to transformed colonies when replated on glass. Fig. 4 shows the orientation characteristics of clones removed after 7 and 13 days in suspension; their size was measured before replating on glass. (These were selected for size distribution and not picked at random.) Despite the preponderance of transformed clones in the larger size groups, it should nevertheless be noted that several of the apparently normal clones grew to at least sixty-four cells.

Four clones with orientated and two with random arrangement were examined for further characteristics which are shown in Table 2. T antigen was detected in complement fixation tests with the two random clones, but the level was low (strongly positive at the highest concentration tested, 10 per cent cell suspension). These clones grew in suspension, and one which was tested rapidly produced tumours when transplanted in hamsters. Four orientated clones contained no T antigen detectable in 10 per cent suspensions. None produced tumours in hamsters. The behaviour of all four in 'Methocel' suspension cultures resembled that of uninfected *BHK*21 cells, including the response to polyoma virus. The clones appearing after infection of clone *NC*9 were further characterized, and fully transformed clones were found in addition to those that reverted to normal.

Emigration Patterns

Although most of the clones of infected cells were normally orientated in arrangement from 2–3 days on glass, the cells which emigrated from the colonies in the first 48 h were somewhat randomly orientated, and at this early stage nearly all clones seemed to be transformed. Later, as already indicated, the cells from most colonies became orientated in parallel. This raised the possibility that the infected cells had temporarily developed another transformed character, namely, random arrangement, in addition to their ability to grow in suspension.

The distinction between random and orientated arrangement cannot yet be decided quantitatively and objectively. A random type emigration pattern might be a feature of any cells, normal or transformed, during emigration from a spherical colony on to a flat surface, and the pattern is clearly influenced by cell density. Emigration patterns from colonies arising from infected cells were, however, compared with those from colonies of established transformed cells and of normal cells growing in suspension. Because normal cells rarely grow in suspension, "colonies" were obtained from two sources: (1) by using rare clones from uninfected cells which grew in suspension; and (2) by making mock colonies using cell aggregates.

Aggregation occurs when suspended cells in fluid medium are prevented from attaching to a rigid surface, for example, by coating the vessel with agar. Aggregates of suitable size of normal and also transformed *BHK*21 cells were allowed to form for 24 h and were then centrifuged out of the medium and resuspended in 'Methocel'. On successive days thereafter they were removed

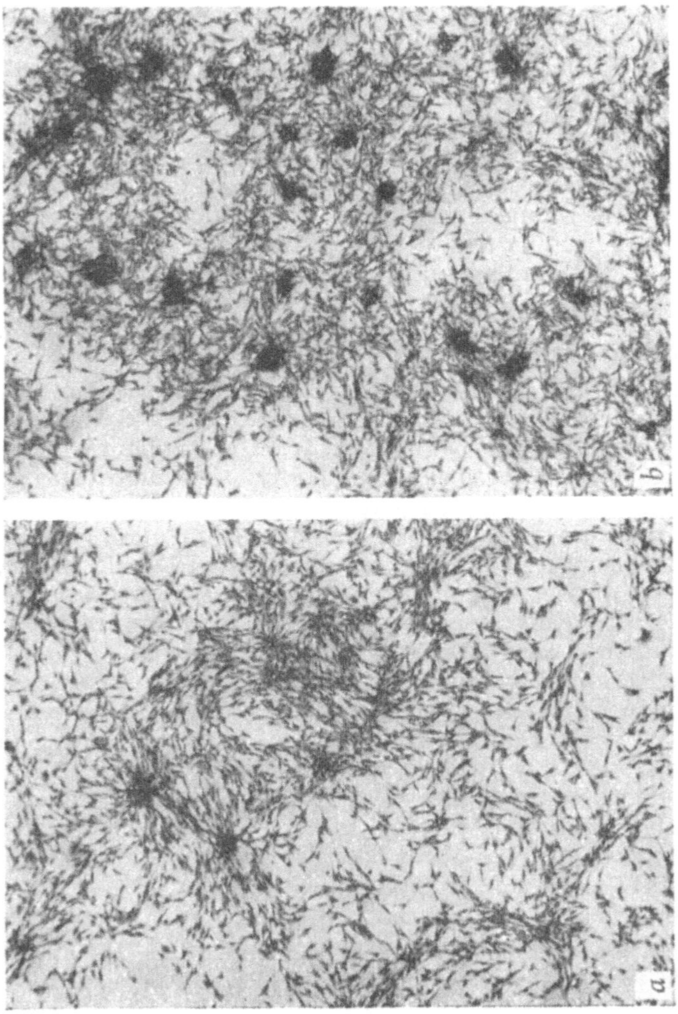

Fig. 5. Emigration patterns on glass 24 h after attachment of colonies previously developed for 72 h in 'Methocel'. (*a*) Colonies of normal cells appearing spontaneously in 'Methocel' suspension culture. (*b*) Colonies arising after polyoma virus infection. (Giemsa, × 25.)

replated in fluid medium on glass coverslips for examination of the emigration patterns.

Cells emigrating from normal "colonies", whether true clones or aggregates, usually showed some ordered directional arrangement soon after flattening, while established transformed cells showed a different, more random, pattern. A high proportion of the clones arising from freshly infected cells showed emigration patterns different from normal cells, and closer to those of established transformed cells. These abnormal patterns were lost after about 48 h on glass from all except the few permanently transformed clones. Typical patterns of emigration from normal colonies are shown in Fig. 5 in comparison with those from colonies arising from recently infected cells.

Significance of Results

Growth of most types of normal cell, and of the *BHK*21 and *NIL* 2 cell lines, is "anchorage dependent". Thus growth of suspended cells stops, or is delayed, in the *G*1 phase of the cell cycle until the inhibition is removed by anchorage to a rigid surface (unpublished observations). Transformation into a tumour cell can result in the permanent loss of this anchorage dependence, with the well known consequence that some transformed cells grow in suspension.

The results reported here show that polyoma virus can initiate up to six consecutive cell divisions in stationary normal cells in suspension cultures. The clones which result then fall into two classes: (1) a large number in which the cells regain the characteristics of untransformed cells, including their incapacity for further growth in suspension, and (2) a few in which the characteristics of transformation are perpetuated indefinitely. The temporary change, like permanent transformation[10], is not initiated by empty virus capsids but requires "full" DNA containing particles of polyoma virus. It is therefore reasonable to assume that it is caused by the viral DNA or possibly host DNA enclosed in viral capsids[11,12].

Thus the infected cells can develop one of the characteristics of transformation, that is, loss of anchorage dependence for a limited period only. From a recent report it is possible that this change is also accompanied by a synthesis of T antigen[13]. The fact that the alteration in behaviour only lasts for a few divisions would make it difficult to identify in conventional culture systems; moreover, it has not been possible to investigate the temporary acquisition of other characteristics of transformation such as increased transplantability and high glycolytic capacity. Nevertheless, there is some evidence from the emigration patterns (admittedly subjective and non-quantitative) that there is a temporary change in cell arrangement during this short period. Because at least one, and possibly two, of the characteristics of transformed cells are gained and then lost, the phenomenon is termed "abortive transformation".

It is interesting to recall that, in their early work on polyoma transformation *in vitro*, Sachs and his colleagues reported that mass cultures of hamster fibroblasts developed a general change in arrangement a few days after exposure to virus. This was difficult to understand at the time because it was known that only a very small proportion of cells were fully transformed[14]. It now seems likely, however, that the change visible in the mass cultures was caused by abortive transformation in a high proportion of cells. Abortive transformation, if accompanied by synthesis of transplantation antigen, might also account in part for the transplantation immunity which develops in animals inoculated with virus.

Polyoma and *SV*40 viruses are known to induce DNA synthesis of the host cell in dense stationary cultures[15], or even X-irradiated cultures[16], of several types of cells. Some studies have involved cells which are destroyed by the viruses, but it has been shown that in human diploid cells and 3*T*3 cells the DNA synthesis may be followed by stimulation of mitosis[16,17]. It has been noted previously that infection with *SV*40 virus increased the efficiency with which a polyoma transformed line of hamster cells[18] formed colonies, and that *SV*40 DNA (but not intact virus) stimulated the growth of *BHK*21 cells[19].

There is a delay of 16 h between infection and the appearance of the enzymes concerned with DNA synthesis[20]. Part of this delay might be explained if the virus affects a control mechanism early in the *G*1 phase of the cell cycle, affected both by density and anchorage. Enzyme and DNA synthesis would then follow with a time lag as a normal progress through the remainder of the *G*1 and *S* phases, leading eventually to cell division, unless prevented by vegetative growth and cell lysis.

In the experiments reported here, it was not possible to induce cell division in more than 33 per cent, or occasionally 50 per cent, of the cells, and this ceiling was not affected by increasing the virus dose. It is possible, however, that some of the cells were damaged and unable to divide in any circumstances. Preliminary experiments do not suggest that DNA synthesis is initiated in a higher proportion of cells than of those that divide.

Though the adsorbed multiplicity is not known the ratio of virus : cell used in these experiments was very high (up to 10^5 particles/cell), and when directly added to the 'Methocel' cultures the virus remained throughout the experiment. Even when excess virus was removed by adsorption and washing before 'Methocel' culture a large amount of cell associated virus would be expected to remain[21]. It is not known at this stage if each cell division cycle is initiated afresh by uncoating of residual virus, or if one initial event will result in several divisions. The proportion of cells which were not growing remained the same on successive days with no late recruits, but this might be caused by cell incompetence. Further experiments are in progress with limiting doses of virus to determine whether growth stimulation occurs repeatedly.

It can be concluded from these results that physiological changes associated with transformation do not depend entirely on the stable integration of the viral genome. Though temporary and unstable integration (for example, by repeated insertion and excision[22]) cannot be ruled out in abortively transformed cells, it is simpler to suppose that the transformed phenotype is caused by viral gene product(s) which may function whether the viral DNA is free or attached to cellular DNA. If so, it is possible that physiological transformation, like bacterial conversion[23], may even be manifest during the vegetative cycle in the short period before cell lysis occurs.

I thank Miss Vivienne Waxman for expert technical assistance. I also thank Dr L. V. Crawford and his colleagues for preparing purified polyoma virus, Dr E. A. C. Follett for particle counts and Dr I. A. Macpherson and his colleagues for undertaking the complement fixation tests. This work was supported by the Medical Research Council.

1 Stoker, M., and Macpherson, I., *Virology*, **14**, 359 (1961).
2 Sachs, L., Medina, D., and Berwald, Y., *Virology*, **17**, 491 (1962).
3 Macpherson, I., and Montagnier, L., *Virology*, **23**, 291 (1964).
4 Macpherson, I., and Stoker, M., *Virology*, **16**, 147 (1962).
5 Diamond, L., *Intern. J. Cancer*, **2**, 143 (1967).
6 Todaro, G. J., and Green, H., *Virology*, **28**, 756 (1966).
7 Black, P. H., *Virology*, **28**, 760 (1966).
8 Crawford, L. V., *Virology*, **18**, 177 (1962).
9 Crawford, L. V., Crawford, E. M., and Watson, D. H., *Virology*, **18**, 170 (1962).
10 Abel, P., and Crawford, L. V., *Virology*, **19**, 470 (1963).
11 Winocour, E., *Virology*, **31**, 15 (1967).
12 Michel, M. R., Hirt, B., and Weil, R., *Proc. US Nat. Acad. Sci.*, **58**, 1381 (1967).
13 Meyer, G., Lherisson, A.-M., and Bonneau, H., *Bull. Cancer*, Paris, 54/1, 79 (1967).
14 Sachs, L., in *Tumour Viruses of Murine Origin*, Ciba Symposium, 1962 (edit. by Wolstenholme, G. E. W., and O'Connor, M.), 380 and discussion 392–3 (Churchill Ltd., London).
15 Dulbecco, R., Hartwell, L. H., and Vogt, M., *Proc. US Nat. Acad. Sci.*, **53**, 403 (1964).
16 Gershon, D., Sachs, L., and Winocour, E., *Proc. US Nat. Acad. Sci.*, **56**, 918 (1966).
17 Sauer, G., and Defendi, V., *Proc. US Nat. Acad. Sci.*, **56**, 452 (1966).
18 Macpherson, I. A., in *Wistar Inst. Bienn. Res. Rep.* 1962–1963, 52.
19 Black, P. H., and Rowe, W. P., *Proc. US Nat. Acad. Sci.*, **54**, 1126 (1965).
20 Hartwell, L. H., Vogt, M., and Dulbecco, R., *Virology*, **27**, 262 (1965).
21 Fraser, K. B., and Gharpure, M., *Virology*, **18**, 505 (1962).

[22] Thomas, R., in *The Molecular Biology of Viruses*, Society of General Microbiology Symposium, 1968 (edit. by Crawford, L. V., and Stoker M. G. P.), 315.

[23] Uetake, H., Luria, S. E., and Burrous, J. W., *Virology*, 5, 68 (1958).

CELL-TRANSFORMING ABILITY OF A TEMPERATURE-SENSITIVE MUTANT OF POLYOMA VIRUS

MICHAEL FRIED

Conditional lethal mutants have been shown to be extremely useful in the study of the physiology of viral reproduction and maturation.[1,2] Since such mutants are distributed over most of the genetic map,[1,2] they could prove to be valuable in the study of tumor viruses. They may be used to "tag" almost all of the viral genes, a property of obvious physiological and genetical implication; they can also be useful for studying the mechanisms by which these viruses transform normal cells into cancer cells. For instance, it may be possible to determine, on one hand, which viral functions are needed for the *initiation* of cell transformation and, on the other hand, which functions are essential for the *maintenance* of the transformed state. Furthermore, these mutants allow the use of selective techniques and thus provide a sensitive tool for performing marker rescue experiments to find out whether any of the genes of the transforming virus persist in the transformed cells.

This article describes the transforming ability of a conditional lethal mutant of polyoma virus which is temperature-sensitive in the cytocidal virus cell interaction.[3]

Materials and Methods.—Virus strains: The parental large plaque strain of polyoma virus (PY) has been described elsewhere.[4] The temperature-sensitive mutant (TS-a) was isolated from a stock of PY after nitrous acid treatment.[3]

Cell culture: BHK/C13 hamster cells[5] were obtained from Dr. Michael Stoker of the Institute of Virology in Glasgow, Scotland, and were always cultivated in reinforced[6] Eagle's medium containing 10% calf serum and 10% tryptone phosphate broth.

Transformation assay: The transformation assay was essentially that described by Montagnier and Macpherson[7,8] which distinguishes the cells transformed by polyoma virus from normal cells by their enhanced ability to form colonies in agar. Hamster cells (BHK/C13) were infected in suspension at a concentration of 10^6 cells/ml in reinforced Eagle's medium for 1 hr at 38.5°C. The infected cells were washed three times with reinforced Eagle's medium to remove the unabsorbed virus and then suspended in 0.33% agar containing 10% calf serum and 10% tryptone phosphate broth in reinforced Eagle's medium at a concentration of 10^4 cells/ml. One and one-half ml of the 0.33% agar cell suspension was plated onto a preset base layer of 5 ml of 0.5% agar in 60-mm plastic Petri dishes. After the agar had hardened, the dishes were incubated at 31.5° and/or 38.5°C in well-humidified incubators containing 7% CO_2.

It should be pointed out that under these conditions the seeding of more than $2–3 \times 10^4$ cells/dish, as originally suggested by Montagnier and Macpherson,[7,8] resulted in the production of many normal cell colonies when incubated at 38.5°C.

Incubation conditions: The assay dishes were divided into four equal groups which were each incubated in one of the following fashions before scoring for transformation: (1) high-temperature incubation—incubation 16–18 days at 38.5°C; (2) "shift up" incubation—incubation first for 4 days at 31.5°C and then 16–18 days at 38.5°C; (3) "shift down" incubation—incubation first for 2 days at 38.5°C and then 32–36 days at 31.5°C; (4) low-temperature incubation—incubation 32–36 days at 31.5°C.

Scoring of transformation: Transformation was usually scored by counting the macroscopic colonies arising after the various incubation periods. Corrections were made for the background of normal colonies by subtracting the number of colonies appearing on the uninfected dishes. In order to verify the transformed nature of a colony, it was picked out of the agar and plated under fluid medium in a plastic Petri dish. After being pipetted free of the agar, the colony usually stuck to the bottom of the dish. Within 1 day at 38.5°C or 2 days at 31.5°C, cells began to migrate out from the colony and started to divide. The transformed colonies grew in a nonorientated crisscross manner in a multilayered mat (loss of contact inhibition).[5] The cells from the normal colonies retained their contact inhibition and grew in an aligned parallel fashion in a single layer. The transformed cells, either as the whole colony or as individual cells, had a great tendency to become dislodged from the substrate under conditions of the slightest agitation. This led to the production of satellite colonies from the parental single colony each time the dish was disturbed. For the most part, the normal colonies did not produce satellite colonies under conditions of slight agitation.

Dishes kept under the low-temperature incubation had a higher background of normal colonies; with these dishes the following procedure was used to estimate the number of transformed colonies. The whole top agar from each dish was removed in medium and transferred to two large plastic Petri dishes (100 mm). The agar was broken up by pipetting in order to free the colonies. These dishes were then incubated at 31.5°C for 3.5 days without being disturbed in order to keep the production of satellite colonies to a minimum. After this incubation period, the colonies were fixed by the gentle addition of formaldehyde (final concentration 4%) directly to the medium in the plate. The colonies were then stained with methylene blue and scored for the number of transformed colonies by microscopic examination.[5] By this method too high an estimate can be obtained by counting colonies derived from single cells or pieces of broken colonies. To avoid counting colonies initiated by single cells, only large colonies were scored. This method can also tend to give a low estimate of transformation due to colonies remaining embedded in the agar. When plating single colonies in individual dishes (see above), if perchance a colony is not adequately freed from the agar at the time of the initial transfer, it can be freed at a later time with no ill consequences. When many colonies are transferred into the same dish, it is important that the colonies not be disturbed until after fixation to avoid the production of satellite colonies. Thus, colonies embedded in the agar after the transfer are lost.

Inoculation of hamsters: One- to three-day-old Syrian hamsters were inoculated either subcutaneously (SQ) or intraperitoneally (IP) with 0.1 or 0.3 ml of virus. The young animals were kept as a litter with the mother until they were 18 days old, at which time they were placed in separate cages. The animals were checked twice a day when kept as a litter, and once a day when separated, for death or the production of tumors. Tumors were detected by macroscopic examination only.

Results.—Transformation in vitro: Hamster cells that were exposed to five different concentrations of TS-a (1000, 100, 10, 1, and 0 PFU/cell) were plated in agar and incubated under the different temperatures as described in *Materials and Methods.* The amount of transformation at these different virus concentrations and incubating conditions is presented in Table 1. Each number in Table 1 represents the total number of transformed colonies on 15 replicate dishes.

TABLE 1

Input multiplicity, PFU per cell	Total Number of Transformed Colonies per 15 Replicate Dishes			
	High-temperature* incubation, 16–18 days at 38.5°C	"Shift up"* incubation, 4 days at 31.5°C, then 16–18 days at 38.5°C	"Shift down"* incubation, 2 days at 38.5°C, then at 31.5°C	Low† temperature, 32–36 days at 31.5°C
1000	5	504	3	415
100	1	296	0	225
10	0	69	0	56.7
1	0	7	0	7.5
0	0	0	0	0

* All the transformed colonies were determined by replating individual colonies under fluid medium as explained in the text.

† All the numbers of transformed colonies were obtained by plating out the whole top agar of 6–8 plates of each set under fluid medium as explained in the text; the numbers are normalized to 15 dishes.

High-temperature and "shift up" incubations: A background of one colony per dish was found by macroscopic examination of the dishes of uninfected cells. In the cases where there were less than 100 colonies per 15 replicate dishes, all the colonies were picked from the agar and tested for their transformed morphology by replating in fluid medium at 38.5°C. In the instances where there were more than 100 colonies per 15 replicate dishes, a representative number (60–100) of colonies were replated in fluid medium at 38.5°C. Almost all the colonies so treated stuck to the bottom of the dishes on replating in fluid medium and could be classified as either normal or transformed. In a few cases, colonies were observed which could not be classified. It is not known whether these are normal cell mutants selected by the agar or transformed cells with an unusual morphology.

"Shift down" incubation: Under this incubation condition, a background of three colonies per dish was observed by macroscopic examination of the dishes of uninfected cells. All the colonies arising on all the dishes were picked from the agar and replated in fluid medium at 31.5°C. Only about 35 per cent of these replated colonies could be classified as either normal or transformed. Some of the colonies did not stick to the dish on replating. In some of the colonies that stuck, the cells did not migrate out of the colony; in others, the cells migrated but did not divide. Nonclassifiable colonies were observed in approximately the same frequency from the uninfected and infected sets of dishes. Therefore, it is most likely that, for the most part, these nonclassifiable colonies were derived from normal cells. But since this was not shown to be the case, the results presented in Table 1 can only be taken as a lower limit of the transformation under the "shift up" incubating conditions. If one assumes that transformation among the nonclassifiable colonies is in the same proportion as it is among the classifiable colonies, or even if all the nonclassifiable colonies are transformed, it is still found that transformation is greatly inhibited as compared to cells kept at the low-temperature or "shift up" incubations.

Low-temperature incubation: By macroscopic examination, there were about 60–80 normal colonies per dish of cells kept at the low-temperature incubation. This high background hampered any accurate count of the transformed colonies. An estimate of the general number of transformed colonies was made by plating out the whole top agar from 6 to 8 dishes of each virus concentration in fluid medium at 31.5°C as described in *Materials and Methods.*

Since it is not known to what effect the errors involved in plating out the whole top agar affected the true number of the transformed colonies, the results presented in Table 1 for the low-temperature incubation can only be taken as approximate.

The validity of these data, however, is shown by the presence of transformed colonies when the cells were infected by one PFU per cell: since at this input multiplicity not more than one transformed colony was recognized per original dish, the number cannot be an overestimate.

Since the numbers obtained at the various input multiplicities show the same general relationship to each other as those obtained in the "shift up" experiment, it seems that the dose response to the virus is similar in the two cases. However, the considerable similarity between the absolute numbers shown by the table may be spurious.

Transformation in vivo: In a preliminary experiment to test the tumor-produc-

TABLE 2

Virus	Inoculation route	Total number of animals inoculated	Number of Animals with— No tumors	Subcutaneous tumors	Internal tumors
TS-a	SQ	19	0	19	3
TS-a	IP	15	5	8	5
PY	SQ	16	0	1	16
PY	IP	13	0	2	13

The total number of tumors at the different sites for each type of inoculation adds up to more animals than were inoculated, since some animals had tumors at more than one site.

ing ability of the mutant, 5 three-day-old hamsters were inoculated subcutaneously with 10^6 PFU of TS-a. All five animals developed subcutaneous fibrosarcomas about the site of inoculation 40–90 days after inoculation. No tumors were found associated with the internal areas of the body. A hemagglutination inhibition titer[9] of 1280 was detected in the sera of one of the tumor-bearing animals (the only animal so tested) 90 days after inoculation. Usually the wild type (PY), when inoculated under similar conditions, causes angiomatous tumors of the liver and lungs and tumors of the internal organs predominantly in the kidneys and heart.[10]

In order to test whether the tumors induced by TS-a were truly restricted to the subcutaneous site, possibly due to the lower temperature in this area, the experiment was repeated on a larger scale, using different routes of inoculation. One-to three-day-old hamsters were inoculated either subcutaneously or intraperitoneally with either 9×10^6 PFU of TS-a or 2.7×10^7 PFU of PY and observed for 70 days.

The results of this experiment are presented in Table 2. All the animals inoculated with PY, either through the SQ or IP route, died from the rupture of angiomatous tumors of the liver or lungs between 12 and 40 days after infection. Very few of the TS-a-inoculated animals died as a direct effect of the virus but were sacrificed when their tumors became too burdensome or at the end of the experiment. Except for some PY-infected animals that were lost previous to the 16th day after inoculation, all had, in addition to the angiomatous tumors, macroscopic tumors of the internal organs mainly in the heart and kidneys. Subcutaneous tumors were only detected in those PY animals which survived for the longest time. On the contrary, only three (two SQ and one IP inoculated) TS-a-infected animals had any indication of angiomatous tumors of the liver at the time of death. Only one TS-a-infected animal had an internal tumor associated with an internal organ (heart). All the other internal tumors were either free or found to be loosely connected to the internal body wall.

The results presented here clearly indicate that there is an inhibition in hamsters infected with TS-a, either by a SQ or IP route, of both the angiomatous tumors of the liver and lungs and the production of tumors of the other internal organs that are found in the early stages of PY infection.

Discussion and Conclusions.—The results presented here show that TS-a is temperature-sensitive in its transforming ability. It appears that after completion of the temperature-sensitive step at one temperature (either 31.5°C or 38.5°C) the effect on the transformation produced, at least for the most part, cannot be reversed by then shifting the infected cells to the other temperature.

Because of technical difficulties involved in determining the transformation rate at the low temperature, the exact degree of inhibition of transformation (transformation inhibition factor) caused by the temperature-sensitive gene at 38.5°C is not known. At 1000 PFU/cell there is a 100-fold decrease in the production of transformed colonies in infected cells kept at the high temperature for the whole incubation period, compared to cells exposed to the "shift up" incubation. Since in the "shift up" incubation the infected cells were kept at 31.5°C for only 4 days, the true transformation inhibition factor created by high temperature at 1000 PFU/cell may even be greater than 100-fold. It is not known whether the transformation produced at 38.5°C is due to leakage, revertants, or is a consequence of just the initial virus input.

The temperature-sensitive gene involved in TS-a is needed for the initiation of the transformed state, but not necessarily for the maintenance of this state. This is true since the cells transformed at 31.5°C retain their transformed characteristics upon cultivation at 38.5°C. This observation is reminiscent of the temperature-sensitive sex factor of *E. coli.*[11] This sex factor is sensitive to high temperature when replicating autonomously but insensitive when integrated with the host genome. But this result is also compatible with a hit-and-run effect of the virus, i.e., once the effect of the virus on the cell producing transformation is completed, no portion of the virus genetic material is needed any longer for the maintenance of the transformed state.

It is interesting to remark that the quantitative aspects of the transformation covered by this mutant in the "shift up" incubation are similar to those covered by a nontemperature-sensitive strain of the virus at 37°C.[7, 8] With TS-a, in fact, approximately 3×10^4 PFU are needed to produce one transformed colony; the dose response curve begins to level off between 10 and 100 PFU per cell; the maximum proportion of transformed cells is at least 0.2 per cent.

The *in vivo* results are consistent with TS-a also being temperature-sensitive in the animal. One could postulate that transformation more readily takes place at the subcutaneous site because of the lower temperature of this area compared with the internal organs. The increased resistance to the neoplastic effect of the virus when inoculated by the IP route (30 per cent of the animals inoculated by this route had no detectable tumors) might be due to the slower growth of the virus because of the higher temperature, allowing the immunological mechanisms of the animal to fight off the oncogenic response.

Summary.—A mutant of polyoma virus which was isolated by its temperature sensitivity in the cytocidal virus cell interaction has also been found to be temperature-sensitive in its ability to convert normal hamster cells to neoplastic cells *in vitro*. The activity of the gene involved has been shown to be required for the initiation, but not for the maintenance of the transformed state.

The author wishes to thank Professor Renato Dulbecco and Dr. Marguerite Vogt for their advice, encouragement, and stimulation in the pursuit of this work, and also for their many helpful discussions and suggestions in the preparation of this paper.

This work was supported in part by a grant (112) from The Nutrition Foundation, and was aided by a USPHS graduate training grant (5T1 GM86).

[1] Edgar, R. S., G. H. Denhardt, and R. H. Epstein, *Genetics*, **49**, 639 (1964).

[2] Epstein, R. H., A. Bolle, C. M. Steinberg, E. Kellenberger, E. Boy de la Tour, R. Chevalley, R. S. Edgar, M. Susman, G. H. Denhardt, and I. Lielausis, in *Synthesis and Structure of Macromolecules*, Cold Spring Harbor Symposia on Quantitative Biology, vol. 28 (1963), p. 375.

[3] Fried, M., submitted for publication.

[4] Dulbecco, R., and M. Vogt, *Virology*, **16**, 41 (1962).
[5] Macpherson, I., and M. Stoker, *Virology*, **16**, 147 (1962).
[6] Vogt, M., and R. Dulbecco, these PROCEEDINGS, **49**, 171 (1963).
[7] Montagnier, L. and I. Macpherson, *Compt. Rend.*, **258**, 4171 (1964).
[8] Macpherson, I., and L. Montagnier, *Virology*, **23**, 291 (1964).
[9] Rowe, W. P., J. W. Hartley, J. D. Estes, and R. J. Huebner, *J. Exptl. Med.*, **109**, 379 (1959).
[10] Stewart, S. E., *Advan. Virus Res.*, **7**, 61 (1960).
[11] Jacob, F., S. Brenner, and F. Cuzin, in *Synthesis and Structure of Macromolecules*, Cold Spring Harbor Symposia on Quantitative Biology, vol. 28 (1963), p. 329.

Abortive Transformation by the Tsa Mutant of Polyoma Virus

RENATO DULBECCO

MICHAEL STOKER

THE Tsa mutant of polyoma virus (PyTsa) is a conditional lethal mutant isolated by Fried[1]. PyTsa grows normally at the permissive temperature (31·5° C), but at the non-permissive temperature (38·5° C) there is a failure of normal viral DNA synthesis, and of vegetative growth in mouse cells. Transformation of hamster cells by PyTsa is also inhibited at 38·5° C; this raises the question, Is the mutant gene required simply for the viral induced change in the cell physiology or is it concerned with the stable propagation of the transformed state? These alternatives can be distinguished by the phenomenon of abortive transformation[2].

When hamster cells of the BHK21 line are exposed to polyoma virus at sufficiently high multiplicities, a large proportion of cells develop a characteristic of transformed cells, namely the ability to grow in suspension. In most of the transformed clones, however, this altered growth character is lost after a few divisions and the cells regain their original state (abortive transformants). Only a small proportion of clones retain the transformed character indefinitely, presumably through stable perpetuation of the viral genome (stable transformation). We therefore investigated the ability of the Tsa mutant of polyoma virus to initiate abortive transformation and stable transformation respectively, in permissive and non-permissive temperature conditions.

BHK21 cells were cultured and infected in methyl-cellulose[2,3]. 10^5 haemagglutinating units (HAU) of PyTsa, partially purified by density gradient centrifugation, were added to 2×10^5 cells in 1 ml. methyl-cellulose suspension cultures in agar lined holes, and incubated at 38·5° C or 31·5° C. After 7 days the cultures were transferred to 37° C. Similar cultures were infected with 10^5 HAU wild type virus, either the large plaque (PyLP)[1] partially purified as given here, or stock small plaque (PySP) virus concentrated[4] but not purified further.

The diameters of at least 200 developing colonies were measured in each culture on successive days, and clone sizes were expressed as single cell volume equivalents[2]. To identify stably transformed clones the 1 ml. methyl-cellulose suspension cultures were diluted with 4 ml. of

Table 1. GROWTH OF BHK21 CELLS IN METHYL-CELLULOSE SUSPENSION CULTURES AFTER INFECTION WITH PyTsa MUTANT AND PySP WILD TYPE POLYOMA VIRUS

Incubation	Virus	Per cent clones of different sizes (cell equivalents)* ⩾8	⩾32	⩾128	Identity of clones† Trans-formed/Total	Per cent
	—	19·6	nt	nt	nt	
7 days at 31·5° C	PyTsa	64	nt	nt	nt	
	PySP	60	nt	nt	nt	
	—	5·5	nt	nt	nt	
7 days at 38·5° C	PyTsa	23·5	nt	nt	nt	
	PySP	65·0	nt	nt	nt	
	—	26	6·1	0·5	0/2	—
7 days at 31·5° C, 1 day at 37° C	PyTsa	59	29	6·7	16/65	28
	PySP	47	13	2·9	8/91	8·8
	—	10	0·5	—	0/10	—
7 days at 38·5° C, 1 day at 37° C	PyTsa	30	5·9	0·5	0/450	< 0·25
	PySP	63	44	17	13/101	13
	—	29	10	0	0/7	—
7 days at 31·5° C, 5 days at 37° C	PyTsa	63	29	7·7	11/27	41
	PySP	83	36	5·7	10/55	18
	—	21	3·4	1·0	0/72	—
7 days at 38·5° C, 5 days at 37° C	PyTsa	43	15	3·0	0/420	< 0·25
	PySP	73	41	28	10/59	17

10^5 haemagglutinating units of virus added to 2×10^5 cells and incubated at permissive (31·5° C) and non-permissive (38·5° C) temperatures for 7 days before transferring to 37° C. * Clone sizes calculated from diameters of at least 200 spherical colonies at times shown. † Clones of more than 100 cells identified from orientated or random arrangement of cells after replating on plastic. Totals not comparable.

nt, Not measured.

growth medium to reduce viscosity. These were cultured in plastic Petri dishes on which the colonies attached and the cells emigrated. After 2 days most attached colonies showed normal parallel orientation of cells. Transformed colonies were identified by the persistent random arrangement of the emigrating cells.

Table 1 shows the results of a typical experiment with PyTsa and PySP (less complete experiments using PyLP as the wild type control gave similar results). In the uninfected controls a small proportion of cells grew as reported previously. This background of normal growth was increased by incubation at 31·5° C, but was not enough to obscure the effect of the virus.

Growth in suspension, above the background found in uninfected controls, was found after infection by both PyTsa and wild type virus at both temperatures. Compared with input virus and response at 31·5° C, the response to PyTsa at 38·5° C was somewhat less than the response to PySP (and also to PyLP in other experiments), but the growth stimulation was none the less obvious. When the resulting colonies were examined further by replating, however, typical transformation was found to follow infection with PyTsa at low temperature only. As expected, no stably transformed colonies were identified after infection at 38·5° C. PySP (and also PyLP in other experiments) gave rise to stable transformation at either temperature. The transformation frequencies appeared to be highest with PyTsa at 31·5° C, but it is doubtful if the method allows accurate comparison.

These results show that, in non-permissive conditions, the Tsa mutant is able to initiate a temporary change in BHK21 cells leading to growth in suspension, that is, abortive transformation, but it is unable to give rise to stable transformation. It is known that PyTsa will also induce host cell DNA synthesis during vegetative growth in mouse cells in non-permissive conditions (personal communication, M. Fried), and this may be caused by the same viral function that induces the initiation of cell division in suspended hamster cells. Abortive transformation by PyTsa is somewhat less efficient than by either SP or LP wild type virus, and this may be caused by lack of normal replication of viral DNA.

The data suggest that the gene which is affected in the Tsa mutant is not concerned with the expression of the transformed phenotype, but rather with the events leading to stable perpetuation of the viral genome. If, as we suppose, this involves the conversion of a free double stranded circular molecule into a linear form which is then inserted into host cell DNA, then the Tsa function might be concerned, for example, with strand scission or ligation. Whatever the defect, it also interferes with production of normal virus DNA during vegetative growth. Finally, these results emphasize the distinction between the temporary expression of the transformed phenotype of the cell and the stable inheritance of the transformed state.

Part of this work was carried out while one of us (M. S.) was a visiting scientist at the Salk Institute for Biological Studies.

[1] Fried, M., *Virology*, **25**, 669 (1965).
[2] Stoker, M., *Nature*, **218**, 234 (1968).
[3] Stoker, M., O'Neill, C., Berryman, S., and Waxman, V., *Int. J. Cancer*, **3**, 686 (1968).
[4] Crawford, L. V., *Virology*, **18**, 177 (1962).

Temperature-Dependent Properties of Cells Transformed by a Thermosensitive Mutant of Polyoma Virus

Renato Dulbecco and Walter Eckhart

Abstract. In cells transformed by ts-3, a thermosensitive mutant of polyoma virus, the loss of inhibition of DNA synthesis by topographical factors (topoinhibition), is rendered temperature-dependent, providing evidence that the viral genome controls this essential aspect of transformation. The expression of two other attributes of transformed cells, growth in agar and serum requirement for initiation of DNA synthesis in a wound in culture, is not made temperature-dependent. In productive infection of BALB-3T3 cells by ts-3, virus-induced stimulation of cellular DNA synthesis and movement is rendered temperature-dependent.

Cells transformed in tissue culture by polyoma virus are distinguishable from normal cells on the basis of several attributes. Transformed cells acquire new morphology and growth patterns, attain a higher maximal density in culture than normal cells, and are much less sensitive to inhibition of cell multiplication by topographical factors (topoinhibition) or by serum limitation.[1]

Cell cultures productively infected by polyoma virus undergo changes that resemble those of transformation. They have a much decreased sensitivity to topoinhibition and are stimulated to synthesize cellular DNA and enzymes involved in DNA synthesis. In addition, cultures that require a high concentration of serum for cell survival, mutiplication, and movement, become less dependent upon serum for all these activities.[2]

Viral DNA becomes part of the genetic material of some transformed cells,[3] suggesting that some of the attributes of transformed cells may be conferred by the expression of viral genes. Temperature-sensitive mutants of polyoma virus have been isolated in several laboratories to test this point, but so far no mutation has been found to affect the characteristics of transformed cells. We now wish to report on a thermosensitive mutant of polyoma virus, ts-3, which renders some of the characteristics of transformed cells temperature-dependent, thus providing evidence for the control of these characteristics by the viral genome.

Materials and Methods. Cell cultures were maintained in plastic tissue culture dishes (Nunc, Denmark) in reinforced Eagle's medium supplemented with calf serum, except where noted. The virus used for infection was polyoma large plaque, or the thermosensitive mutant ts-3 derived from it (by Dr. Marguerite Vogt), as described.[4] (The properties of the mutant ts-3 will be described in detail elsewhere.) Virus preparations purified by equilibrium sedimentation in cesium chloride density gradients gave the same results as unpurified lysates.

DNA synthesis was measured by labeling cultures with tritiated thymidine; the proportion of cells incorporating thymidine was determined by radioautography. Total incorporation per culture was determined after selective extraction of DNA by the method of Hirt.[5]

Topoinhibition and wound serum requirement were determined[1] in cell cultures that had been wounded. Topoinhibition, which measures the effect of topographical factors inhibiting DNA synthesis, is defined as TI = (WP − LP)/WP, where WP is the proportion of cells incorporating thymidine in the wound, and LP is the proportion incorporating thymidine in the layer, measured *at or near zero serum concentration* where the ratio is maximal. Wound serum requirement, which measures the serum requirements of cells for DNA synthesis in the absence of topographical factors, is defined as WSR = (MxWP − MnWP)/MxWP, where MxWP is the proportion of cells incorporating thymidine in the wound at high serum concentration, and MnWP is the proportion incorporating thymidine in the wound without serum. Movement was determined as described.[2]

Transformed cells were derived by infecting BHK cells in suspension with wild type polyoma, or with ts-3, at a multiplicity of about 100 plaque-forming units per cell, and incubating the infected cells in 0.34% agar at 31.5°C. Clones arising in the agar were picked and reseeded in liquid medium. WTC1-1A and ts-3C1-7C were recloned once after being transferred from agar to liquid; ts-3C1-1 was not recloned. The cultures were routinely carried at 31.5°C and transferred to 39°C at least 2 days prior to any measurements made at 39°C.

Results. Induction of cellular DNA synthesis after infection of BALB-3T3 cells by wild type polyoma, or by ts-3, was measured by uptake of tritiated thymidine and radioautography. The results are presented in Fig. 1 and Table 1. There is some induction of cellular DNA synthesis in ts-3-infected cells at 39°C, but the amount of thymidine incorporated into cellular DNA is much lower at 39°C than at 32°C. The proportion of ts-3-infected cells incorporating thymidine at 39°C is also lower than the proportion at 32°C. This difference is not caused by a difference in synthesis of viral DNA at the two temperatures, because the amount of thymidine incorporated into viral DNA is much less than the amount incorporated into cellular DNA, and because ts-a, another thermosensitive mutant, induces incorporation in an equal proportion of infected cells at high and low temperature, even though viral DNA synthesis is defective at high temperature (ref. 6 and unpublished data).

Ts-3 is thermosensitive in productive infection with viral DNA, indicating that the thermosensitive defect does not involve adsorption or uncoating of the virus (unpublished results).

Induction of cell movement was measured after infection of BALB-3T3 cells by wild type polyoma, or by ts-3. The results are presented in Fig. 2 and Table 2. There is some induction of movement by ts-3 at 39°C, but the amount is less at 39°C than at 31.5°C, whereas induction of movement by wild type infection is greater at 39°C than at 31.5°C (Table 2). At late times after infection, there is some dissociation of cell movement from DNA synthesis: movement is reduced less than DNA synthesis in ts-3-infected cultures at 39°C in comparison to the values for wild type infected cultures (Fig. 2). A similar dissociation has been observed with 3T3 cells infected by wild type polyoma virus in the absence of serum, where movement took place without DNA synthesis.[2]

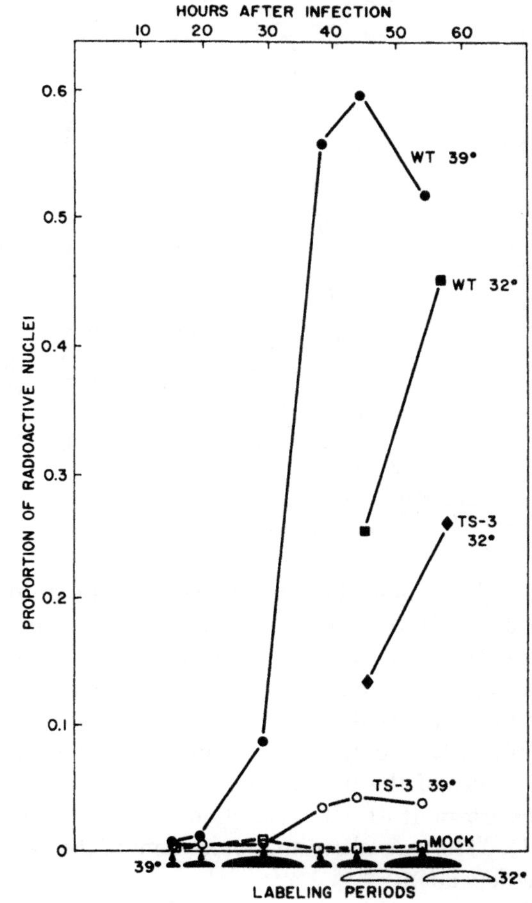

FIG. 1. Induction of cellular DNA synthesis after infection of BALB-3T3 cells by wild type or ts-3 virus. BALB-3T3 cells were grown to confluency in medium containing 10% calf serum. The cells were washed with buffer and infected with CsCl-purified preparations of wild type polyoma or ts-3 at approximately 10^7 PF units/plate in buffer containing 2% horse serum. Mock-infection was performed with buffer containing 2% horse serum. After infection, the cultures were covered with 5 ml of Eagle's medium, making the final serum concentration 0.05%. Tritiated thymidine was added to each culture at 2 μCi/ml at a final concentration of 10^{-6} M thymidine. Cultures labeled for the times indicated were fixed by washing with cold 5% trichloroacetic acid, followed by 95% ethanol, for radioautography. Kodak AR10 stripping film was added directly to the plates.

TABLE 1. *Induction of cellular DNA synthesis after infection.*

Culture	Labeling period, hours after infection	39°C —cpm incorporated into— Supernatant	39°C Pellet	32°C —cpm incorporated into— Supernatant	32°C Pellet
Mock-infected	30–40 hr	2.5×10^3	11×10^3		
	40–50 hr	3.5×10^3	28×10^3		
	43–55 hr			1.2×10^3	9.0×10^3
	54–66 hr			0.9×10^3	9.5×10^3
Ts-3-infected	30–40 hr	2.5×10^3	18×10^3		
	40–50 hr	4.0×10^3	36×10^3		
	43–55 hr			20×10^3	210×10^3
	54–66 hr			24×10^3	260×10^3
Wild type-infected	30–40 hr	15×10^3	160×10^3		
	40–50 hr	15×10^3	160×10^3		
	43–55 hr			37×10^3	480×10^3
	54–66 hr			18×10^3	240×10^3

BALB-3T3 cells were infected as described in Fig. 1, except that after infection the cultures were covered with fresh Eagle's medium to which had been added enough depleted medium containing 10% serum to make the final serum concentration 2%. The cultures were labeled for the times indicated, and harvested by the Hirt extraction. (Cellular DNA is in pellet; viral DNA in supernatant.)

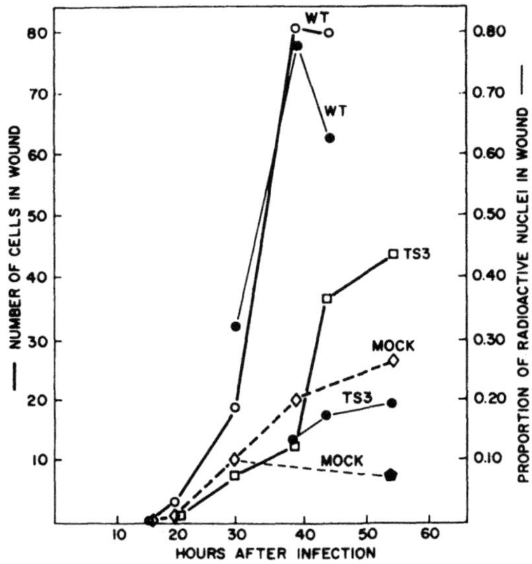

FIG. 2. Induction of movement and of DNA synthesis in BALB-3T3 cells in wounds after infection at 39°C. Confluent cultures of BALB-3T3 cells were wounded, and then infected as described in the legend to Fig. 1. Tritiated thymidine was added 10 hr after infection, and was present continuously thereafter. At various times after infection, cell movement was determined by counting the number of cells that had migrated into the area of the wound. The proportion of radioactive nuclei in the wound was determined by radioautography.

Topoinhibition (TI) and wound serum requirement (WSR) were measured in cultures of BHK cells transformed by wild type polyoma, or by ts-3. Two independently isolated, ts-3 transformed lines were used. The growth parameters measured in two experiments, in cultures grown at 39°C and at 31.5°C, are presented in Table 3. The values were calculated from the data shown in Fig. 3 (*left, center, right*). TI is temperature-dependent for the ts-3 transformed cells, the values of the parameter being closer to those for transformed BHK in cultures grown at 31.5°C, and closer to those for normal BHK in cultures grown at 39°C. The wound serum requirement of the ts-3 transformed cells is not temperature-dependent.

The somewhat lower values of TI for the ts-3 transformed cells grown at 39°C in the second, as compared to the first, experiment may have arisen from incomplete reversal of the TI parameter since the cells were grown for a shorter time at 39°C before being used in the second experiment. The lower values of TI may also have arisen from the accumulation of irreversibly-transformed cells in the cultures that were carried for some length of time.

TABLE 2. *Induction of cell movement after infection of BALB-3T3 cells.*

	Movement	
Infection	39°C*	31.5°C†
Mock	0.50	0.49
Wild type	1.22	0.82
Ts-3	0.72	0.88

Confluent cultures of BALB-3T3 cells were wounded and then infected with unpurified virus at approximately 1.5 × 10⁷ plaque forming units (PF units)/plate. Cell movement was measured by determining the changes in the gradient of cell concentration at either edge of the wound. The value for movement is calculated as the area bounded by the curve defining the gradient of cell concentration at the edge of the wound, normalized to the cell concentration in an undisturbed area of the cell layer.[2]

* 24 hr after infection.

† 48 hr after infection.

TABLE 3. *Growth parameters of ts-3-BHK and py-BHK cells.*

		39°C		32°C	
Expt.	Clone	WSR	TI	WSR	TI
1	1A (wild type)	0.13	0.01	0.51	0.05
		0.10	0.03	0.50–0.60	0.09
		0.30	~0	0.60	0.09
	1(ts-3)	0.57	0.37	0.52	0.05
		0.60	0.62	0.54	~0
	7C(ts-3)	0.51	0.63	0.32	0.17
		0.56	0.82		
2	1A (wild type)	0.16	0.10	0.10	0.06
	1(ts-3)	0.35	0.38	0.40	0.10
	7C(ts-3)	0.32	0.35	0.09	0.08
	* Normal BHK	0.8–1.0	0.75–0.85		

* Previous result (see ref. 1).
Confluent cultures of wild type-transformed BHK and ts-3-transformed clones grown at 39 or 32°C were wounded, and the values of topoinhibition and wound serum requirement were calculated from the data shown in Fig. 3. Each value represents a determination made on a single plate. Expt. 2 was performed about 3 weeks after Expt. 1.

Growth in agar: Each of the transformed cell lines was seeded in 0.34% agar to determine the efficiency with which it would form colonies at 39 and 31.5°C. The results are shown in Table 4. The property of growth in agar is not temperature-dependent in ts-3 transformed cells.

Discussion. In BHK cells transformed by the thermosensitive mutant ts-3, topoinhibition, which measures the inhibition of DNA synthesis by topographical factors, is temperature-dependent. This implies that the loss of topoinhibition,

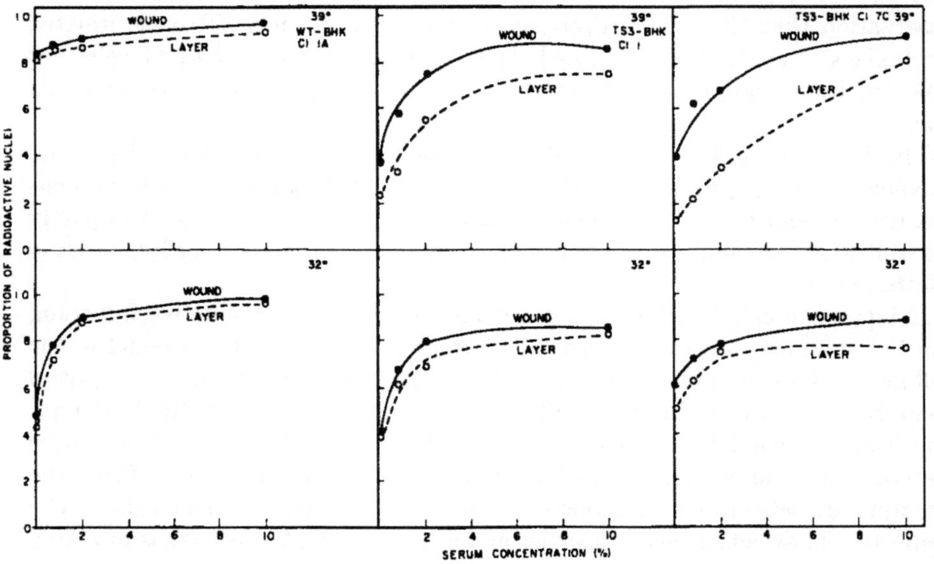

FIG. 3. (*Left, Center, Right*) The proportion of radioactive nuclei in cells in wounds. Wild type and ts-3 transformed cultures were wounded, incubated in media containing different amounts of serum, and labeled continuously with tritiated thymidine from 8 hr after wounding. Cultures were fixed for radioautography 30 hr after wounding at 39°C, and 48 hr after wounding at 32°C.

TABLE 4. *Growth of transformed cells in agar.*

Clone	Proportion of cells forming colonies in agar 39°C	32°C
1A (wild type)	0.04	0.12
1(ts-3)	0.40	0.20
7C(ts-3)	0.04	0.02

Wild type or ts-3 transformed cells grown at 39 or 32°C were removed from dishes by treatment with EDTA or trypsin. The cells were seeded in 0.34% agar containing 10% calf serum, and the proportion of cells forming colonies at each temperature was determined.

which is characteristic of transformed cells, is under the control of a viral gene in polyoma-transformed cells.

Two other attributes of ts-3 transformed cells, ability to grow in agar and serum requirement for initiation of DNA synthesis in a wound, are not temperature-dependent. Two possible explanations could be offered for the lack of temperature-dependence of these attributes. One possibility is that the expression of growth in agar and wound serum requirement does not require a functional ts-3 gene product. The other possibility is that the expression of these attributes requires a functional ts-3 gene product, but requires it in smaller amounts than do other attributes for full expression. If the ts-3 mutation reduces, but does not abolish, the function of the gene, the amount of functional gene product might be such that the expression of some attributes of transformation is reduced, whereas that of others is still almost complete.

It is clear that transformation involves changes in more than one cellular system, requiring the action of one or more viral genes. The involvement of a number of independent cellular systems in transformation is supported by the observation that the "flat" revertants of SV40-transformed 3T3 cells isolated by Pollack[7] show a maximum cell density characteristic of normal cells but topoinhibition and wound serum requirements characteristic of transformed cells.[1]

The ts-3 mutant has reduced ability to induce cellular DNA synthesis and movement during productive infection of BALB-3T3 cultures at the nonpermissive temperature. Movement is less affected, perhaps because it is partly controlled by a cellular system not affected by the ts-3 mutation, or less severely affected by it.

Induction of cellular DNA synthesis and movement in productive infection, and topoinhibition in transformed cells, can be controlled by the product of the ts-3 gene. This product could control these functions directly, or could control them indirectly by modifying a cellular structure involved in all the functions. The latter mechanism seems more likely because, according to evidence to be presented in a subsequent communication, the ts-3 mutation also affects the structure of cell-surface membranes in infected and transformed cells. The connection between surface properties and control of DNA synthesis is probably part of a fundamental regulatory mechanism in animal cells.

We gratefully acknowledge the excellent technical assistance of M. Bajak and M. Dulbecco.

Supported by the U.S. Public Health Service, National Cancer Institute grant CA-07592 and National Cancer Institute Contract no. 67-1147.

[1] Dulbecco, R., *Nature*, **227**, 802 (1970).
[2] Dulbecco, R., *Proc. Nat. Acad. Sci. USA*, **67**, 1214 (1970).
[3] Sambrook, J., H. Westphal, P. R. Srinivasan, and R. Dulbecco, *Proc. Nat. Acad. Sci. USA*, **60**, 1288 (1968).
[4] Eckhart, W., *Virology*, **38**, 120 (1969).
[5] Hirt, B., *J. Mol. Biol.*, **26**, 365 (1967).
[6] Eckhart, W., *Trans. Roy. Soc. London* in press (1970).
[7] Pollack, R., H. Green, and G. J. Todaro, *Proc. Nat. Acad. Sci. USA*, **60**, 126 (1968).

VIII
Further Problems in the Biology of Small DNA Viruses

GROWTH CONTROL IN CULTURED CELLS: SELECTION OF SUBLINES WITH INCREASED SENSITIVITY TO CONTACT INHIBITION AND DECREASED TUMOR-PRODUCING ABILITY

Robert E. Pollack, Howard Green, and George J. Todaro

As growing cultures of mammalian cells reach a high density, their growth rate decreases and in some cases a substantially nondividing population may result. This arrest of growth requires cell contact,[1,2] though soluble substances are also involved.[3–5] The phenomenon has usually been called contact inhibition of division.

Transformants and tumors are relatively insensitive to growth inhibition, but the present experiments have shown that, even in such populations, it is possible to select for sensitive cells by exposure to 5-fluoro-2-deoxyuridine (FUdR) and to obtain stable variants greatly altered in their culture behavior and in tumor-producing ability.

Materials and Methods.—All cultures were maintained at 36.5°C in 20-cm^2 plastic Petri dishes in Dulbecco and Vogt's modification of Eagle's medium[6] supplemented with 10% calf serum. The medium was changed twice weekly.

The cells used in these experiments were established fibroblast lines originating in this laboratory. 3T3-4 and 3T6-7 are established lines of mouse embryo origin.[7] Py11 and SV101 are transformants of 3T3 produced by polyoma virus and SV40, respectively.[8] PyH 75 is a hamster established line originating from a polyoma virus-induced tumor.[9] The lines were cloned twice before they were used in selection experiments. All variants selected are designated as Fl (flat) with superscripts indicating the number of FUdR treatments preceding their isolation.

FUdR[10] was obtained from Hoffmann-LaRoche, Nutley, New Jersey. Concentrated stock solutions were kept frozen until used. To remove FUdR before transferring surviving cells, cultures were incubated in fresh medium for 15–30 min before washing and trypsinization. In order to prevent any fluorouracil, produced by breakdown of the deoxyriboside, from interfering with RNA synthesis, a 10-fold greater quantity of uridine was added with the FUdR.[11]

The capacity of a cell line to produce tumors was tested in 21-day-old Syrian hamsters. Known numbers of cells were injected subcutaneously over the back, and the animals examined for tumors 3 times weekly. A nodule exceeding 1 cm in diameter was considered positive. All tumors grew progressively thereafter.

Results.—*Sensitivity of different lines to killing by FUdR:* Antimetabolites that interfere with DNA synthesis affect the viability of cells only if they are in the *S* period of the division cycle, i.e., synthesizing DNA. For example, nongrowing cells are unaffected by concentrations of iododeoxyuridine lethal to growing cells.[12]

The susceptibility of a number of lines to FUdR was tested by exposing exponentially growing cultures to FUdR for 48 hours, removing the FUdR, transferring the cultures at suitable dilutions, and counting colonies produced by the survivors 10–14 days later. Figure 1 shows the results obtained with PyH75, SV101, Py11, and 3T3-4. In all cases, the number of surviving cells could be reduced to 10^{-5} or less. This was accomplished at a concentration of 0.1

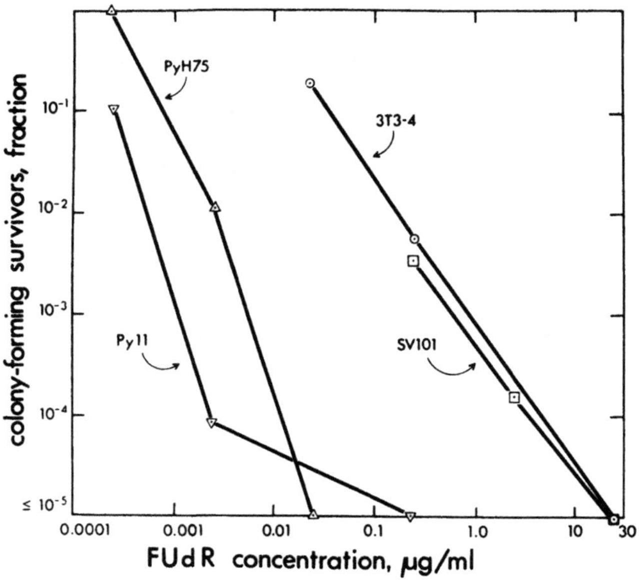

FIG. 1.—Fraction of cells forming colonies after 48-hr exposure to FUdR at low cell density.

µg/ml for two polyoma virus-transformed fibroblast lines (PyH75 and Py11) and 25 µg/ml for SV101 and 3T3-4. Other lines examined were about as sensitive as SV101, so that the polyoma lines may be unusually sensitive to FUdR.

Survival of cells treated with FUdR at different cell densities and relation to contact sensitivity: Lines 3T3-4 and 3T6-7 differ greatly in their saturation densities.[7] Cultures of each line were inoculated at low densities with the cells distributed as evenly as possible (Table 1). At intervals, a plate was trypsinized, and the cells were counted and plated to determine the number of colony-formers. FUdR was then added to a sister culture and allowed to remain there for two days. The cells were then trypsinized, counted, and the number of colony-forming survivors was determined. No increase in total cell number occurred in the presence of FUdR.

In Figure 2*A*, the number of colony-forming survivors expressed as a fraction of the control is plotted on a logarithmic scale against cell number per unit area at the time the FUdR was added. At a sufficiently low density, fewer than 10^{-4} cells survived. As the density increased, an increasing proportion of the cells survived, and the number of survivors always rose much faster than the cell density. For example, the number of 3T3 survivors rose from 10^{-3} to 1.0, whereas the cell density rose from 0.4 to $4 \times 10^4/cm^2$. At the same density, only 1.8×10^{-3} of the 3T6-7 population survived FUdR, and only at a density of $40 \times 10^4/cm^2$ (its saturation density) did the 3T6-7 survivor fraction reach 1.0. At a low density, the killing curves converge and might be expected to meet at a density sufficiently low that no cell contact occurs.

Properties of FUdR survivors: Colonies were isolated from cultures treated with FUdR at densities at which most of the cells were killed. These colonies grew as rapidly as the parent line in sparse culture, but many were found to have lower saturation densities. They thus appear to be variants having greater contact sensitivity. Several of these variants were examined for sensitivity to FUdR while growing at very low cell densities and were found to leave fewer than 10^{-4} survivors. They did not therefore have increased FUdR resistance while synthesizing DNA. However, at sufficiently high cell densities their greater sensitivity to contact inhibition made them more resistant to FUdR than the parent population.

TABLE 1. *Saturation density and plating efficiency on 3T3 monolayers of various lines and their flat derivatives.*

	Saturation density (cells/cm² × 10^{-4})	RPE*
3T6-7	40.0	1.0
Fl¹-1 3T6-7	7.0	0.003
3T3-4	4.0	0
SV101	46.0	0.6
Fl²-1 SV101	7.0	0.06
Py11	37.5	0.1
Fl¹-1 Py11	7.0	0.01
PyH75	>75	1.1
Fl²-1 PyH75	50	0.9
Fl²-21 PyH75	45	0.2†

* RPE: relative plating efficiency = $\frac{\text{colonies on 3T3 monolayer}}{\text{colonies on bare Petri dish}}$.

† This value does not include a small number of microcolonies formed on 3T3. These colonies are visible several days after plating but do not continue to grow.

One variant colony of 3T6-7, Fl¹-3T6-7, was isolated, grown, and recloned. It was found to reach a saturation density of 7 × 10^4/cm². When cultures at different cell densities were again treated with FUdR, a survival curve was obtained (Fig. 2*A*) that was very much displaced from that of the parent line and very similar to that of 3T3-4.

In contrast, the same selection procedure applied to 3T3-4 did not yield variants with appreciably lower saturation densities nor with displaced FUdR killing curves. Presumably 3T3 is already close to the maximum sensitivity to growth inhibition.

Selection of flat variants from virus-transformed cell lines: The FUdR killing curve of SV101, a cloned SV40 transformant of 3T3 that saturates at 47 × 10^4/cm², is shown in Figure 2*B*. It can be seen that the number of survivors is several orders of magnitude below that of 3T3-4 treated at the same density. Only at densities higher than the saturation density of 3T3-4 are survivors detectable. The surviving fraction never reaches 0.1, even at the highest cell density. This probably means that even at the nominal saturation density the greater part of the culture is not resting but contains dividing and dying cells. Consistent with this, cells in mitosis can usually be seen even in very thick cultures.

From the surviving colonies of this first FUdR selection, a relatively flat colony was isolated and grown to mass culture. It had a saturation density of 15–20 ×

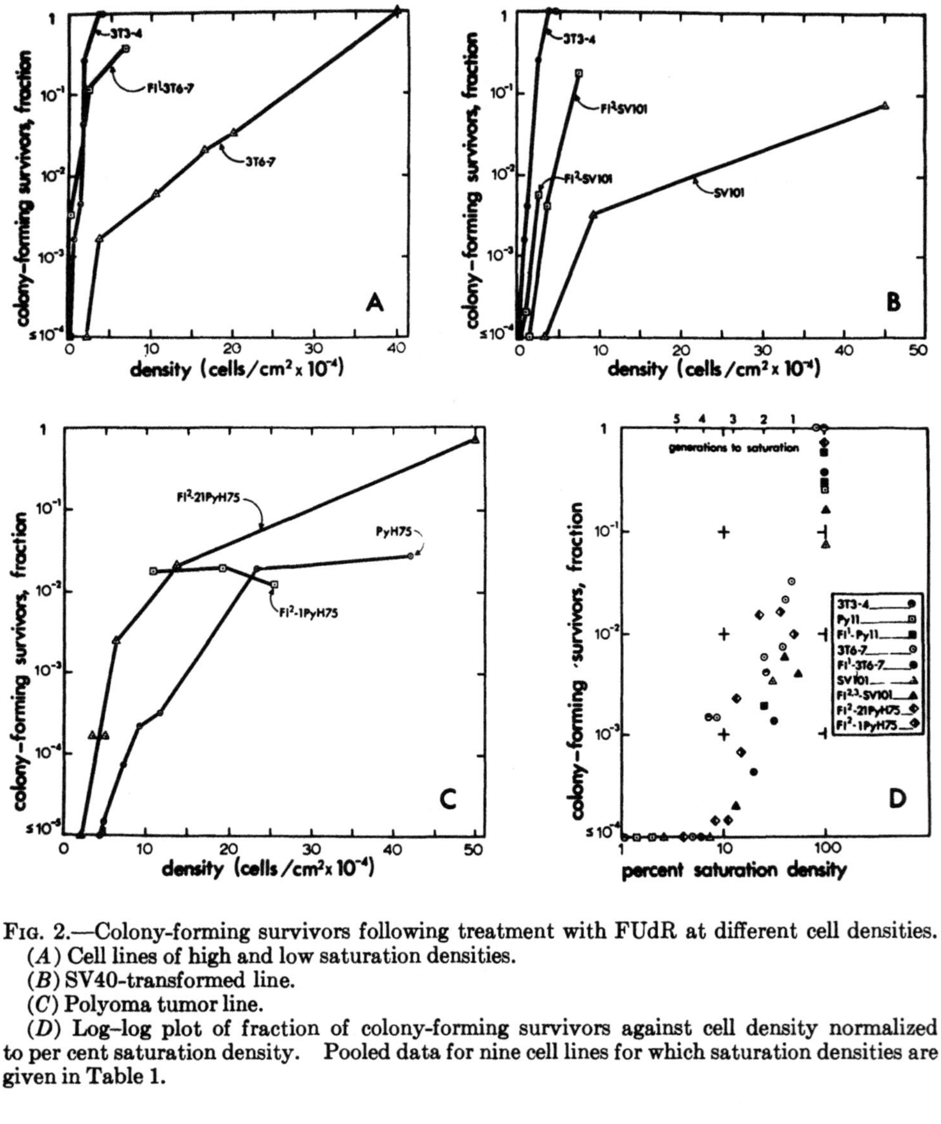

FIG. 2.—Colony-forming survivors following treatment with FUdR at different cell densities.
(*A*) Cell lines of high and low saturation densities.
(*B*) SV40-transformed line.
(*C*) Polyoma tumor line.
(*D*) Log–log plot of fraction of colony-forming survivors against cell density normalized to per cent saturation density. Pooled data for nine cell lines for which saturation densities are given in Table 1.

$10^4/cm^2$. This line (Fl^1-SV101) was carried through a second FUdR cycle, and the surviving colonies were more generally flat. One clone (Fl^2-SV101) was found to reach a saturation density of about $7 \times 10^4/cm^2$. A cycle of FUdR treatment on this clone (Fig. 2*B*) showed marked displacement of the killing curve. It may be seen that at a density of $5 \times 10^4/cm^2$ fewer than 10^{-4} of SV101 survived, whereas survivors of Fl^2-SV101 reached 10^{-2}. This behavior is still clearly distinguishable from that of 3T3-4, which was completely protected at this density. A further FUdR selection on a line originating from a surviving colony did not result in variants with improved survival in FUdR (Fig. 2*B*, Fl^3-SV101). Similar results were obtained with polyoma virus-transformant

Pyll. In this case, a single cycle of FUdR was sufficient to produce colonies of cells that had saturation density not much higher than 3T3-4 (Table 1).

Selection of flat variants from a virus-induced hamster tumor line: PyH75 is a tumor line produced by the injection of polyoma virus into newborn hamsters. Following its culture evolution into an established line, PyH75 retained its ability to produce tumors.[9] The clone isolated from this line and used here grows in a very loose net characteristic of polyoma transformants.[13] The cells are fusiform, and since they attach to the dish over a relatively narrow surface, they have a refractile appearance. As the cell density increases, nodules of more densely packed cells develop, perhaps as a result of retraction, and these break out of the plane of the cell layer, forming balls of largely unattached cell masses. As a result, no stable saturation density is attained though attached cell densities of $75 \times 10^4/\text{cm}^2$ have been measured.

When PyH75 was treated with FUdR, fewer than 10^{-5} of the cells survived when the density at the time of exposure was less than $4.5 \times 10^4/\text{cm}^2$ (Fig. 2*C*). At a fivefold higher density, the surviving fraction was about 10^{-2}. A culture grown up from a mixture of several surviving colonies was taken through a second cycle of FUdR. Of the colonies surviving the second cycle, some grew very slowly even at low densities and failed to do well on transfer. These colonies were discarded. Others grew at the usual rate, but the cells were obviously more adherent to the culture vessel than the parent line and were packed more tightly together. Two such colonies, Fl^2-1 PyH75 and Fl^2-21 PyH75, were grown up and subjected to third cycles. Figure 2*C* shows that at each density the survival was 10–100 times better than that of the parent. Both lines had quite high saturation densities (Table 1), but the cell layers were stable and, at least in the case of Fl^2-21, the FUdR killing curve indicates that there was not much cell turnover at the saturation density.

When the data for the nine lines with different stable saturation densities (Table 1) were pooled in a log–log plot (Fig. 2*D*) in which cell density was normalized to per cent of saturation density, all points clustered about a single curve. Thus all lines displayed a common response to increasing density. During the last 3.5 cell generations preceding saturation, the surviving fraction increased roughly as the fourth power of the density.

Colony-forming ability on a monolayer of 3T3: The ability of one cell type to inhibit the growth of another has been studied in various systems.[14–17] Viral transformants are often able to form colonies very well on a confluent lawn of the parent cell type. For example, SV40 transformants of 3T3 or of human fibroblasts will grow very well when inoculated on a resting monolayer of either 3T3 or human diploid cells.

The ability of the flat variant lines to grow on a 3T3 monolayer was compared to their ability to grow on the bare surface of the Petri dish (Table 1). All flat derivatives showed a marked drop in colony-forming ability on 3T3 when compared with the parent line.

Tumor-producing ability of FUdR-selected variants: In general, a relation has been thought to exist between insensitivity to contact inhibition in culture and ability to produce tumors in animals. However, it has not been possible up to

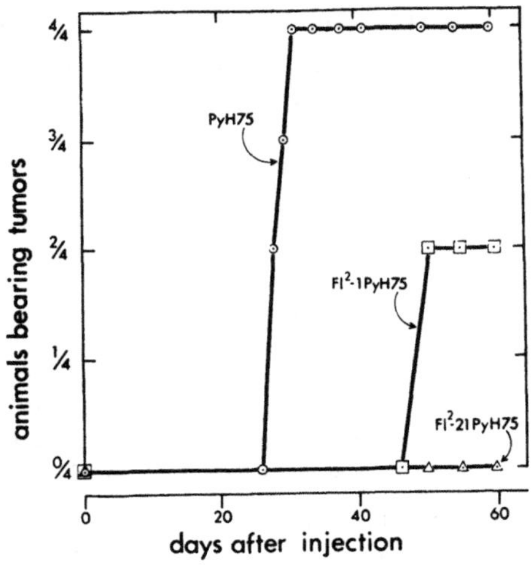

FIG. 3.—Tumor incidence following injection of PyH75 and its flat variants into 21-day hamsters (5×10^4 cells/animal). With 5×10^5 cells/animal, one of four hamsters injected with Fl^2-21 PyH75 cells developed a tumor 35 days after injection; the other three animals remained tumor-free.

now to select at will for sensitivity to contact inhibition of growth and to systematically correlate tumor-producing power with culture behavior. The PyH75 variants have made it possible to do this, and the results are shown in Figure 3.

Weanling hamsters were injected with 5×10^4 cells of PyH75, and two Fl^2 sublines, Fl^2-1 PyH75 and Fl^2-21 PyH75, obtained from the second cycle of FUdR treatment described above. PyH75 cells produced tumors of 1-cm diameter in all the injected animals within 30 days. Fl^2-1 PyH75 cells produced tumors in only half the injected animals within 50 days. The same number of Fl^2-21 PyH75 cells produced no tumors during the observation period of 60 days. In a further experiment, when a group of four animals was injected with cells of this line in ten times higher number (5×10^5 cells), one animal developed a tumor 35 days after injection. It is therefore clear, even on the basis of this small experiment, that FUdR-selected variants of a polyoma virus-induced tumor have reduced tumor-producing power.

Significance of flat variants for the study of transformation: When tumors are serially transplanted in animal hosts, the selective conditions favor those cells with the highest degree of neoplastic character; transplanted tumors evolve in this direction.[18] Similarly, when established cell lines are serially cultivated, the culture conditions select in favor of variants with the ability to grow under conditions in which the growth of the rest of the population is arrested. This is probably why many long-term established lines do not reach any stable saturation density and have neoplastic character. Culture behavior and tumor-producing power for cells of similar origin have also been related by the observation that multilayer, but not monolayer, clones of polyoma virus-transformed cells are tumorigenic in hamsters.[13]

The present experiments show that the variation also occurs in the opposite direction and that, with FUdR, it is possible to select in culture for cells with increased sensitivity to growth inhibition. The cells so selected are stably different from the original population since their descendants show: (1) altered morphology including greater flattening and adherence to the culture vessel; (2) lower saturation density; (3) reduced ability to form colonies on a 3T3 monolayer; (4) altered FUdR survival curve; and (5) reduced tumor-producing capacity. Since all of the original lines were cloned twice before the selection experiments were begun, the variants selected must have developed during the period of growth of the cloned lines preceding the selection experiment (20–40 cell generations). A fluctuation experiment[19, 20] has confirmed the pre-existence of flat variants in an SV40-transformed 3T3 population at a frequency of approximately one in 10^5.

The type of variation described here also occurred in lines that have never been infected with viruses. It could therefore be a cellular property independent of the virus. However, it seemed possible that since the virus transformants in general have low sensitivity to contact inhibition of growth, a powerful selection in favor of sensitive cells might disclose the presence of "revertants" that have lost the viral genome. This possibility was tested on flat variants of SV101 and PyH75.

The two tested variant clones that arose from SV101, like the original line, contained the SV40 T-antigen, and the two clones derived from the polyoma-transformed hamster line PyH75 still contained the polyoma T-antigen.[21] Tests for the SV40 viral genome by cocultivation with monkey kidney cells in the presence of irradiated Sendai virus[22] yielded SV40 from both SV101 and Fl^2-SV101. This type of variation therefore occurs in cells expressing viral functions. However, examination of a larger number of flat variants may disclose whether or not revertants do occur, as in the case of Rous virus hamster transformants.[23]

The relation between contact inhibition in culture and tumor-producing capacity has been clouded by the ambiguity of the criteria used and especially by the fact that some cell lines, though quite sensitive to contact inhibition of movement, are not sensitive to contact inhibition of growth.[3, 24, 25] It should therefore be emphasized that the present studies are concerned only with effects on growth and that comparisons have been made between lines and their proximate clonal descendants.

Summary.—When cloned populations of cultured mammalian cells were treated with FUdR at different cell densities, the growing cells were killed, but the non-growing cells were unaffected and their progeny could be obtained selectively. Among these progeny were variants stably different from the original population in that they were more sensitive to contact inhibition of division and for this reason were protected from FUdR killing. Such variants could be isolated from established lines and from transformants produced by oncogenic viruses. They had a number of common properties in culture and greatly reduced tumor-producing ability *in vivo.*

The results give strong support to the hypothesis that contact inhibition of division in culture is related to growth control *in vivo.* Variation in the direction of greater sensitivity to both evidently occurs spontaneously even in viral transformants and in tumors in which some viral functions continue to be present.

The FUdR selection method may be of general applicability for the detection and isolation of cells with increased sensitivity to any form of growth control.

This work was aided by grants, fellowships, and awards from the U.S. Public Health Service: CA 06793; F2AI 33, 222 (R.E.P.); 5K6-CA1181 (H.G.); and 1-K3-CA-21230 (G.J.T.).

[1] Fisher, H. W., and J. Yeh, *Science*, **155**, 581 (1967).
[2] Schutz, L., and P. T. Mora, *J. Cell Physiol.*, in press.
[3] Stoker, M. G. P., in *Current Topics in Developmental Biology* (New York: Academic Press, in press), vol. 2.
[4] Kruse, P. F., and E. Miedema, *J. Cell Biol.*, **27**, 273 (1965).
[5] Todaro, G. J., Y. Matsuya, S. Bloom, A. Robbins, and H. Green, *Wistar Inst. Symp. Monogr.*, **7**, 89 (1967).
[6] Eagle, H., V. Oyama, M. Levy, and A. Freeman, *J. Biol. Chem.*, **226**, 191 (1957).
[7] Todaro, G. J., and H. Green, *J. Cell Biol.*, **17**, 299 (1963).
[8] Todaro, G. J., H. Green, and B. Goldberg, these PROCEEDINGS, **51**, 66 (1964).
[9] Todaro, G. J., K. Nilausen, and H. Green, *Cancer Res.*, **23**, 825 (1963).
[10] Heidelberger, C., *Progr. Nucleic Acid Res. Mol. Biol.*, **4**, 1 (1965).
[11] Rueckert, R. R., and G. Mueller, *Cancer Res.*, **20**, 1584 (1960).
[12] Todaro, G. J., and H. Green, *Virology*, **24**, 393 (1964).
[13] Vogt, M., and R. Dulbecco, in *Cold Spring Harbor Symposia on Quantitative Biology*, vol. 27 (1962), p. 367.
[14] Eagle, H., and E. Levine, *Nature*, **213**, 1102 (1967).
[15] Todaro, G. J., and H. Green, these PROCEEDINGS, **55**, 302 (1966).
[16] Green, H., and G. J. Todaro, in *Molecular Biology of Viruses*, ed. J. S. Colter (New York: Academic Press, 1967), p. 667.
[17] Stoker, M. G. P., M. Shearer, and C. O'Neill, *J. Cell Sci.*, **1**, 297 (1966).
[18] Sanford, K. K., G. D. Likely, and W. Earle, *J. Natl. Cancer Inst.*, **15**, 215 (1954).
[19] Luria, S., and M. Delbrück, *Genetics*, **28**, 491 (1943).
[20] Pollack, R., unpublished experiments.
[21] Assays for polyoma-specific T-antigen were kindly performed by Mr. J. Lehman and Dr. V. Defendi.
[22] Koprowski, H., F. C. Jensen, and Z. Steplewski, these PROCEEDINGS, **58**, 127 (1967).
[23] MacPherson, I., "Malignant transformation by viruses," *Recent Results Cancer Res.*, **6**, 1 (1966).
[24] Stoker, M. G. P., and I. MacPherson, *Nature*, **203**, 1355 (1964).
[25] Defendi, V., J. Lehman, and P. Kraemer, *Virology*, **19**, 592 (1963).

Mutation Causing Temperature-Sensitive Expression of Cell Transformation by a Tumor Virus

Hartmut C. Renger and Claudio Basilico

A procedure has been devised to isolate 3T3 mouse fibroblasts transformed by simian virus 40 (SV40) that express their transformed phenotype at low (32°C) but not at high (39°C) temperature. Three parameters typical of malignant growth *in vitro:* (*a*) high saturation density in culture, (*b*) ability to form colonies on monolayers of normal 3T3 cells, and (*c*) lack of contact inhibition of DNA synthesis, are temperature sensitive. These phenotypic changes are fully reversible. The serum requirement for growth appears to be largely unchanged by temperature. These cells seem to owe their behavior to a cellular, rather than to a viral, alteration since after fusion of the temperature-sensitive transformed cells with permissive monkey cells, a procedure that leads to rescue (i.e., multiplication of the virus), wild-type SV40 virus is produced.

Several lines of evidence demonstrate that the viral genome, or at least a portion of it, becomes part of the genetic material of SV40 or polyoma virus-transformed cells (1, 2). Because of this fact, and because viral functions are necessary for the initial establishment of transformation (3), it has generally been assumed that the continuing expression of viral gene(s) was also responsible for maintaining the cell in a transformed state. This hypothesis has recently found experimental support with the isolation of a temperature-sensitive mutant of polyoma virus that renders some of the properties of transformed cells temperature dependent (4, 5). Most of the characteristics of Rous sarcoma virus-transformed cells also appear to be under the control of viral genes (6).

Abbreviations: ts, temperature-sensitive; SV3T3, SV40-transformed 3T3 cells.

To gain more information on the problem of which characteristics of SV40-transformed cells are under the control of viral or cellular genes, we devised a method of selecting for SV40-transformed cells whose transformed phenotype was temperature dependent. It took advantage of the fact that, *in vitro*, SV40-transformed 3T3 cells will continue to divide even under crowded culture conditions, while normal 3T3 cells do not (contact inhibition of growth). By shifting 3T3 cells that had been transformed at 32°C to a temperature of 39°C under conditions in which normal cells are contact inhibited, and then exposing the cell population to 5-fluoro-2-deoxyuridine (FdU), which is lethal to dividing cells (9, 10), one could selectively kill all the wild-type transformants, but not the normal 3T3 cells. It was assumed that, under such conditions, possible temperature-sensitive transformants would have escaped killing, since at the high temperature they should have behaved like normal, untransformed cells. The selection was applied to cells transformed by mutagenized virus. Since SV40 can easily be rescued from transformed cells (7, 8), this procedure should enhance the chances of obtaining viral mutants affected in the genes for transformation.

In this paper we report the isolation and characterization of several SV40-transformed 3T3 cell lines, which by most *in vitro* criteria express their transformed phenotype at 32°C, but not at 39°C. However, the virus rescued from these transformed cells seems to be completely normal with respect to both growth and transforming ability. This result suggests that the mutants described here are host-cell, rather than virus, mutants.

MATERIALS AND METHODS

Cells and Medium. The established cell lines 3T3 (mouse) and BSC-1 (monkey) were grown in Dulbecco's modification of Eagle's medium, supplemented with 5 or 10% calf serum. Medium was changed regularly on growing or resting cells, usually every 3 days at 39°C and every 4 days at 32°C. Gammaglobulin-free serum was purchased from Grand Island Biologicals. Cells were grown in plastic Petri dishes (Falcon).

Virus. Small-plaque SV40 was used for all experiments. It was propagated in BSC-1 cells at low multiplicities of infection [about 1 plaque-forming unit (PFU)/cell]. Purified high-titer preparations of virus were obtained by density gradient centrifugation. Viral infectivity was determined by plaque assay on monolayers of BSC-1 cells.

Mutagenized stocks of SV40 were obtained by adding different concentrations of *N*-methyl-*N*-nitroso-*N*-nitroguanidine (nitrosoguanidine) to BSC-1 cultures 24 hr after infection with about 10 PFU/cell. The cultures were harvested about 50 hr later. Concentrations of 5, 10, and 20 μg/ml of nitrosoguanidine reduced the yield of infectious virus 55, 80, and 96%, respectively. Each mutagenized stock to be further used was passaged once at low multiplicities of infection (about 0.1 PFU/cell) in BSC-1 cells at 32°C to eliminate defective mutants. Stocks treated with 10 and 20 μg/ml of nitrosoguanidine were used for all the experiments to be described.

Transformation Assay. SV40 transformation was assayed essentially as described by Todaro and Green (11): 3T3 cells were infected at 37°C and then incubated at the desired temperature in medium containing 10% calf serum. When transformed foci were visible by eye, the plates were fixed and stained (about 2 weeks at 39°C, 3 weeks at 32°C).

Virus Rescue. Virus was rescued from SV40-transformed 3T3 cells by fusion with BSC-1 monkey cells (7, 8). An equal number (about 10^6 cells) of both BSC-1 and transformed cells were mixed at 0°C in isotonic Tris·HCl buffer (about 1 ml total volume). 1 ml of isotonic Tris·HCl buffer, containing about 10^4 hemagglutination units of β-propiolactone-inactivated Sendai virus was then added to enhance cell fusion. After 5 min in the cold, the mixture was transferred to 37°C for 30 min. After gentle resuspension the cells were diluted in medium containing 5% fetal calf serum, and incubated at the desired temperature.

T-Antigen Assay. Cells synthesizing SV40 T-antigen were detected (12) by immunofluorescence with hamster antiserum to SV40 T-antigen and fluorescein-labeled anti-hamster globulin.

Chromosomes were prepared as described (12).

DNA synthesis was determined from the incorporation of tritium-labeled thymidine into acid-insoluble material. At

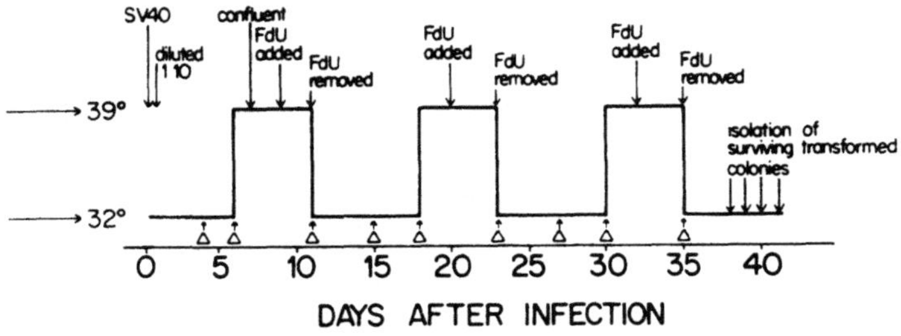

FIG. 1. Schematic outline of the selection procedure used to isolate ts SV3T3 cells. Medium containing 10% calf serum was changed at the times indicated (Δ).

the end of the labeling period, the cultures were washed twice with cold isotonic Tris·HCl buffer and lysed in 1% sodium dodecyl sulfate. After 10 min at room temperature, the lysate was scraped off with a rubber policeman. The viscosity was lowered and high molecular weight compounds were sheared by strong pipetting. Aliquots were then taken, precipitated with trichloroacetic acid, filtered on Millipore filters, dried, and counted in a Beckman LS-250 scintillation spectrometer.

RESULTS

Isolation of temperature-sensitive transformants

Fig. 1 outlines the experimental procedure used to select for temperature-sensitive SV40-transformed 3T3 cells (ts transformants, ts SV3T3). 3T3 cells, subconfluent at 37°C, were infected with 1 PFU/cell of the mutagenized SV40 (which resulted in a transformation frequency of about 0.1%). About 5 hr later, the cells were trypsinized, plated at a 1:10 dilution, and incubated at 32°C. When the cells had almost reached confluence, they were transferred to 39°C. 2 or 3 days later FdU was added at a concentration of 30 μg/ml. FdU blocks the enzyme deoxythymidylate synthetase (9), resulting in inhibition of DNA synthesis and eventual death of growing cells. However, cells that are in a resting state because of contact inhibition, remain unaffected by the drug. A 5-fold excess of uridine was always present to prevent FdU from interfering with RNA synthesis due to its possible breakdown into fluorouracil.

After 2 or 3 days, the cells were washed free of the drug by successive medium changes, and transferred to 32°C. Control plates indicated that more than 90% of the transformed colonies growing at 39°C were killed by the FdU treatment. The FdU treatment was usually repeated two more times. Of the few transformed colonies that survived all three FdU cycles, the ones that were large and seemed least affected were considered as possible ts transformants and isolated. Preliminary testing for temperature-sensitive behavior involved the measurement of their respective saturation densities at the two temperatures. The majority of the isolates did not turn out to be satisfactory, showing either very little or no difference. They were therefore discarded. Cell lines that exhibited a significantly higher saturation density at 32°C than at 39°C were propagated at 32°C and examined further. The frequency at which they were obtained was on the order of 10^{-4}.

Properties of the ts transformants

Although numerous ts SV3T3 cell lines have been isolated, only those that have been fully characterized will be described. These lines are identified by a letter, followed by a number (e.g., H1). Subclones are indicated by the line name, followed by a dash and a number (e.g., H6-15). Several lines have been used as the control wild-type SV3T3: one is a line called SV101 (10) that has been in culture for over 1 year; others are SV40-transformed lines isolated in the course of transformation experiments performed at 32°C. All gave identical results under the conditions of our experiments and will, therefore, be referred to as wild-type SV3T3.

Saturation Density. Transformed cells are able to grow efficiently in multilayers, while normal cells generally stop dividing after formation of a monolayer. Measurement of the saturation density may, therefore, give an indication of the degree of malignancy of a given cell line.This parameter is especially useful in 3T3 cells, which, unless transformed, have a very low saturation density (13).

We thus determined the saturation densities of our cell lines at 39, and 32°C. When an equal number of these cells was allowed to grow at different temperatures, a confluent monolayer always formed earlier in the cultures at the high temperature. This observation was in agreement with the

TABLE 1. *Saturation density and T-antigen synthesis of ts SV3T3 cells at 39 and 32°C*

Cells	Saturation density (cells × 10^{-5}/cm² of surface area)		T-Antigen*	
	39°C	32°C	39°C	32°C
3T3	0.8	1.0	—	—
Wild-type SV3T3	9.0	10.0	+	+
ts H6	1.25	10.5	+	+
ts H6-7	1.2	8.0	+	+
ts H6-15	1.0	8.5	+	+
ts H1	1.5	8.5	+	+
ts R10-3	1.75	8.0	+	+

* (+) more than 99% of the cells were positive; (—) no cells positive.

generation times determined during exponential growth (about 18 hr at 39°C and 24 hr at 32°C for both H6 and H1 cells). This result indicated that these cells were not temperature sensitive for growth.

Soon after reaching confluence, however, the cells stopped increasing in number at 39°C, whereas at 32°C growth continued well beyond that point. The cell number was determined between 10 and 14 days after formation of a monolayer. Table 1 shows that at 39°C the cell lines ts H6, ts H6-7, ts H6-15, ts H1, and ts R10-3 compare favorably with their normal parent 3T3 cell line, whereas at 32°C they grow to high saturation densities and compare well with wild-type SV3T3. Hence, it appears that contact inhibition is rather pronounced at 39°C. At 32°C, on the other hand, the cells grow like typical transformants. This temperature effect was fully reversible. When cells that had been maintained at 39°C for as long as two weeks were shifted to 32°C, they eventually reached the same saturation density as cells that had not been exposed to 39°C.

Tumor Antigen. Fluorescent antibody staining of the ts SV3T3 cells grown at both 32 and 39°C revealed the presence of SV40 T-Antigen (14) in all the lines tested (Table 1). No significant difference in the intensity of T-antigen appearance could be observed between the two temperatures. It seems clear, therefore, that these cells are transformed by SV40, and that the appearance of this virus-specific antigen is not affected by temperature.

Colony-Forming Ability on 3T3 Monolayers. If normal 3T3 cells are plated on top of a contact-inhibited monolayer of 3T3 cells, they will not form colonies. SV40 or polyoma-transformed cells, on the other hand, will grow above 3T3 cells and form dense colonies (15). We determined the colony-forming ability of our ts transformants on 3T3 monolayers at different temperatures: cells were plated on fully confluent 3T3 monolayers, and incubated at 39 or 32°C. At the time when colonies had grown to be visible by eye, the cells were fixed and stained for colony counting. Table 2 shows the results. Whereas the colony-forming ability of SV40-transformed 3T3 control lines is about equal at both temperatures, the mutant lines have a much higher colony-forming ability at 32°C than at 39°C. The few colonies given by the mutants at 39°C

TABLE 2. *Colony-forming ability of ts SV3T3 cells on 3T3 monolayers**

Cells	Percent of cells plated yielding colonies at 39°C	Percent of cells plated yielding colonies at 32°C	Ratio 39°C/32°C
3T3	0	0	—
Wild-type SV3T3	53	51	1.04
ts H6	0.47	25.5	0.018
ts H6-7	0.75	32.5	0.023
ts H6-15	0.75	78	0.009
ts H1	0.2	12.5	0.016
ts R10-3	2	56.5	0.035

* 200 and 2000 cells were plated on fully confluent monolayers of 3T3 cells in 50-mm petri dishes and incubated at 39 and 32°C. Medium was changed every 3 days at 39°C and every 4 days at 32°C. About 9 days later at 39°C, and 14 days later at 32°C, the cells were fixed and stained for colony counting.

TABLE 3. *Growth of ts SV3T3 cells in medium containing normal or gammaglobulin-free serum**

		39°C		32°C	
Cells	Serum	N_5/N_0	AG/G (%)†	N_7/N_0	AG/G (%)†
3T3	Normal	10		12	
			<10		8
	Gammaglobulin-free	<1		1	
Wild-type SV3T3	Normal	95		90	
			95		78
	Gammaglobulin-free	90		70	
ts H1	Normal	18		23	
			46		56
	Gammaglobulin-free	8.4		13	
ts H6	Normal	21		49	
			57		86
	Gammaglobulin-free	12		42	

* 5×10^4 cells were plated in 50-mm petri dishes in medium containing either 6% normal calf serum or 6% gammaglobulin-free calf serum and incubated at 39 and 32°C. 12 hr later, the cell number per plate was determined (N_0). 5 Days (39°C) or 7 days later (32°C), the cells were trypsinized and counted (N_5 and N_7, respectively).

† Ratio of increase in cell number in medium containing gammaglobulin-free serum over increase in cell number in medium containing normal serum.

were generally less dense than the colonies given by the wild type, or by the ts transformants at 32°C (see *Discussion*).

The standard plating efficiency (on bare plates) was not affected by temperature, again indicating that these cells are temperature sensitive with respect to growth control, rather than growth *per se*.

Serum Dependence. Normal 3T3 cells are strongly dependent on serum to grow, whereas SV40-transformed 3T3 cells grow well even in low concentrations of serum (16, 17). In addition, a gammaglobulin-associated factor was shown to be necessary for the growth of normal, but not of transformed, 3T3 cells (17). Therefore we tested the ability of the mutant cell lines to grow at the two temperatures in medium containing gammaglobulin-free serum and in medium containing whole calf-serum. Growth was determined by measurement of the increase in cell number after a few days of incubation at 39 or 32°C. Table 3 shows that both ts H6 and ts H1 appear to be slightly more serum dependent at 39°C than at 32°C. However, these transformants are definitely much less dependent on serum factors than the normal 3T3 cells, even at high temperature.

DNA Synthesis. In normal 3T3 cells, contact inhibition of growth is accompanied by inhibition of DNA synthesis (16).

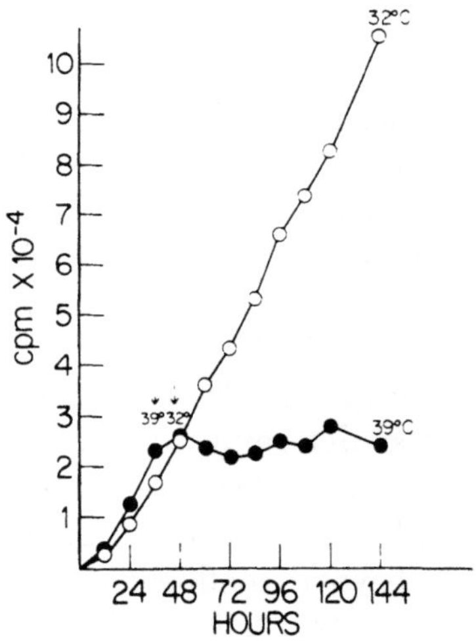

Fig. 2. DNA synthesis in ts SV3T3 cells. About 2 × 10^5 ts H6-15 cells were plated in 30-mm petri dishes in medium containing [^{3}H]dT (1.6 μCi, 8 μg/ml) and incubated at 39°C and at 32°C (time 0). Medium was changed 72 hr later. The amount of DNA synthesized was determined at the times indicated. The *ordinate* gives the radioactivity incorporated into DNA in equal aliquots of the cultures. The *arrows* indicate the time at which visible confluence was reached at the two temperatures.

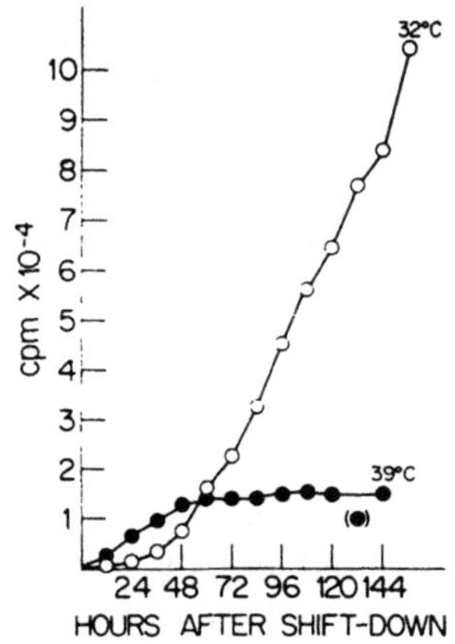

FIG. 3. DNA synthesis in ts SV3T3 cells after shift-down. ts H6-15 cells were plated at 39°C and allowed to reach confluence. 2 days later, the medium was replaced with fresh medium containing [^{3}H]dT (1 μCi, 5 μg/ml) and half of the cultures was shifted to 32°C. The amount of DNA synthesized was determined at the times indicated. The *ordinate* gives the radioactivity incorporated into DNA in equal aliquots of the cultures.

These cells stop synthesizing DNA once confluence is reached. Transformation by SV40 or lytic infection by polyoma virus overcomes this inhibition (4, 16, 18). As a result, DNA synthesis proceeds even in dense cultures. We examined the effect of cell density on the DNA-synthesizing activity of the ts transformants at 39 and 32°C. Cells of the ts H6 line were plated in medium containing [^{3}H]thymidine. The amount of DNA synthesized was determined at 12-hr intervals. Fig. 2 shows that DNA synthesis decreases greatly at 39°C after confluence is reached. At 32°C, however, the cells continue to synthesize DNA beyond confluence. A similar conclusion was reached when the rate of DNA synthesis was determined by pulse labeling the cells for 2 hr with [^{14}C]dT at various times before and after they reached confluence.

To determine the extent to which the effect of high temperature is reversible, the following experiment was done: H6-15 cells were incubated at 39°C. 2 days after visible confluence was reached, the old medium was removed and new medium containing [^{3}H]dT was added. Half of the cultures were then shifted down to 32°C, while half were kept at 39°C.

The amount of DNA synthesized was determined at intervals. Fig. 3 shows that cells kept at 39°C initially synthesized DNA at a slow rate, which approached zero by about 48 hr. At 32°C the cells synthesized DNA at an increasing, but initially slower, rate than at 39°C for the first 36 hr. By 48 hr, however, DNA synthesis had reached a constant rate, which was directly comparable to the rate observed in cells that never were exposed to 39°C. This observation indicates a full reversibility of the effect of high temperature that is complete by about 48 hr after shift down. Similar results were obtained with ts H1 cells.

Chromosomes. It has been reported that a reduction in the tumorigenicity, and a partial reversion of cells originally transformed by SV40 or polyoma virus, is accompanied by a significant change in chromosome number (19, 20). We therefore checked the chromosome number of ts H6 cells, growing at 39°C and at 32°C. A median number of 67 and 68 chromosomes, respectively, was found (all telo- or acrocentric). These values are close to the number of chromosomes of the parent 3T3 line. It seems, therefore, that no significant change in chromosome number occurred in the ts transformants.

Properties of the transforming viruses

Virus can be rescued from most SV40-transformed cells by means of fusion with permissive monkey cells (7, 8). ts H6, ts H6-7, ts H1, and ts R10-3 were tested several times and shown to be free of infectious virus. These cells were fused with permissive BSC-1 cells and the virus was allowed to grow for about 10 days at 32°C. Virus was rescuable from all the lines tested, but the yield varied greatly, depending on the line used. The ability of each rescued virus to multiply, establish transformation, and maintain the transformed state was then tested at both 39 and 32°C and compared with that of wild-type SV40.

Lytic Growth. The plaque assay was used to test lytic growth. As compared to wild-type SV40, the four rescued virus stocks gave a similar number of plaques at both temperatures (Table 4).

Transformation. If the rescued virus was unable to establish or maintain the transformed state at 39°C, then no

TABLE 4. *Properties of virus rescued from ts SV3T3 cells*

Experiment*	Virus†	Plaque-forming ability (PFU at 39°C/PFU at 32°C)	Ratio of transforming ability at 39°C to 32°C	Transforming ability‡ (shift to 39°C from 32°C)
1	Wild-type SV40	1.9	2.1	2.5
	SV40 H6	1.0	3.5	1.5
	SV40 H6-7	0.86	1.8	1.5
2	Wild-type SV40	1.4	0.73	1.17
	SV40 H6	2.0	0.52	0.79
	SV40 H1	2.1	0.53	0.89
	SV40 R10-3	2.4	0.56	0.83

* The viruses used in 1 and 2 were obtained from two independent fusion experiments.

† SV40 H6 designates virus rescued from ts H6 cells. The other viruses rescued were designated accordingly.

‡ Ratio: transformation frequency obtained by infecting cells at 32°C and shifting to 39°C 4 days later to transformation frequency at 32°C.

transformed colonies would be expected when the transforming ability was tested at 39°C. If, instead, the virus was only impaired in its ability to *establish* transformation at 39°C, transformed colonies should appear in plates that were kept at 32°C and then shifted to 39°C after a suitable period of time. When tested for these properties, however, the four rescued virus stocks appeared completely normal, giving rise to about the same relative number of transformed colonies at both temperatures as wild-type SV40 (Table 4).

DISCUSSION

The results clearly indicate that the cell lines ts H6, ts H6-7, ts H6-15, ts H1, and ts R10-3 are transformed by SV40, but that these cells grow in a typically transformed manner only at low temperature. At high temperature, they are subject to growth control and resemble more closely their untransformed 3T3 parents. Furthermore, this temperature effect is fully reversible.

It is important to note that these cells are not temperature sensitive for growth. When growing exponentially they have faster generation times at the high temperature, and their efficiency of plating (under standard conditions) is unchanged by temperature. When, however, we consider the parameters used to determine *in vitro* whether a cell behaves in a transformed or normal manner, one can see that the saturation density reached by these cell lines at 32°C is almost 10-times higher than at 39°C, indicating a substantial difference in the degree of growth control between the two temperatures. This difference is emphasized if one compares the corresponding amounts of DNA synthesized.

The ability to form colonies on a confluent monolayer of contact-inhibited cells has been used to determine normal or malignant growth behavior of different cell lines (15). SV40-transformed 3T3 cells plate very efficiently on top of normal 3T3 cells at both 32 and 39°C. Our ts transformants, however, grow in a comparable manner only at 32°C; the colony-forming ability is greatly reduced at 39°C. The few colonies that grow at 39°C bring about the question of reversion. When some of these colonies were isolated and tested for their colony-forming ability on 3T3 monolayers, they seemed to have maintained their ts behavior. This result indicated that those cells were not true revertants. True revertants do, however, arise in these cell lines. These cells grow equally on top of 3T3 monolayers at 39 and 32°C, and are capable of quickly overgrowing the rest of the cell population at 39°C. The rate at which they arise has not yet been determined, but their frequency is rather low.

The fact that our ts transformants appear quite independent of the requirement for serum at both temperatures is reminiscent of the observation that the phenotypic "flat" revertants of SV40-transformed cells that have regained contact inhibition of growth, still exhibit independence of the requirement for serum (16, 17). Also, the polyoma ts3-transformed cells are independent of serum requirement at the nonpermissive temperature (4). In our cells, some difference between the degree of dependence is apparent between the two temperatures, but it is not substantial. This phenomenon could be due to a differential suppression of transformed growth characteristics. Alternatively, these characteristics may be under independent control.

Transformed cells, due to surface alterations, show enhanced agglutination by wheat-germ agglutinin or concanavalin A (21). This degree of agglutination cannot be observed with normal cells or revertants of SV40-transformed cells that do not grow to high cell densities (22). We are indebted to Dr. Max Burger for testing two of our lines (ts H1 and ts H6) with respect to their agglutinability by wheat-germ agglutinin. Preliminary results indicate that these cells exhibit a significantly higher degree of agglutination when growing at 32° C than at 39° C. This observation suggests that these cells undergo surface changes when exposed to different temperatures.

In view of the fact that mutagenized SV40 was used for the infection, it was somewhat surprising to find that the virus rescued from these lines was fully wild type with respect to infectivity or transforming ability. It would seem, therefore, that the ts SV3T3 cells tested owe their ts behavior to a spontaneous cellular, rather than a viral, mutation. We are aware of the possible objection that these cells may have been multiply transformed, although we took precautions against this by using low multiplicities in the original infection. If cells were multiply transformed, fusion might have yielded only normal virus, which could, for example, be integrated at a site of the cellular genome where rescue is possible, but where the virus cannot exert any control over the cell. This hypothesis finds some support in the fact that the presence of viral genetic material in cells that have completely normal growth characteristics has recently been shown by means of nucleic acid hybridization (23). In our ts transformants, the ts virus may be integrated at a functional site, thus exerting control over the cell. However, this ts virus may have lost a function necessary for rescue by fusion. Although it cannot be completely discounted, this possibility does not seem likely. We have obtained supporting evidence that the ts transformants are cellular mutants. An attempt to retransform ts H6 cells, growing at 39° C, by the use of high multiplicities of infection (about 200 PFU/cell) was not successful. This is consistent with the interpretation that ts H6 carries a cellular mutation that prevents the expression of transformation at 39° C.

The failure to obtain ts transformation mutants of SV40 may have the following explanation. The only indication we have for the existence of viral transformation mutants among the oncogenic DNA viruses is the ts3 polyoma mutant (4, 5). It is possible that such mutants occur with very low frequency. Therefore, they cannot be isolated by our selection method because of the relatively high number of spontaneously occurring cellular ts transformants. We are trying to isolate additional ts transformants in an attempt to answer this question.

If our ts transformants are indeed host mutants, the temperature sensitivity of the phenotypic change suggests the involvement of a specific protein of the host-cell. As for the mechanism of the postulated change, in the ts transformants a reversible event must occur at 39°C that effectively blocks the interaction of viral and cellular functions that ordinarily result in malignant cell growth. If the role of the viral gene(s) responsible for transformation lies in switching off some of the normal cellular controls, one could think of a repressor–inducer type of interaction, the viral product being the inducer, which can bind to the cellular repressor at 32°C, but not at 39°C, because of altered repressor characteristics. Alternatively, the ts transformed cells might have developed a way of suppressing the expression of viral functions at 39°C. The fact that T-antigen synthesis in our cells is not affected by temperature, and viral mRNA synthesis seems also to proceed at both temperatures (unpublished results) makes us favor the first type of interpretation.

This investigation was supported by PHS grant CA11893 from the National Cancer Institute and by grant E-610 of the American Cancer Society. We thank Dr. Severo Ochoa for his helpful criticism of the manuscript. H.C.R. was a fellow of the Jane Coffin Childs Memorial Fund for Medical Research; C.B. is a scholar of the Leukemia Society.

1. Sambrook, J., Westphal, H. Srinivasan, P. R. & Dulbecco, R. (1968) *Proc. Nat. Acad. Sci. USA* **60,** 1288–1295.
2. Gelb, L. D., Kohne, D. E. & Martin, M. A. (1971) *J. Mol. Biol.* **57,** 129–145.
3. Fried, M. (1965) *Proc. Nat. Acad. Sci. USA*, **53,** 486–491.
4. Dulbecco, R. & Eckhart, W. (1970) *Proc. Nat. Acad. Sci. USA* **67,** 1775–1781.
5. Eckhart, W., Dulbecco, R. & Burger, M. M. (1971) *Proc. Nat. Acad. Sci. USA* **68,** 283–286.
6. Martin, G. S. (1970) *Nature* **227,** 1021–1023.

7. Koprowski, H., Jensen, F. C. & Steplewski, Z. (1967) *Proc. Nat. Acad. Sci. USA* **58**, 127–133.
8. Watkins, J. F. & Dulbecco, R. (1967) *Proc. Nat. Acad. Sci. USA* **58**, 1396–1403.
9. Heidelberger, C. (1965) *Progr. Nucleic Acid Res. Mol. Biol.* **4**, 1–50.
10. Pollack, R. E., Green, H. & Todaro, G. J. (1968) *Proc. Nat. Acad. Sci. USA* **60**, 126–133.
11. Todaro, G. J. & Green, H. (1964) *Virology* **23**, 117–119.
12. Basilico, C., Matsuya, Y. & Green, H (1970) *Virology* **41**, 295–305.
13. Todaro, G. J., & Green, H. (1963) *J. Cell Biol.* **17**, 299–313.
14. Habel, K. (1967) *Curr. Top. Microbiol. Immunol.* **41**, 85–99.
15. Aaronson, S. A., Todaro, G. J. & Freeman, A. E. (1970) *Exp. Cell Res.* **61**, 1–5.
16. Dulbecco, R. (1970) *Nature* **227**, 802–806.
17. Jainchill, J. L. & Todaro, G. J. (1970) *Exp. Cell Res.* **59**, 137–146.
18. Basilico, C., Matsuya, Y. & Green, H. (1969) *J. Virol.* **3**, 140–145.
19. Rabinowitz, Z. & Sachs, L. (1970) *Nature* **225**, 136–139.
20. Pollack, R., Wolman, S. & Vogel, A. (1970) *Nature* **228**, 938–970.
21. Burger, M. M. & Goldberg, A. R. (1967) *Proc. Nat. Acad. Sci. USA* **57**, 359–366.
22. Pollack, R. E. & Burger, M. M. (1969) *Proc. Nat. Acad. Sci. USA* **62**, 1074–1076.
23. Smith, H. F., Gelb, L. D. & Martin, M. A. *Proc. Nat. Acad. Sci. USA* **69**, in press.

AUTHOR INDEX

SUBJECT INDEX

Single numbers indicate pages on which consideration of topic begins; number ranges refer to entire papers.

GPSR Compliance

The European Union's (EU) General Product Safety Regulation (GPSR) is a set of rules that requires consumer products to be safe and our obligations to ensure this.

If you have any concerns about our products, you can contact us on ProductSafety@springernature.com

In case Publisher is established outside the EU, the EU authorized representative is:

Springer Nature Customer Service Center GmbH
Europaplatz 3
69115 Heidelberg, Germany

Batch number: 10615115

Printed by Printforce, the Netherlands